Two-Dimensional
Carbon

edited by
Yihong Wu
Zexiang Shen
Ting Yu

Two-Dimensional
Carbon

Fundamental Properties, Synthesis, Characterization, and Applications

Published by

Pan Stanford Publishing Pte. Ltd.
Penthouse Level, Suntec Tower 3
8 Temasek Boulevard
Singapore 038988

Email: editorial@panstanford.com
Web: www.panstanford.com

British Library Cataloguing-in-Publication Data
A catalogue record for this book is available from the British Library.

Two-Dimensional Carbon: Fundamental Properties, Synthesis, Characterization, and Applications
Copyright © 2014 by Pan Stanford Publishing Pte. Ltd.
All rights reserved. This book, or parts thereof, may not be reproduced in any form or by any means, electronic or mechanical, including photocopying, recording or any information storage and retrieval system now known or to be invented, without written permission from the publisher.

For photocopying of material in this volume, please pay a copying fee through the Copyright Clearance Center, Inc., 222 Rosewood Drive, Danvers, MA 01923, USA. In this case permission to photocopy is not required from the publisher.

ISBN 978-981-4411-94-3 (Hardcover)
ISBN 978-981-4411-95-0 (eBook)

Printed in the USA

Contents

Preface		xi

1. Introduction   **1**
Yihong Wu

1.1	Carbon Allotropes	1
1.2	Two-Dimensional Carbon	3
1.3	Scope of This Book	4

2. Electronic Band Structure and Properties of Graphene   **11**
Yihong Wu

2.1	Lattice structure		11
2.2	Electronic Band Structure		12
	2.2.1	Tight-Binding Model	12
	2.2.2	Low-Energy Electronic Spectrum	15
	2.2.3	Effect of Magnetic Field	19
	2.2.4	Quantum Confinement and Tunneling	20
2.3	Electrical Transport Properties		23
	2.3.1	Weak (Weak Anti-) Localization	24
	2.3.2	Electrical Conductivity and Mobility	26
2.4	Summary		29

3. Growth of Epitaxial Graphene on SiC   **35**
Xiaosong Wu

3.1	Introduction		36
	3.1.1	Graphene	36
	3.1.2	Epitaxial Graphene on SiC	37
3.2	Growth on Si-Face		39
	3.2.1	Growth Mechanism	39
	3.2.2	Control the Growth	42
	3.2.3	Beyond One Layer	46
	3.2.4	More Than Just Si Desorption	47
3.3	Growth on C-Face		49
3.4	Conclusion		56

4. Chemical Vapor Deposition of Large-Area Graphene on Metallic Substrates 67

Wei Wu and Qingkai Yu

4.1	Introduction	67
4.2	Graphene Synthesis Methods	68
	4.2.1 Mechanical Exfoliation	68
	4.2.2 Graphite Oxide Reduction	69
	4.2.3 Thermal Decomposition of SiC	69
	4.2.4 CVD of Hydrocarbons on Metals	70
4.3	CVD Graphene on Nickel	70
4.4	CVD Graphene on Copper	78
4.5	Single-Crystal Graphene	84
4.6	Summary	88

5. Growth and Electrical Characterization of Carbon Nanowalls 95

Yihong Wu

5.1	Introduction	95
5.2	Synthesis of Carbon Nanowalls by MPECVD	96
	5.2.1 Growth Apparatus and Surface Morphology	96
	5.2.2 Growth Mechanism	99
5.3	Electrical Transport Properties	100
	5.3.1 Electrical Transport Across the CNW Junctions	102
	5.3.2 Electrical Transport in Carbon–Metal Point-Contact	106
5.4	Magnetic Properties	114
5.5	Summary	116

6. Structural Characterization of Carbon Nanowalls and Their Potential Applications in Energy Devices 121

Masaru Tachibana

6.1	Introduction	122
6.2	Synthesis of CNWs	123
6.3	Structural Characterization of CNWS	124
	6.3.1 Raman Spectroscopy	125
	6.3.2 Transmission Electron Microscopy	129
6.4	Potential Applications of CNWs to Energy Devices	134

Contents | vii

		6.4.1	CNWs as Negative Electrode in Lithium Ion Battery	134
			6.4.1.1 Preparation of electrode	135
			6.4.1.2 Cyclic voltammogram	135
			6.4.1.3 Charge/discharge curve	137
			6.4.1.4 High rate property	138
			6.4.1.5 CNWs as catalyst support in fuel cell	140
			6.4.1.6 Preparation of Pt/CNW	140
			6.4.1.7 Physical characterization of Pt/CNW	141
			6.4.1.8 Electrochemical characterization of Pt/CNW	144
	6.5	Summary		146

7. Raman and Infrared Spectroscopic Characterization of Graphene   153

Da Zhan and Zexiang Shen

	7.1	Introduction		153
	7.2	Raman Spectroscopic Features of Graphene		154
		7.2.1	Identify the Thickness and Stacking Geometry of Graphene	155
		7.2.2	Probing Doped Charges in Graphene	159
		7.2.3	Defects in Graphene Probed by Raman Spectroscopy	164
		7.2.4	Graphene Edge Chirality Identified by Raman Spectroscopy	169
		7.2.5	Raman Spectroscopy Probes Strain in Graphene	173
	7.3	Chemical Functional Groups and Energy Gaps of Graphene Probed by Infrared Spectroscopy		174
	7.4	Summary		176

8. Graphene-Based Materials for Electrochemical Energy Storage   183

Jintao Zhang and Xiu Song Zhao

	8.1	Introduction		183
		8.1.1	A Family of Graphene-Based Materials	184
		8.1.2	Electrochemical Energy Storage Systems	187

8.2		Synthesis of Graphene-Based Materials	190
	8.2.1	Synthesis of Graphene	190
	8.2.2	Synthesis of Chemically Modified Graphene	195
	8.2.3	Synthesis of Graphene-Based Composite Materials	202
8.3		Graphene-Based Materials as Supercapacitor Electrodes	207
	8.3.1	Energy Storage in Supercapacitor	207
	8.3.2	Basic Principles and Techniques for Evaluation of Supercapacitor	210
	8.3.3	Graphene and Reduced Graphene Oxide as Supercapacitor Electrodes	216
	8.3.4	Graphene Based Composite Materials as Supercapacitor Electrodes	223
8.4		Summary and Perspectives	232
8.5		Acknowledgements	233

9. Chemical Synthesis of Graphene and Its Applications in Batteries   **247**

Xufeng Zhou and Zhaoping Liu

9.1		Introduction	247
9.2		Chemical Synthesis of Graphene	248
	9.2.1	Direct Exfoliation of Graphite	248
	9.2.2	Graphene from Graphite Oxide	249
	9.2.3	Graphene from Graphite Intercalated Compounds	254
9.3		Applications of Graphene in Batteries	256
	9.3.1	Li ion Batteries	256
	9.3.2	Supercapacitors	264
	9.3.3	Other types of Batteries	269
9.4		Summary	271

10. Photonic Properties of Graphene Device   **279**

Hua-Min Li and Won Jong Yoo

10.1	Introduction	280
10.2	Energy Band Structure of Graphene	283
10.3	Photonic Absorption of Graphene	284
10.4	Photocurrent Generation in Graphene	286
10.5	Technology for Performance Improvement	289

		10.5.1 Asymmetric Metallization	290
		10.5.2 Graphene Stack Channel	290
		10.5.3 Graphene Plasmonics	291
	10.6	Summary	292

11. Graphene Oxides and Reduced Graphene Oxide Sheets: Synthesis, Characterization, Fundamental Properties, and Applications   **297**

Shixin Wu and Hua Zhang

	11.1	Introduction	297
	11.2	Methods of Production of Graphene	298
	11.3	Introduction to Graphene Oxide	299
	11.4	Methods to Produce Stable Dispersion of Reduced Graphene Oxide	300
	11.5	Characterizations and Fundamental Properties of Reduced Graphene Oxide	302
		11.5.1 Characterization with Atomic Force Microscopy	302
		11.5.2 Characterization with X-Ray Diffraction	303
		11.5.3 Characterization with X-Ray Photoelectron Spectroscopy (XPS)	304
		11.5.4 Characterization with Raman Spectroscopy	305
		11.5.5 Conductivity	305
	11.6	Sensing Applications of Reduced Graphene Oxide	306
		11.6.1 Field-Effect Transistor Sensors	306
		11.6.2 Electrochemical Sensors	308
		11.6.3 Matrices for Mass Spectrometry	309
	11.7	Device Applications of Reduced Graphene Oxide	310
		11.7.1 Memory Devices	310
		11.7.2 Solar Cells	310
	11.8	Applications of Graphene Oxide	313
		11.8.1 Fluorescence Sensors	313
		11.8.2 Solar Cells	315
	11.9	Conclusion	316

Index   329

Preface

Graphene in the ideal form is a single layer of carbon atoms arranged in a honeycomb lattice, consisting of two interpenetrating Bravais sublattices. It is this unique lattice structure that gives graphene a range of peculiar properties that most metals and semiconductors lack. As far as electronic applications are concerned, its gapless and linear energy spectrum, high carrier mobility, frequency-independent absorption, and long spin diffusion length make it a material of choice for a variety of electronic, photonic, and spintronic devices. Apart from these applications, owing to its unique electronic properties, graphene has also attracted tremendous attention for applications that are due to primarily its unique shape and surface morphology, and low-cost production of related materials such as few-layer graphene sheets and graphene oxides. These graphene derivatives are more attractive and viable than single-layer graphene for applications that require a large quantity of materials with low cost and that rely less on graphene's electronic properties. These different types of graphene-based carbon nanostructures are referred to as two-dimensional (2D) carbon in this book, which include but are not limited to single layer graphene, few-layer graphene, vertically aligned few-layer graphene sheets (or carbon nanowalls), reduced graphene oxide and graphene oxide, etc. As far as large-scale applications are concerned, we feel that these graphene-related materials may be a step closer to reality than their pure graphene counterpart, in particular, in energy storage–related applications. This has motivated us to pull together a team of researchers who are doing frontier research in the respective fields to discuss fundamental properties of graphene, synthesis and characterization of graphene and related 2D carbon structures, and associated applications in an edited book.

The book is organized into 11 chapters. Following the introduction in Chapter 1, Yihong Wu gives a brief overview of electronic band structure and properties of graphene in Chapter 2. In addition to the description of band structure based on the tight-binding model, several unique electron transport properties of graphene are discussed. Chapters 3 and 4 cover the growth of graphene on

SiC substrates by Xiaosong Wu and on metallic substrates by Wei Wu and Qingkai Yu, respectively. The former discusses the growth mechanism of graphene on both Si-face and C-face of SiC, while the latter deals with the growth of graphene on nickel and copper substrates using chemical vapor deposition. Chapter 5 discusses the growth and electrical transport properties of carbon nanowalls on different types of substrates. Emphases are placed on how to design and form different types of electrical contacts that allow for the study of electrical transport properties of material structures with an unusual surface morphology. This is then followed by Chapter 6, in which Masaru Tachibana writes about the structural characterization of carbon nanowalls using Raman spectroscopy and transmission electron microscopy, and their potential applications in energy storage such as lithium ion batteries and fuel cells. In Chapter 7, Zexiang Shen and Da Zhan discuss the structural properties of 2D carbon based on Raman spectroscopy studies. Chapters 8 and 9 are devoted to the energy storage applications of graphene obtained by the chemical reduction route, which is more cost effective compared with other vapor deposition–based techniques. Xiu Song Zhao and Jintao Zhang focus on the applications of 2D carbon in supercapacitor in Chapter 8, followed by Zhaoping Liu and Xufeng Zhou dealing with battery applications in Chapter 9. The photonic properties of graphene are discussed by Won Jong Yoo and Hua-Min Li in Chapter 10. In Chapter 11, Hua Zhang and Shixin Wu discuss another important material derived from graphene, graphene oxide, and its potential applications in sensor and memory devices.

Owing to the very competitive environment of graphene research, many researchers would put their priority in doing research and writing papers rather than contributing to book chapters. In this context, we would like to thank all the contributing authors for their excellent chapters; without their extra efforts, we would not have the book in the present form. We would like to thank Prof. Andrew Thye Shen Wee of National University of Singapore for giving the opportunity to edit this book. Finally, we would like to thank Mr. Stanford Chong and his team at Pan Stanford Publishing for their help on this project.

Yihong Wu
Zexiang Shen
Ting Yu

Chapter 1

Introduction

Yihong Wu
Department of Electrical and Computer Engineering,
National University of Singapore, 4 Engineering Drive 3,
Singapore 117576
elewuyh@nus.edu.sg

1.1   Carbon Allotropes

The properties of a material at the mesoscopic scale are determined not only by the nature of its chemical bonds but also by its dimensionality and shape. This is particularly true for carbon-based materials. In the ground state, the carbon atom has four valence electrons, two in the 2s sub-shell and two in the 2p sub-shell. When a large number of carbon atoms come together to form materials under appropriate conditions, an individual carbon atom will promote one of its 2s electrons into its empty 2p orbital and then form bonds with other carbon atoms via sp hybrid orbitals. Depending on the number of p orbitals (1 to 3) mixing with the s orbital, it will lead to the formation of three kinds of sp hybrid orbitals called sp, sp^2, and sp^3. Carbon atoms with sp^2 and sp^3 hybrid orbitals are able to form

---

Two-Dimensional Carbon: Fundamental Properties, Synthesis, Characterization, and Applications
Edited by Yihong Wu, Zexiang Shen, and Ting Yu
Copyright © 2014 Pan Stanford Publishing Pte. Ltd.
ISBN 978-981-4411-94-3 (Hardcover), 978-981-4411-95-0 (eBook)
www.panstanford.com

three and four bonds with neighboring carbon atoms, respectively, which form the bases of graphene and diamond.

An ideal graphene is a monatomic layer of carbon atoms arranged in a honeycomb lattice; therefore, graphene is a perfect two-dimensional (2D) material in the ideal case. As ideal 2D crystals in free state are unstable at finite temperature [1], graphene tends to evolve into other types of structures with enhanced structural stability, such as graphite, fullerene, nanotubes, and their derivatives [2]. Graphite is formed by the layering of a large number of graphene layers mediated by the van der Waals force; therefore, from the point of view of physics, it falls into the category of three-dimensional (3D) systems. Under appropriate conditions, a single- or multiple-layer graphene can also roll up along certain directions to form a tubular structure called carbon nanotubes (CNTs) [3]. The CNTs, which can be in the form of single-walled, double-walled, and multiple-walled structures, are considered one-dimensional (1D) objects as far as their physical properties are concerned [4]. With the introduction of pentagons, graphene can also be wrapped up to form zero-dimensional (0D) fullerenes [5]. In addition to cylindrical CNTs and spherical fullerenes, there also exist intermediate carbon nanoforms, such as nanocones with different stacking structures [6]. Although ideal graphene is unstable, it may become stable through the introduction of local curvatures, as discussed in [7], or through the support of foreign materials. Macroscopic single-layer graphene was successfully isolated from graphite through mechanical exfoliation in 2004 and was found to be stable on a foreign substrate, highly crystalline, and chemically inert under ambient conditions [8–10], albeit with local roughness and ripples [11]. This discovery has led to an explosive interest in 2D carbon nanostructures, which also earned K. S. Novoselov and A. K. Geim the 2010 Nobel Prize in Physics. In addition to crystalline carbon allotropes, there are also amorphous carbons and carbons with mixed phases, such as activated carbon and diamond-like carbon (DLC). As far as large-scale industrial applications are concerned, bulk carbons are still dominant, although nanocarbons are expected to play an increasingly important role in future. A pictorial summary of different forms of single-phase carbon, or allotropes of carbon, is presented in Fig. 1.1. For a detailed discussion on nomenclature of sp^2 carbon nanoforms, the reader may refer to [6].

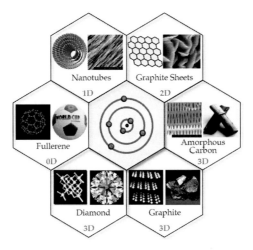

Figure 1.1 Major allotropes of carbon.

1.2 Two-Dimensional Carbon

Among all carbon allotropes, graphene stands out because of its unique lattice structure: a monatomic honeycomb lattice with a perfect 2D dimensionality. The specific lattice structure in combination with the valence electron configuration of carbon atoms gives rise to peculiar electronic band structures, which distinguish graphene from other allotropes. The quasi-particles (or electrons and holes) in graphene behave like massless relativistic particles, or Dirac fermions, with the electrons and holes degenerated at the Dirac points [12–16]. This gives rise to a number of peculiar physical properties that are either not found or superior to those found in other carbon allotropes [17]. Some of the unique physical phenomena that have been observed or explored so far include unconventional integer quantum Hall effect (IQHE) [9, 10], Klein tunneling [18–20], valley polarization [21, 22], universal (non-universal) minimum conductivity [23–26], weak (weak anti-) localization [23, 27–31], ultrahigh mobility [23, 32–34], specular Andreev reflection at the graphene–superconductor interface [35, 36], exceptional thermal conductivity [37, 38], and superior mechanical properties [39].

Since the discovery of single-layer graphene, tremendous progress has been made in the development/redevelopment of various types of techniques for synthesizing both single-layer graphene

(SLG) and few-layer graphene (FLG) sheets, such as epitaxial growth on both SiC and metallic substrates [40–44], reduction from graphite oxide [45], chemical vapor deposition (CVD) [7, 46–48], and electrical discharge [49]. It is worth noting that most of these techniques are not new and they have been used to grow various types of 2D graphitic materials before the discovery of graphene. Depending on the synthesis techniques and conditions, in addition to pure graphene, various secondary forms of graphene can also be formed. These carbon nanostructures are typically multilayer graphene with a varying degree of curvature, defects, and morphology. Although different terminologies have been introduced to describe these nano-carbon forms [6], in general, they can all be referred to as two-dimensional carbon, which is the focus of this book. Just like diamond and graphite, perfect crystalline materials are always desirable, but they are more difficult to produce and thus often too expensive for large-scale applications; on the contrary, partially perfect carbons such as synthetic graphite/diamond, activated carbon, and DLC are more widely used in industry. The same scenario may also happen to graphene, which warrants a book to discuss 2D carbon in a more inclusive manner instead of purely on graphene.

1.3   Scope of This Book

This book is not intended to focus on the fascinating properties of graphene that have already been covered by other books. Instead, after a brief introduction of the band structure and electronic properties of graphene, we focus more on the synthesis and characterization of 2D carbons in general and the associated applications, in particular, in the area of energy storage. Based on this spirit, this book is organized into 11 chapters. Following the introduction, Yihong Wu gives a brief overview of electronic band structure and properties of graphene in Chapter 2. In addition to the description of band structure based on the tight-binding model, several unique electron transport properties of graphene are discussed, including quantum Hall effect, weak (weak anti-) localization, and electrical conductivity and mobility. Chapters 3 and 4 cover the growth of graphene on SiC substrates by Xiaosong Wu and on metallic substrates by Qingkai Yu, respectively. The former discusses the growth mechanism of graphene on both Si-face and C-face of SiC, while the latter deals with the growth of graphene on nickel and copper substrates using CVD. Chapter 5 discusses the growth and electrical transport properties of carbon nanowalls on different types of substrates. Emphasis is placed on how to design and form different types of electrical contacts that allow for the study of electrical transport properties of material structures with an unusual surface morphology. This is then followed by Chapter 6, in which Masaru Tachibana writes about the structural characterization of carbon nanowalls using Raman spectroscopy and transmission electron microscopy, and their potential applications in energy storage such as lithium ion batteries and fuel cells. In Chapter 7, Zexiang Shen discusses the structural properties of 2D carbon based on Raman spectroscopy studies. Chapters 8 and 9 are devoted to the energy storage applications of graphene obtained by the chemical reduction route, which is more cost effective compared with other vapor deposition-based techniques. X. S. Zhao focuses on the applications of 2D carbon in supercapacitor in Chapter 8, followed by Zhaoping Liu dealing with battery applications in Chapter 9. The photonic properties of graphene are discussed by Yoo Won Jong in Chapter 10. In Chapter 11, Hua Zhang discusses another important material derived from graphene—graphene oxide—and its potential applications in sensor and memory devices.

The current interest in graphene is phenomenal, as evidenced by the large number of publications published in the last few years. Many reviews have been written on graphene, covering various aspects from fundamental physics and electronic properties [16, 23, 50–55] to material synthesis [40–45, 56, 57] and applications [58–62]. Several books with different emphases are already available. It is not possible for any book to cover all the relevant topics on graphene and related nanostructures. By extending the coverage to both flat and vertically aligned graphene sheets, it is hoped that this book can serve as a good reference for research on 2D carbon in general rather than graphene only.

References

1. Marder M.P. (2010) *Condensed Matter Physics*, 2nd ed, JohnWiley, Hoboken, New Jersey.
2. Dresselhaus M.S., Dresselhaus G., Eklund P.C. (1996) *Science of Fullerenes and Carbon Nanotubes*, Academic Press, San Diego.
3. Iijima S. (1991) Helical microtubules of graphitic carbon, *Nature*, **354**, 56–58.

4. Saito R., Dresselhaus G., Dresselhaus M.S. (1998) *Physical Properties of Carbon Nanotubes*, Imperial College Press, London.

5. Kroto H.W., Heath J.R., Obrien S.C., Curl R.F., Smalley R.E. (1985) C-60: Buckminsterfullerene, *Nature*, **318**, 162–163.

6. Suarez-Martinez I., Grobert N., Ewels C.P. (2012) Nomenclature of sp^2 carbon nanoforms, *Carbon*, **50**, 741–747.

7. Wu Y.H., Yang B.J., Zong B.Y., et al.(2004) Carbon nanowalls and related materials, *J Mater Chem*, **14**, 469–477.

8. Novoselov K.S., Geim A.K., Morozov S.V., et al. (2004) Electric field effect in atomically thin carbon films, *Science*, **306**, 666–669.

9. Novoselov K.S., Geim A.K., Morozov S.V., et al.(2005) Two-dimensional gas of massless Dirac fermions in graphene, *Nature*, **438**, 197–200.

10. Zhang Y.B., Tan Y.W., Stormer H.L., Kim P. (2005) Experimental observation of the quantum Hall effect and Berry's phase in graphene, *Nature*, **438**, 201–204.

11. Fasolino A., Los J.H., Katsnelson M.I. (2007) Intrinsic ripples in graphene, *Nat Mater*, **6**, 858–861.

12. Wallace P.R. (1947) The band theory of graphite, *Phys Rev*, **71**, 622.

13. McClure J.W. (1957) Band structure of graphite and de Haas-van Alphen effect, *Phys Rev*, **108**, 612.

14. Slonczewski J.C. and WeissP.R. (1958) Band structure of graphite, *Phys Rev*, **109**, 272.

15. Semenoff G.W. (1984) Condensed-matter simulation of a three-dimensional anomaly, *Phys Rev Lett*, **53**, 2449.

16. Neto A.H.C., Guinea F., Peres N.M.R., Novoselov K.S., Geim A.K. (2009) The electronic properties of graphene, *Rev Mod Phys*, **81**, 109–162.

17. Ando T., Fowler A.B., Stern F. (1982) Electronic properties of two-dimensional systems, *Rev Mod Phys*, **54**, 437–672.

18. Katsnelson M.I., Novoselov K.S., Geim A.K. (2006) Chiral tunnelling and the klein paradox in graphene, *Nat Phys*, **2**, 620–625.

19. Beenakker C.W.J. (2008) Colloquium: Andreev reflection and klein tunneling in graphene, *Rev Mod Phys*, **80**, 1337–1354.

20. Stander N., Huard B., Goldhaber-Gordon D. (2009) Evidence for klein tunneling in graphene p-n junctions, *Phys Rev Lett*, **102**, 026807.

21. Rycerz A., Tworzydlo J. Beenakker C.W.J. (2007) Valley filter and valley valve in graphene, *Nat Phys*, **3**, 172–175.

22. Cresti A., Grosso G., Parravicini G.P. (2008) Valley-valve effect and even-odd chain parity in p-n graphene junctions, *Phys Rev B*, **77**, 233402.

23. Geim A.K., Novoselov K.S. (2007) The rise of graphene, *Nat Mater*, **6**, 183–191.

24. Ando T., Zheng Y.S., Suzuura H. (2002) Dynamical conductivity and zero-mode anomaly in honeycomb lattices, *J Phys Soc Jpn*, **71**, 1318–1324.

25. Ziegler K. (2007) Minimal conductivity of graphene: Nonuniversal values from the kubo formula, *Phys Rev B*, **75**, 233407.

26. Adam S., Hwang E.H., Galitski V.M., Das Sarma S. (2007) A self-consistent theory for graphene transport, *Proc Natl Acad Sci USA*, **104**, 18392–18397.

27. Suzuura H., Ando T. (2002) Crossover from symplectic to orthogonal class in a two-dimensional honeycomb lattice, *Phys Rev Lett*, **89**, 266603.

28. McCann E., Kechedzhi K., Fal'ko V.I., Suzuura H., Ando T., Altshuler B.L. (2006) Weak-localization magnetoresistance and valley symmetry in graphene, *Phys Rev Lett*, **97**, 146805.

29. Morozov S.V., Novoselov K.S., Katsnelson M.I., et al. (2006) Strong suppression of weak localization in graphene, *Phys Rev Lett*, **97**, 016801.

30. Tikhonenko F.V., Horsell D.W., Gorbachev R.V. Savchenko A.K. (2008) Weak localization in graphene flakes, *Phys Rev Lett*, **100**, 056802.

31. Wu X.S., Li X.B., Song Z.M., Berger C., de Heer W.A. (2007) Weak antilocalization in epitaxial graphene: Evidence for chiral electrons, *Phys Rev Lett*, **98**, 136801.

32. Bolotin K.I., Sikes K.J., Jiang Z., et al. (2008) Ultrahigh electron mobility in suspended graphene, *Solid State Commun*, **146**, 351–355.

33. Orlita M., Faugeras C., Plochocka P., et al. (2008) Approaching the Dirac point in high-mobility multilayer epitaxial graphene, *Phys Rev Lett*, **101**, 267601.

34. Du X., Skachko I., Barker A. Andrei E.Y. (2008) Approaching ballistic transport in suspended graphene, *Nat Nanotechnol*, **3**, 491–495.

35. Beenakker C.W.J. (2006) Specular andreev reflection in graphene, *Phys Rev Lett*, **97**, 067007.

36. Zhang Q.Y., Fu D.Y., Wang B.G., Zhang R., Xing D.Y. (2008) Signals for specular andreev reflection, *Phys Rev Lett*, **101**, 047005.

37. Balandin A.A., Ghosh S., Bao W., et al. (2008) Superior thermal conductivity of single-layer graphene, *Nano Lett*, **8**, 902–907.

38. Balandin A.A. (2011) Thermal properties of graphene and nanostructured carbon materials, *Nat Mater*, **10**, 569–581.

39. Lee C., Wei X., Kysar J.W., Hone J. (2008) Measurement of the elastic properties and intrinsic strength of monolayer graphene, *Science*, **321**, 385–388.

40. de Heer W.A., Berger C., Wu X.S., et al. (2007) Epitaxial graphene, *Solid State Commun*, **143**, 92–100.

41. Hass J., de Heer W.A., Conrad E.H. (2008) The growth and morphology of epitaxial multilayer graphene, *J Phys: Condens Matter*, **20**, 323202.

42. Gall N.R., Rut'kov E.V., Tontegode A.Y. (1997) Two dimensional graphite films on metals and their intercalation, *Int J Mod Phys B*, **11**, 1865–1911.

43. Oshima C., Nagashima A. (1997) Ultra-thin epitaxial films of graphite and hexagonal boron nitride on solid surfaces, *J Phys: Condens Matter*, **9**, 1–20.

44. Wintterlin J., Bocquet M.L. (2009) Graphene on metal surfaces, *Surf Sci*, **603**, 1841–1852.

45. Park S.J., Ruoff R.S. (2009) Chemical methods for the production of graphenes, *Nat Nanotechnol*, **4**, 217.

46. Wu Y.H., Qiao P.W., Chong T.C., Shen Z.X. (2002) Carbon nanowalls grown by microwave plasma enhanced chemical vapor deposition, *Adv Mater*, **14**, 64–67.

47. Wang J.J., Zhu M.Y., Outlaw R.A., et al. (2004) Free-standing subnanometer graphite sheets, *Appl Phys Lett*, **85**, 1265–1267.

48. Hiramatsu M., Shiji K., Amano H., Hori M. (2004) Fabrication of vertically aligned carbon nanowalls using capacitively coupled plasma-enhanced chemical vapor deposition assisted by hydrogen radical injection, *Appl Phys Lett*, **84**, 4708–4710.

49. Ebbesen T.W., Ajayan P.M. (1992) Large-scale synthesis of carbon nanotubes, *Nature*, **358**, 220–222.

50. Geim A.K., MacDonald A.H. (2007) Graphene: Exploring carbon flatland, *Phys Today*, **60**, 35–41.

51. Katsnelson M.I. (2007) Graphene: Carbon in two dimensions, *Materials Today*, **10**, 20–27.

52. Ando T. (2007) Exotic electronic and transport properties of graphene, *Physica E: Low-Dimensional Syst Nanostruct*, **40**, 213–227.

53. Ferrari A.C. (2007) Raman spectroscopy of graphene and graphite: Disorder, electron–phonon coupling, doping and nonadiabatic effects, *Solid State Commun*, **143**, 47–57.

54. Malard L.M., Pimenta M.A., Dresselhaus G., Dresselhaus M.S. (2009) Raman spectroscopy in graphene, *Phys Rep*, **473**, 51–87.

55. Molitor F., Guttinger J., Stampfer C., et al. (2011) Electronic properties of graphene nanostructures, *J Phys Condens Matter*, **23**, 243201.

56. Singh V., Joung D., Zhai L., Das S., Khondaker S.I., Seal S. (2011) Graphene based materials: Past, present and future, *Prog Mater Sci*, **56**, 1178–1271.

57. Wu Y.H., Yu T., Shen Z.X. (2010) Two-dimensional carbon nanostructures: Fundamental properties, synthesis, characterization, and potential applications, *J Appl Phys*, **108**, 071301.

58. Grande L., Chundi V.T., Wei D., Bower C., Andrew P., Ryhanen T. (2012) Graphene for energy harvesting/storage devices and printed electronics, *Particuology*, **10**, 1–8.

59. Huang X., Qi X.Y., Boey F., Zhang H. (2012) Graphene-based composites, *Chem Soc Rev*, **41**, 666–686.

60. Machado B.F., Serp P. (2012) Graphene-based materials for catalysis, *Catal Sci Technol*, **2**, 54–75.

61. Kim K., Choi J.Y., Kim T., Cho S.H., Chung H.J. (2011) A role for graphene in silicon-based semiconductor devices, *Nature*, **479**, 338–344.

62. Jo G., Choe M., Lee S., Park W., Kahng Y.H., Lee T. (2012) The application of graphene as electrodes in electrical and optical devices, *Nanotechnology*, **23**, 112001.

Chapter 2

Electronic Band Structure and Properties of Graphene

Yihong Wu

Department of Electrical and Computer Engineering, National University of Singapore,
4 Engineering Drive 3, Singapore 117576
elewuyh@nus.edu.sg

2.1   Lattice structure

The peculiar electronic properties of graphene originate from its unique lattice structure. Graphene has a single layer of carbon atoms arranged in a honeycomb lattice, as shown in Fig. 2.1a. The primitive cell spanned by the following two lattice vectors

$$\vec{a}_1 = \left(\frac{3}{2}a, -\frac{\sqrt{3}}{2}a \right), \quad \vec{a}_2 = \left(\frac{3}{2}a, \frac{\sqrt{3}}{2}a \right) \tag{2.1}$$

contains two atoms, one of type A and the other of type B, which represent the two triangular lattices. Here, $a = 0.142$ nm is the carbon bond length. Type A atoms occupy the lattice sites $\vec{R} = m\vec{a}_1 + n\vec{a}_2$, where m and n are integers, and the B atoms are shifted with respect to the A atoms in each primitive cell by $\vec{\tau} = (\vec{a}_1 + \vec{a}_2)/3$.

Two-Dimensional Carbon: Fundamental Properties, Synthesis, Characterization, and Applications
Edited by Yihong Wu, Zexiang Shen, and Ting Yu
Copyright © 2014 Pan Stanford Publishing Pte. Ltd.
ISBN 978-981-4411-94-3 (Hardcover), 978-981-4411-95-0 (eBook)
www.panstanford.com

Electronic Band Structure and Properties of Graphene

The corresponding reciprocal lattice vectors are given by

$$\vec{g}_1 = \frac{4\pi}{3\sqrt{3}a}\left(\frac{\sqrt{3}}{2}, -\frac{3}{2}\right), \quad \vec{g}_2 = \frac{4\pi}{3\sqrt{3}a}\left(\frac{\sqrt{3}}{2}, \frac{3}{2}\right) \tag{2.2}$$

which also form a honeycomb lattice, but appears to be rotated by 30° when compared with the real lattice. The first Brillouin zone (BZ) is a hexagon with a side length of $\dfrac{4\pi}{3\sqrt{3}a}$. Inside the first BZ, points

$$\vec{K} = \left(\frac{2\pi}{3a}, \frac{2\pi}{3\sqrt{3}a}\right) \text{ and } \vec{K}' = \left(\frac{2\pi}{3a}, -\frac{2\pi}{3\sqrt{3}a}\right) \text{ are of particular interest,}$$

where, as it will become clear later, the A and B lattices decouple, forming the so-called Dirac point.

2.2 Electronic Band Structure

2.2.1 Tight-Binding Model

In the ground state, each carbon has four valence electrons, two in the 2s sub-shell and two in the 2p sub-shell. When forming bonds with other carbon atoms, an individual carbon atom will first promote one of its 2s electrons into its empty 2p orbital and then form bonds with other carbon atoms via sp hybrid orbitals. In case of graphene, two 2p orbitals (p_x and p_y) hybridize with one 2s orbital to form three sp^2 hybrid orbitals, and during this process the other 2s electron is promoted to the $2p_z$ orbital. The hybrid orbits are given by

$$\phi_1^i = \frac{1}{\sqrt{3}}s^i + \frac{\sqrt{2}}{\sqrt{3}}p_x^i$$

$$\phi_2^i = \frac{1}{\sqrt{3}}s^i - \frac{1}{\sqrt{6}}p_x^i + \frac{1}{\sqrt{2}}p_y^i$$

$$\phi_3^i = \frac{1}{\sqrt{3}}s^i - \frac{1}{\sqrt{6}}p_x^i - \frac{1}{\sqrt{2}}p_y^i$$

$$\phi_4^i = p_z^i,$$

$$\tag{2.3}$$

where s, p_x, p_y, and p_z are the valence orbitals before hybridization and $i = $ A and B, indicating the A and B atoms in the honeycomb lattice. The three sp^2 hybrid orbitals lie in the xy plane and form an angle of 120^0 with one another. In contrast, the $2p_z$ orbital is perpendicular to the xy plane. Due to the strong directionality of the sp^2 hybrid orbitals, they subsequently form the so-called σ bonds with the three nearest neighbor carbon atoms in the honeycomb lattice. The σ bonds are energetically stable and localized; therefore, they do not contribute to the electrical conduction. The overlap of the $2p_z$ orbitals of neighboring carbon atoms leads to the formation of π-bonds and anti-bonds (π^*), which are responsible for the high electrical conductivity of graphene.

The band structure of graphene can be calculated using the tight-binding approximation by taking into account only the p_z obitals [1, 2]. The calculation involves the construction of a wave function which is the linear combination of Bloch wave functions for A and B atoms and the use of variational principle to obtain the eigenfunctions and eigenstates. Under this framework, the single-particle electron wavefunction in a crystal can be written as

$$\Psi_{\vec{k}}^{(n)} = \sum_{\alpha i} C_{\alpha i}^{(n)} \chi_{\vec{k}\alpha i}(\vec{r}), \tag{2.4}$$

where n is the band index, $\vec{k}$ is the wave vector, α is the index of orbitals for each atom, i is the index of atoms in a primitive cell, $C_{\alpha i}^{(n)}$ is the coefficient to be determined, and $\chi_{\vec{k}\alpha i}(\vec{r})$ is a linear combination of atomic orbitals which satisfies the Bloch's theorem:

$$\chi_{\vec{k}\alpha i}(\vec{r}) = \frac{1}{\sqrt{N}} \sum_{\vec{R}_l} e^{i\vec{k}\cdot\vec{R}_l} \phi_\alpha(\vec{r} - \vec{t}_i - \vec{R}_l) \tag{2.5}$$

with the summation running over all primitive cells (N in this case) of the crystal. Here, $\vec{r}$ is the position vector, $\vec{t}_i$ is the position of atom i in a specific primitive cell, and $\vec{R}_l$ is the position of the lth primitive cell. In case of graphene, $i = $ A and B and α can be omitted because we are only interested in the p_z orbital; therefore, $\chi_{\vec{k}\alpha i}(\vec{r})$ can be written as

$$\chi_{\vec{k}A}(\vec{r}) = \frac{1}{\sqrt{N}} \sum_{\vec{R}_l} e^{i\vec{k}\cdot\vec{R}_l} p_z(\vec{r} - \vec{t}_A - \vec{R}_l) \tag{2.6}$$

for atom A and

$$\chi_{\vec{k}B}(\vec{r}) = \frac{1}{\sqrt{N}} \sum_{\vec{R}_l} e^{i\vec{k}\cdot\vec{R}_l} p_z(\vec{r} - \vec{t}_B - \vec{R}_l) \tag{2.7}$$

for atom B. The linear combination of these two gives

$$\Psi_{\vec{k}}^{(n)} = C_A^{(n)} \chi_{\vec{k}A}(\vec{r}) + C_B^{(n)} \chi_{\vec{k}B}(\vec{r}). \tag{2.8}$$

By substituting Eq. (2.8) into Schrödinger equation and minimizing the energy, one obtains the following secular equation:

$$\begin{bmatrix} H_{11} - \varepsilon_{\vec{k}}^{(n)} & H_{12} \\ H_{21} & H_{22} - \varepsilon_{\vec{k}}^{(n)} \end{bmatrix} \begin{bmatrix} C_{\vec{k}A}^{(n)} \\ C_{\vec{k}B}^{(n)} \end{bmatrix} = 0. \tag{2.9}$$

Here, $\varepsilon_{\vec{k}}^n$ is the energy and $H_{\alpha\beta}$ (α, β = 1, 2) are the interaction matrix elements. The latter can be readily obtained by taking into account the nearest neighbor interactions:

$$H_{11} = \langle \chi_{\vec{k}A} | H | \chi_{\vec{k}A} \rangle = \sum_{\vec{R}} e^{i\vec{k}\cdot\vec{R}} \langle p_z(\vec{r} - \vec{t}_A) | H | p_z(\vec{r} - \vec{t}_A - \vec{R}) \rangle = \varepsilon_{zA}$$

$$H_{22} = \langle \chi_{\vec{k}B} | H | \chi_{\vec{k}B} \rangle = \sum_{\vec{R}} e^{i\vec{k}\cdot\vec{R}} \langle p_z(\vec{r} - \vec{t}_B) | H | p_z(\vec{r} - \vec{t}_B - \vec{R}) \rangle = \varepsilon_{zB}$$

$$H_{12} = \langle \chi_{\vec{k}A} | H | \chi_{\vec{k}B} \rangle = \sum_{\vec{R}} e^{i\vec{k}\cdot\vec{R}} \langle p_z(\vec{r} - \vec{t}_A) | H | p_z(\vec{r} - \vec{t}_B - \vec{R}) \rangle$$

$$= -\gamma_0 - \gamma_0 e^{-i\vec{k}\cdot\vec{a}_1} - \gamma_0 e^{-i\vec{k}\cdot\vec{a}_2} \tag{2.10}$$

$$H_{21} = \langle \chi_{\vec{k}B} | H | \chi_{\vec{k}A} \rangle = \sum_{\vec{R}} e^{i\vec{k}\cdot\vec{R}} \langle p_z(\vec{r} - \vec{t}_B) | H | p_z(\vec{r} - \vec{t}_A - \vec{R}) \rangle$$

$$= -\gamma_0 - \gamma_0 e^{i\vec{k}\cdot\vec{a}_1} - \gamma_0 e^{i\vec{k}\cdot\vec{a}_2}.$$

Here, ε_{zA} (ε_{zB}) is the energy of p_z orbital for atom A (B) after the hybridization but without the formation of bonds with neighboring atoms and γ_0 is the hopping energy between nearest neighbor atoms. As we are only interested in the excitation spectrum, we may set ε_{zA} (ε_{zB}) = 0. The electronic band structure can be obtained by letting the determinant be zero, i.e.,

$$\det \begin{bmatrix} H_{11} - \varepsilon_{\vec{k}}^{(n)} & H_{12} \\ H_{21} & H_{22} - \varepsilon_{\vec{k}}^{(n)} \end{bmatrix} = 0. \tag{2.11}$$

Solving Eq. (2.11) gives the energy dispersion:

$$E(\vec{k}) = \pm\gamma_0 \sqrt{1 + 4\cos\left(\frac{3}{2}k_x a\right)\cos\left(\frac{\sqrt{3}}{2}k_y a\right) + 4\cos^2\left(\frac{\sqrt{3}}{2}k_y a\right)}. \tag{2.12}$$

Note that, for clarity, we have replaced $\varepsilon_{\vec{k}}^n$ by $E(k)$ since we have only two bands which are corresponding to the conduction and the valence band of electrons. In Eq. (2.12), k_x and k_y are the components of $\vec{k}$ in the (k_x, k_y) plane, $\gamma_0 = 2.75\,\text{eV}$ is the nearest-neighbor hopping energy, and plus (minus) sign refers to the upper (π^*) and lower (π) band. Figure 2.1c shows the three-dimensional electronic dispersion (left) and energy contour lines (right) in the k-space. Near the K and K' points, the energy dispersion has a circular cone shape which, to a first order approximation, is given by

$$E(\vec{k}) = \pm\hbar v_F |\vec{k}|. \tag{2.13}$$

Here $v_F = \dfrac{3\gamma_0 a}{2\hbar} \approx 10^6\,\text{ms}^{-1}$ is the Fermi velocity. Note that in Eq. (4) the wave vector $\vec{k}$ is measured from the K and K' points, respectively. This kind of energy dispersion is distinct from that of non-relativistic electrons, i.e., $E(k) = \dfrac{\hbar^2 k^2}{2m}$, where m is the mass of electrons. The linear dispersion becomes "distorted" with increasing k away from the K and K' points due to a second order term with a threefold symmetry; this is known as the trigonal warping of the electronic spectrum in literature [3–5]. The peculiarity of electrons in graphene near the K (K') points can be intuitively understood as follows. The $2p_z$ orbital of each carbon atom in the A sub-lattice interacts with the three nearest neighboring atoms in the B sub-lattice (and vice versa) to form energy bands. Although the interaction between the two atoms is strong (as manifested by the large hopping energy), the net interaction with the three nearest neighboring atoms diminishes as $\vec{k}$ approaches the K (K') points. This can be readily verified by substituting the $K(0, 4\pi/3\sqrt{3}a, 0)$ and $K'(2\pi/3a, 2\pi/3\sqrt{3}a, 0)$ points into Eqs. (2.10) and (2.12). The strong interaction with individual neighboring atoms makes it possible for electrons to move at a fast speed in graphene and the diminishing net interaction at Fermi level leads to a zero band gap. This result indicates that any honeycomb lattice consisting of same atoms will exhibit similar energy dispersion curves, and it is not necessary that one must have a carbon lattice.

2.2.2 Low-Energy Electronic Spectrum

Although the electronic band structure of graphene can be calculated by the tight-binding model, the salient features of low-energy electron dynamics in graphene are better understood by modeling the electrons as relativistic Weyl fermions (within the $\vec{k} \cdot \vec{p}$ approximation), which satisfy the 2D Dirac equations [2, 6, 7].

$$-i\hbar v_F \sigma \cdot \nabla \psi = E\psi \quad \text{(around K point)}$$
$$-i\hbar v_F \sigma^* \cdot \nabla \psi' = E\psi' \quad \text{(around K' point)}$$

(2.14)

where $\sigma = (\sigma_x, \sigma_y)$, $\sigma^* = (\sigma_x, -\sigma_y)$, $\sigma_x = \begin{bmatrix} 0 & 1 \\ 1 & 0 \end{bmatrix}$, $\sigma_y = \begin{bmatrix} 0 & -i \\ i & 0 \end{bmatrix}$,

$\psi = (\psi_A, \psi_B)$, and $\psi' = (\psi'_A, \psi'_B)$. Equation (2.14) can be solved to obtain the eigenvalues and eigenfunctions (envelope functions) as follows:

$$E_\alpha = \alpha \hbar v_F (k_x^2 + k_y^2)^{1/2}$$
$$\psi_{\alpha\beta}(\vec{k}) = \frac{1}{\sqrt{2}} \begin{pmatrix} e^{-i\beta\theta_{\vec{k}}/2} \\ \alpha e^{i\beta\theta_{\vec{k}}/2} \end{pmatrix}$$

(2.15)

where $\alpha = 1$ (-1) corresponds to the conduction and valence bands, $\beta = 1$ (-1) refers to the K and K' valley, and $\theta_k = \tan^{-1}(k_y / k_x)$ is determined by the direction of the wave vector in the k-space. Therefore, for both the valleys, the rotation of $\vec{k}$ in the (k_x, k_y) plane (surrounding K or K' point) by 2π will result in a phase change of π of the wave function (so-called Berry phase) [8, 9]. The Berry phase of π has important implications to electron transport properties, which will be discussed shortly.

The eigenfunctions are two-component spinors; therefore, low-energy electrons in graphene possess a pseudospin (with $\alpha = +(-) 1$ corresponding to the up (down) pseudospin) [10]. It is worth stressing that the pseudospin has nothing to do with the real electronic spin; the latter is an intrinsic property of electron with quantum mechanical origin, while the former is a mathematical convenience to deal with A and B atoms in graphene, which represent two intervened triangular lattices. The spinors are also the eigenfunctions of the helicity operator $\hat{h} = \frac{1}{2}\sigma \cdot \frac{\vec{p}}{|\vec{p}|}$. It is straightforward to show that $\hat{h}\psi_{\alpha\beta} = \alpha\beta\frac{1}{2}\psi_{\alpha\beta}$. Taking $\vec{n}$ as the unit vector in the momentum direction, one has $\vec{n} \cdot \sigma = 1$ for electrons and $\vec{n} \cdot \sigma = -1$ for holes, for the K valley, and the opposite applies to the K' valley [11].

The unique band structure near the K point is also accompanied by a unique energy-dependence of density of states (DOS). For a 2D system with dimension $L \times L$, each electron state occupies an area of $2\pi / L^2$ in the k-space. Therefore, the low-energy DOS of graphene can readily be found as $\dfrac{g_s g_v |E|}{2\pi \hbar^2 v_F^2}$, where g_s and g_v are the spin and valley degeneracy, respectively [1, 7, 11]. The linear energy dependence of DOS holds up to $E \approx 0.3\gamma_0$, beyond which the DOS increases sharply due to trigonal warping of the band structure at higher energy [11]. Figure 2.1 compares the basic features of the electronic band structure of graphene with that of conventional 2D electron gas system [12]. In the latter case, the electron is confined in the z direction by electrostatic potentials, leading to the quantization of k_z and thus discrete energy steps. As k_x and k_y still remain as continuous, associated with each energy step is a sub-band with a parabolic energy dispersion curve. Due to energy quantization, the DOS is now given by a sum of step functions, and between the neighboring steps the DOS is constant. In contrast, graphene is a "perfect" 2D system; therefore, there are no sub-bands emerged from the confinement in the z direction. Furthermore, the single band has a linear energy dispersion in the (k_x, k_y) plane instead of a parabolic shape as it is in the case of conventional 2D system. Note that quantum wells with a well thickness of one atomic layer have been realized in several material systems; but these systems are fundamentally different from graphene. In addition to single-layer quantum wells, ultrathin 2D sheets have also been realized in many other material systems [13]. However, these nanosheets are fundamentally different from graphene either in lattice structure or in the constituent elements. Although the linear energy dispersion or Dirac points are also found to exist in some bulk materials, in most cases, they do not play a dominant role in electrical transport; therefore, it is difficult to study electron behavior in these materials directly through electrical transport measurements.

18 | *Electronic Band Structure and Properties of Graphene*

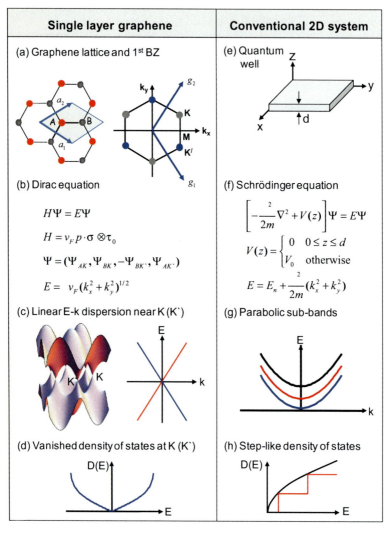

Figure 2.1 Comparison of graphene (a–d) and conventional 2D electron systems (e–h). (a) Lattice structure and first BZ; (b) Dirac equations; (c) 3D (left) and 2D (right) energy dispersions; (d) DOS as a function of energy; (e) schematic representation of a conventional 2DEG confined by electrostatic potentials in the z direction; (f) Schrödinger equation; (g) $E-K$ dispersion curves; (h) DOS as a function of energy. Adapted and modified from Wu et al. [14].

2.2.3 Effect of Magnetic Field

The difference in the behavior of graphene and particles with a parabolic spectrum is manifested when an external magnetic field is applied perpendicularly to the plane. We first look at the case of conventional 2D electron gas system (2DEGs) [12]. Let the magnetic vector potential be $\vec{A} = (-By, 0, 0)$ (Landau gauge), the Schrödinger equation is given by

$$\left(\frac{(\hat{p}_x - eB\hat{y})^2}{2m_e} + \frac{\hat{p}_y^2}{2m_e} + \frac{\hat{p}_z^2}{2m_e} + V_0(z) \right) \psi = E\psi \tag{2.16}$$

where $V_0(z)$ is the confinement electrostatic potential in z direction and m_e is the electron mass. By substituting the wave function $\psi = e^{i(k_x x + k_z z)} \phi(y)$ into Eq. (2.16), one obtains

$$\left(\frac{\hat{p}_y^2}{2m_e} + \frac{1}{2} m_e \omega_c^2 (y - y_0)^2 \right) \phi = (E - E_{zn}) \phi \mathbf{1}$$

where E_{zn} is quantized energy due to confinement in z direction and $y_0 = \frac{-\hbar k_x}{eB}$. The total quantized energy levels, or Landau levels (LLs), are given by

$$E_{nl} = \left(l + \frac{1}{2} \right) \hbar \omega_c + E_{zn} \tag{2.17}$$

where $\omega_c = eB / m_e$ is the cyclotron frequency, n (= 1, 2, 3, ...), and l (= 0, 1, 2, 3, ...) are integers and are the indices for quantization in the z direction and LLs, respectively. The area between two neighboring LLs is $\pi(k_{l+1}^2 - k_l^2) = 2m_e \pi \omega_c / \hbar$; therefore, the degeneracy of one LL is

$$p = \frac{g_s m_e \omega_c L^2}{2\pi \hbar}, \tag{2.18}$$

where g_s = 2 is the spin degeneracy. In the presence of disorder, the Hall conductivity of 2DEGs exhibits plateaus at $lh/2eB$ and is quantized as $\sigma_{xy} = \pm l \frac{2e^2}{h}$ [12], leading to the integer quantum Hall effect (IQHE) [15, 16].

Electronic Band Structure and Properties of Graphene

On the other hand, the low-energy electronic spectrum of electrons in graphene with the presence of perpendicular field is governed by

$$\hbar v_F \sigma \cdot (-i\nabla + e\vec{A}/c)\psi = E\psi \quad \text{(around K point)},$$
$$\hbar v_F \sigma^* \cdot (-i\nabla + e\vec{A}/c)\psi = E\psi \quad \text{(around K' point)}. \tag{2.19}$$

The energy of LLs has been calculated by McClure and is given by [17, 18]

$$E_l = \mathbf{sgn}(l)v_F\sqrt{2e\hbar B|l|}. \tag{2.20}$$

Here, $|l|$ = 0, 1, 2, 3, ... is the Landau index and B is the magnetic field applied perpendicular to the graphene plane. The LLs are doubly degenerate for the K and K' points. Compared with conventional 2DEGs, of particular interest is the presence of a zero-energy state at l = 0, which is shared equally by the electrons and the holes. This has led to the observation of the so-called anomalous integer quantum Hall effect, in which the Hall conductivity is given by [19, 20]

$$\sigma_{xy} = \pm 2(2l+1)\frac{e^2}{h}. \tag{2.21}$$

The measurement by Novoselov et al. [19] was performed at B = 14 T and temperature of 4 K. Instead of a plateau, a finite conductivity of $\pm 2e^2/h$ appeared at the zero-energy. The plateaus at higher energies occurred at half integers of $4e^2/h$. The result agrees well with Eq. (2.21). The l = 0 LL has also been observed in Shubnikov–de Haas oscillations at low field [19, 20], infrared spectroscopy [21, 22], and scanning tunneling spectroscopy [23–25].

2.2.4 Quantum Confinement and Tunneling

The difference in behavior between graphene and normal 2D electron system is also manifested in their response to lateral confinement by electrostatic potentials. A further confinement of 2DEGs from one of the lateral directions leads to the formation of quantum wires. For a quantum wire of size L_z and L_y in the z and y directions, the quantized energy levels are given by

$$E_{n_y, n_z} = \frac{(\hbar k_x)^2}{2m^*} + \frac{\hbar^2}{2m^*}\left(\frac{n_y\pi}{L_y}\right)^2 + \frac{\hbar^2}{2m^*}\left(\frac{n_z\pi}{L_z}\right)^2, \tag{2.22}$$

Electronic Band Structure | 21

where m^* is the effective mass, k_x is the wave vector in x direction, and n_y and n_z are integers. The corresponding density of states is given by

$$\rho(E) = \frac{\sqrt{2m^*}}{\pi\hbar} \sum_{i,j} \frac{H(E - E_{n_y,n_z})}{\sqrt{E - E_{n_y,n_z}}},$$

(2.23)

where H is the Heaviside function.

The counterpart of nanowire in graphene is the so-called graphene nanoribbon (GNR). In addition to the width, the electronic spectrum of GNR also depends on the nature of its edges, i.e., whether it has an armchair or a zigzag shape [26]. The energy dispersion of GNR can be calculated using the tight-binding method [26–29], Dirac equation [30, 31], or first principles calculations [32, 33]. All these models lead to the same general results, i.e., GNRs with armchair edges can be either metallic or semiconducting depending on their width, while GNRs with zigzag edges are metallic with peculiar edge or surface states. For GNRs with their edges parallel to x-axis and located at $y = 0$ and $y = L$, the energy spectra can be obtained by solving Eq. (2.19) with the boundary conditions: $\psi_B(y=0)=0$, $\psi_A(y=L)=0$ at point K and $\psi'_B(y=0)=0$, $\psi'_A(y=L)=0$ at point K' for zigzag ribbons and $\psi_A(y=0)=\psi_B(y=0)=\psi_A(y=L)=\psi_B(y=L)=0$ at point K and $\psi'_A(y=0)=\psi'_B(y=0)=\psi'_A(y=L)=\psi'_B(y=L)=0$ at point K' for armchair ribbons. The eigenvalue equations of the zigzag ribbons near the K point are given by [30]

$$e^{-2\alpha L} = \frac{k_x - \alpha}{k_x + \alpha} \quad \text{and} \quad k_x = \frac{k_n}{\tan(k_n L)}$$

(2.24)

where $\alpha^2 = (\hbar v_F k_x)^2 - \varepsilon^2$ for real α and $\alpha = ik_n$ for pure imaginary α, ε is the energy calculated from the Fermi level of graphene. The first equation has a real solution for α when $k_x > 1/L$, which defines a localized edge state [30]. The solution of the second equation corresponds to confined modes due to finite width of the ribbon. The eigenvalues near the K' point can be obtained by replacement, $k_x \rightarrow -k_x$ [11]. The localized edge state induces a large density of state at K and K' which are expected to play a crucial role in determining the electronic and magnetic properties of zigzag nanoribbons [26–28, 34]. In contrast, there are no localized edge states in armchair GNRs. The wave vector across the ribbon width direction is quantized by

$$k_n = \frac{n\pi}{L} - \frac{4\pi}{3\sqrt{3}a}$$ and the energy is given by $\varepsilon = \pm \hbar v_F \left[k_x^2 + k_n^2 \right]^{1/2}$

[11]. Here, n is an integer. The armchair nanoribbons will be metallic when $L = 3\sqrt{3}na/4$ and semiconducting in other cases.

Although the chiral electrons in graphene can be effectively confined in nanoribbons through the boundaries, they cannot be confined effectively by electrostatic potential barriers in the same graphene. For a one-dimensional potential barrier of height V_0 and width D in the x direction, the transmission coefficient of quasi-particles in graphene is given by [11, 35]

$$T(\phi) = \frac{\cos^2(\theta)\cos^2(\phi)}{[\cos(Dq_x)\cos\phi\cos\theta]^2 + \sin^2(Dq_x)(1 - ss'\sin\phi\sin\theta)^2} \quad (2.25)$$

where $q_x = \sqrt{(V_0 - E)^2/(\hbar v_F)^2 - k_y^2}$, E is the energy, k_y is the wave vector in y direction, $\phi = \tan^{-1}\frac{k_y}{k_x}$, and $\theta = \tan^{-1}\frac{k_y}{q_x}$. The transmission coefficient becomes unity (i) when $Dq_x = n\pi$ with n as an integer, independent of the incident angle and (ii) at normal incidence, i.e., $\phi = 0$. In these two cases, the barrier becomes completely transparent, which is the manifestation of Klein tunneling [6, 35]. Stander et al. have found evidence of Klein tunneling in a steep gate-induced potential step, which is in quantitative agreement with the theoretical predictions [36]. Signature of perfect transmission of carriers normally incident on an extremely narrow potential barrier in graphene was also observed by Young and Kim [37]. Very recently, Klein tunneling was also observed in ultraclean carbon nanotubes with a small bandgap [38]. On the other hand, Dragoman has shown that both the transmission and reflection coefficients at a graphene step barrier are positive and less than unity [39]; therefore it does not support the particle–antiparticle pair creation mechanism predicted by the theory. Further concrete evidence is required to verify the Klein paradox in graphene system.

Figure 2.2 summarizes graphene and normal electron systems under an external magnetic field, in ribbon and wire form. and with a 1D potential barrier. The fundamental properties of graphene summarized in Figs. 2.1 and 2.2 lead to the peculiar electronic, magnetic, and optical properties. In the following text, we give an overview of electrical transport properties that have more experimental results to support the theoretical predictions.

Electrical Transport Properties | 23

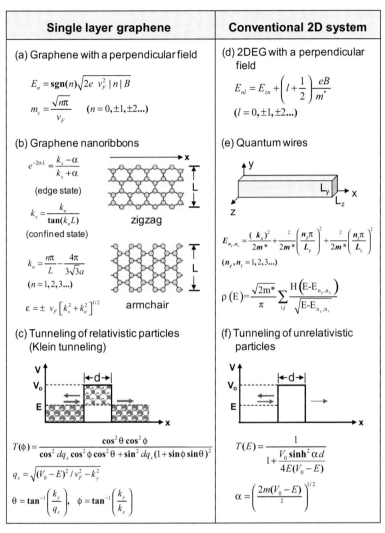

Figure 2.2 Comparison of graphene and normal electron systems under an external magnetic field (a and d), in ribbon and wire form (b and e), and with a 1D potential barrier (c and f).

2.3 Electrical Transport Properties

Due to its unique band structure (linear E–k dispersion, point-shaped Fermi surface, and chiral nature of electrons and holes), graphene exhibits several peculiar electronic properties that are absent in conventional 2DEGs [11, 12]. Among those which have been investigated most intensively include weak (weak anti-) localization [8, 19, 40–43], minimum conductivity [19, 20, 43–46], carrier density dependence of conductivity [47–51], and Klein tunneling [6, 35, 36]. In what follows, we briefly discuss some of these aspects.

2.3.1 Weak (Weak Anti-) Localization

In a weakly disordered system, there are generally two types of scattering events that affect the electron transport processes: elastic and inelastic scattering. In the former case, the electron energy does not change; therefore, its phase evolvement can be traced. In the latter case, however, the electron "forgets" its phase after scattering. The probability for electron to lose its phase memory is the inverse of the phase relaxation time τ_ϕ. When $\tau_\phi \gg \tau$, where τ is the momentum relaxation time, quantum interference between self-returned and multiply scattered paths of electrons on the scale of phase coherence length, $L_\phi = v_F \tau_\phi$, leads to quantum interference corrections (QIC) to the electrical resistance, which manifests itself in the form of weak localization (WL) [52, 53]. In 2D disordered metals, the quantum correction to conductivity is given by $\Delta\sigma_{2D} = -\dfrac{2e^2}{h}\ln\left(\dfrac{L_\phi}{\lambda}\right)$, where λ is the mean-free path. An applied magnetic field starts to break the WL at $B > B_\phi = \dfrac{\hbar}{eL_\phi^2}$ due to the additional loop area dependent phase acquired by the electrons traveling in different directions. Therefore, the WL is usually accompanied with a negative magnetoresistance (MR) effect. In addition to an external magnetic field, the WL can also be destroyed by scattering with magnetic impurities and strong spin–orbit coupling, which flips the spins along the path of electron transport.

Due to the relativistic and chiral nature of electrons in graphene, the WL in this perfect 2D system is expected to be affected by not only inelastic and spin-flip processes but also by a number of elastic scattering processes [40, 41, 43]. In graphene, the envelope wave function of electrons around the K point is given by Eq. (2.15), i.e., $\psi(\vec{k}) = \dfrac{1}{\sqrt{2}}\begin{pmatrix} e^{-i\theta_{\vec{k}}/2} \\ e^{i\theta_{\vec{k}}/2} \end{pmatrix}$, where,

$\theta_{\vec{k}} = \tan^{-1}\left(\dfrac{k_y}{k_x}\right)$. The overlapping between wave functions

$\theta_{\vec{k}} = \tan^{-1}\left(\dfrac{k_y}{k_x}\right)$ and $\psi(0)$ is $\left|\langle\psi(\theta_{\vec{k}})|\psi(0)\rangle\right|^2 = \cos^2(\theta_{\vec{k}}/2)$,

leading to a suppression of intra-valley backscattering (long-range scatters) or the appearance of weak (weak anti-) localization (WAL) [8, 40] The WL will be restored by both inter-valley and intra-valley scatterings. If the former is dominant, whether a WL or a WAL will be observed in an actual graphene sample depends strongly on the ratio between two characteristic times: τ_ϕ and the inter-valley scattering time, τ_{iv}. The WL is expected to occur when $\tau_\phi \gg \tau_{iv}$, and WAL occurs when $\tau_\phi \ll \tau_{iv}$ [40]. The inter-valley scattering can be induced by atomically sharp defects or edges in narrow ribbons. As it has been shown by McCann et al. [41] and Morpurgo and Guinea [43], the phase coherence time τ_ϕ and inter-valley scattering time τ_{iv} are not the only parameters that determine the quantum transport in graphene. The quantum interference within each valley can be affected by trigonal warping and scattering that breaks the chirality of electrons. Such scattering centers include long-range distortions induced by lattice disclinations and dislocations, non-planarity of the graphene layers, and slowly varying random electrostatic potentials that break the symmetry between the two sub-lattices of graphene. All these types of defects are realistically present in real graphene samples; therefore, large differences in the quantum correction to the conductivity measured on different samples should be expected [43]. Yan and Ting studied the weak localization effect on graphene under the presence of charged impurities using the self-consistent Born approximation [54]. This model is considered more realistic than the zero-range potential model. It was found that the QIC to conductivity is dependent on sample size, carrier concentration, and temperature. The WL is present in large size samples at finite carrier doping and its strength becomes weakened or quenched in a wide temperature range when the sample is below a certain critical size (about a few μm at low temperature). Near the zero-doping region, the QIC becomes mostly positive regardless of the sample size, indicating that the electrons become delocalized.

The suppression of weak localization was observed in the very first experiment on graphene by Novoselov et al. [19]. Subsequently, Morozov et al. measured the MR of single-layer graphene flakes of several microns in size placed on the top of SiO_2 (300 nm)/ Si substrate. The negative MR measured was typically 2 orders of magnitude smaller than that expected for metallic samples having a similar range of resistivity, indicating a strong suppression of WL. The authors ruled out both the short phase-breaking length and the magnetic impurities as possible mechanisms for the WL suppression, and instead they attributed the unexpected behavior to the existence of mesoscopic corrugations in graphene sheets that induce a nominal random magnetic field. Wu et al. have observed WAL in epitaxial graphene grown on the carbon rich SiC (0001) surface [55].

Tikhonenko et al. have shown that the WL in graphene exists in a large range of carrier density, including the Dirac region [56]. The authors attributed this to the significant inter-valley scattering. It is argued that total suppression of WL is only possible in experiments where inter-valley scattering is negligible, i.e., in very large samples without sharp defects in the bulk. Similar results have also been observed in bilayer graphene, i.e., the WL is observed at different carrier densities including the Dirac point [57]. In a recent paper from the same group, it was shown that transition between WL and WAL can occur in the same sample depending on the measurement conditions. The WAL prevails over the WL at high temperature and low carrier density. The results are in good agreement with the theoretical predictions [41]. Pezzini et al. studied WL and WAL in graphene at temperatures between 0.3 K and 15 K. A transition from WL to WAL with increasing magnetic field was observed at low temperature, while at high carrier density the WAL was suppressed due to trigonal warping of the conical energy bands [58].

2.3.2 Electrical Conductivity and Mobility

Although the density of states of graphene at the Dirac point is zero, it exhibits a minimum conductivity of the order of e^2/h even at the lowest temperature possible [19, 20]. Away from the Dirac point, it was found that the graphene conductivity is linear in the concentration of carriers (subtracting the residual carriers at half filling) [19, 20]. Miao et al. measured the conductivity of graphene at the Dirac point on samples with different width (W) to length (L) ratio and surface areas (A) [44]. It was found that for devices with relatively large length ($L > \sim 1$ μm) and large area ($A > 3$ μm^2), the values of minimum conductivity were geometry-independent and relatively constant: ~3.3 to $4.7{\times}4e^2/\pi h$. For "small" devices with $L < 500$ nm and $A < 0.2~\mu m^2$, a qualitatively different behavior was observed, depending on the aspect ratio W/L. The minimal conductivity decreased from ~$4{\times}4e^2/\pi h$ at $W/L = 1$ to ~$4e^2/\pi h$ at $W/L = 4$, beyond which it saturated at this value. Similar results have also been observed by Danneau et al. on samples with large W/L ratios and small L (= 200 nm) [45]. In addition, a finite and gate dependent Fano factor reaching the universal value of 1/3 was also observed at the Dirac point, which supports the transport via evanescent waves theory [59].

These results agree well with the theoretical predictions that, in the ballistic regime, the minimal conductivity depends on the graphene's geometry and the microscopic details of the edges, approaching the value of $4e^2/\pi h$ when boundary effects are negligible, i.e., in samples with a large W/L ratio [59]. These theoretical models predict that in perfect graphene (i.e., at the clean limit) and at the Dirac point, the electrical conduction occurs only via evanescent waves, i.e., via tunneling between the electrical contacts [59, 60]. As it has been stated recently by Ziegler [46], depending on whether the Kubo formula or Landauer formula or both of them are used, the theoretically calculated value of minimal conductivity varies from $e^2/\pi h$ [59–66] to $\pi e^2/8h$ [61, 64] and $\pi e^2/4h$ [67, 68] per valley and per spin channel. Ziegler showed that all these values can be obtained from the standard Kubo formula of nearly ballistic quasiparticles by taking limits in different order [46]. Various models have been proposed to account for the difference between theoretical and experimental values of minimal conductivity.

For samples which are not at the clean limit, the minimal conductivity is affected by scattering associated with several different types of scattering centers such as impurities, defects, and phonons [47]. In addition to these conventional scattering centers, ripples also affect electrical transport in graphene. Both the ripples and the charged impurities in the substrate on which the graphene is placed are known to induce electron–hole puddles at low carrier concentration [47, 48]. These puddles have been observed experimentally for graphene samples on SiO_2/Si substrates with a characteristic dimension of approximately 20 −30 nm [69, 70]. From Einstein relation between conductivity and compressibility, a minimal conductivity of the order of $4e^2/h$ is deduced at the Dirac point, which is π times higher than that of the minimal conductivity at the clean limit. Chen et al. have investigated the effect of doping on the conductivity of graphene through controlled doping of potassium in ultrahigh vacuum [71]. It was found that the minimal conductivity decreases only slightly with increasing the doping concentration, although there is a significant decrease in mobility. These results suggest that charge inhomogeneity is responsible for the minimal conductivity obtained experimentally. The former is considered being caused by the charged impurities either inside the substrate or in the vicinity of graphene.

The charged impurities are also responsible for the linear dependence of conductivity on the carrier concentration away from half-filling [47–51]. Ostrovsky et al. showed that the transport properties of a system depend strongly on the character of disorder; both the strength and the type of disorder play an important role in determining the conductivity [65]. Away from the Dirac point, the conductivity exhibits a linear relationship with the carrier concentration in case of strong scatters, while a logarithmic relationship is found for case of weak scatters. Ando demonstrated that the conductivity of graphene limited by charged-impurity scattering increases linearly with the electron concentration and the mobility remains independent of the Fermi energy [49]. It is also shown that the increase of screening with temperature at sufficiently high temperatures leads to the mobility increase proportional to the square of temperature. Hwang et al. developed a detailed microscopic transport theory for graphene by assuming that charged impurities in the substrate are the dominant source of scattering [48]. It was shown that away from the Dirac point and at high carrier density, the electrical transport can be accounted for well by the Boltzmann transport theory, which results in a conductivity that scales linearly with n/n_i, where n is the carrier density and n_i is the impurity distributed randomly near the graphene/substrate interface. For samples with either a large carrier density or low charged impurity concentration, short range scattering by point defects or dislocations would dominate the transport, which leads to sublinear σ–n curves. The theoretical models explain well most of the experimental observations [19, 48, 71, 72].

Removing substrate or using high-κ dielectrics are two possible ways to reduce the scattering from charged impurities [73–78]. From a suspended graphene sheet, Du et al. obtained a mobility value as high as 200,000 $cm^2V^{-1}s^{-1}$ for carrier densities below 5×10^9 cm^{-2}

[73]. The minimum conductivity at low temperature was found to be $1.7 \times 4e^2/\pi h$ for a sample with $L = 0.5$ µm and $W = 1.4$ µm, which is higher than the theoretical value of $4e^2/\pi h$ for ballistic transport at the clean limit. Nevertheless, the sharp change of conductivity with bias voltage suggests that the electrical transport in short and suspended graphene sheets approaches the ballistic regime. Bolotin et al. have investigated the effect of impurity absorbed on the surface of suspended graphene on its electrical transport properties [74]. It was found that, for "dirty" samples, the mobility was low ($28,000$ cm^{2}V^{-1}s^{-1}) even when it was suspended from the substrate. However, the mobility increased significantly after the sample was cleaned in situ in UHV so as to obtain an ultraclean graphene. For these samples, a mobility as high as $170,000$ cm^{2}V^{-1}s^{-1} has been obtained below 5 K. The resistivity of ultraclean graphene is found to be strongly dependent on temperature in the temperature range of 5–240 K. At large carrier densities, $n > 0.5 \times 10^{11}$ cm^2, the resistivity increases with increasing the temperature and becomes linear with temperature above 50 K, suggesting that scattering from acoustic phonons dominates the electrical transport in ultraclean samples. One can estimate the carrier density inhomogeneity from the temperature-dependence of non-universal conductivity at the charge neutral point, which turned out to be $<10^8$ cm^{-2}.

If the enhancement of mobility in suspended graphene is due to the removal of charged impurities from the substrate, different values of mobility would be obtained by replacing SiO_2 with other dielectrics. To this end, Ponomarenko et al. have studied graphene devices placed on a number of different substrates, including SiO_2, polymethylmethacrylate, spin-on glass, bismuth strontium calcium copper oxide, mica, and boron nitride [77]. But the mobility found was almost the same as that of typical graphene devices placed on SiO_2. Similarly, only a small change in mobility ($<30\%$) has been obtained by covering the device with glycerol ($\kappa \approx 45$), ethanol ($\kappa \approx 25$), or water ($\kappa \approx 80$). Further studies are required to understand the different results obtained in suspended samples and samples with different dielectric environment.

2.4 Summary

In this chapter, we have briefly discussed the electronic band structure of graphene by comparing it with that of conventional two-dimensional electron gases. The key characteristics which distinguish graphene from conventional 2DEGs are its point-shaped Fermi surface (Dirac points), linear energy dispersion near the Dirac points, and chirality of the quasi-particles. All these contribute to the appearance of peculiar electronic properties that are not found in other material systems.

References

1. Wallace P.R. (1947) The band theory of graphite, *Phys Rev*, **71**, 622.
2. Slonczewski J.C., Weiss P.R. (1958) Band structure of graphite, *Phys Rev*, **109**, 272.
3. Sugihara K., Ono S., Oshima H., Kawamura K., Tsuzuku T. (1982) Hall-effect in graphite and its relation to the trigonal warping of the energy-bands. 2. Theoretical, *J Phys Soc Jpn*, **51**, 1900–1903.
4. Ajiki H., Ando T. (1996) Energy bands of carbon nanotubes in magnetic fields, *J Phys Soc Jpn*, **65**, 505–514.
5. Saito R., Dresselhaus G., Dresselhaus M.S. (2000) Trigonal warping effect of carbon nanotubes, *Phys Rev B*, **61**, 2981–2990.
6. Beenakker C.W.J. (2008) Colloquium: Andreev reflection and Klein tunneling in graphene, *Rev Mod Phys*, **80**, 1337–1354.
7. Ando T. (2005) Theory of electronic states and transport in carbon nanotubes, *J Phys Soc Jpn*, **74**, 777–817.
8. Ando T., Nakanishi T., Saito R. (1998) Berry's phase and absence of back scattering in carbon nanotubes, *J Phys Soc Jpn*, **67**, 2857–2862.
9. Berry M.V. (1984) Quantal phase factors accompanying adiabatic changes, *Proc R Soc London Ser A—Math Phys Eng Sci*, **392**, pp. 45–57.
10. McEuen P.L., Bockrath M., Cobden D.H., Yoon Y.G., Louie S.G. (1999) Disorder, pseudospins, and backscattering in carbon nanotubes, *Phys Rev Lett*, **83**, 5098–5101.
11. Neto A.H.C., Guinea F., Peres N.M.R., Novoselov K.S., Geim A.K. (2009) The electronic properties of graphene, *Rev Mod Phys*, **81**, 109–162.
12. Ando T., Fowler A.B., Stern F. (1982) Electronic properties of two-dimensional systems, *Rev Mod Phys*, **54**, 437–672.
13. Osada M., Sasaki T. (2009) Exfoliated oxide nanosheets: New solution to nanoelectronics, *J Mater Chem*, **19**, 2503–2511.
14. Wu Y.H., Wang Y., Wang J.Y., et al. (2012) Electrical transport across metal/two-dimensional carbon junctions: Edge versus side contacts, *AIP Advances*, **2**, 012132.

15. Vonklitzing K., Dorda G., Pepper M. (1980) New method for high-accuracy determination of the fine-structure constant based on quantized hall resistance, *Phys Rev Lett*, **45**, 494–497.

16. von Klitzing K. (1986) The quantized hall effect, *Rev Mod Phys*, **58**, 519.

17. McClure J.W. (1960) Theory of diamagnetism of graphite, *Phys Rev*, **119**, 606.

18. McClure J.W. (1956) Diamagnetism of graphite, *Phys Rev*, **104**, 666.

19. Novoselov K.S., Geim A.K., Morozov S.V., et al. (2005) Two-dimensional gas of massless Dirac fermions in graphene, *Nature*, **438**, 197–200.

20. Zhang Y.B., Tan Y.W., Stormer H.L., Kim P. (2005) Experimental observation of the quantum Hall effect and Berry's phase in graphene, *Nature*, **438**, 201–204.

21. Deacon R.S., Chuang K.C., Nicholas R.J., Novoselov K.S., Geim A.K. (2007) Cyclotron resonance study of the electron and hole velocity in graphene monolayers, *Phys Rev B*, **76**, 081406.

22. Sadowski M.L., Martinez G., Potemski M., Berger C., de Heer, W.A. (2006) Landau level spectroscopy of ultrathin graphite layers, *Phys Rev Lett*, **97**, 266405.

23. Li G., Andrei E.Y. (2007) Observation of Landau levels of Dirac fermions in graphite, *Nat Phys*, **3**, 623–627.

24. Miller D.L., Kubista K.D., Rutter G.M., et al. (2009) Observing the quantization of zero mass carriers in graphene, *Science*, **324**, 924–927.

25. Li G.H., Luican A., Andrei E.Y. (2009) Scanning tunneling spectroscopy of graphene on graphite, *Phys Rev Lett*, **102**, 176804.

26. Nakada K., Fujita M., Dresselhaus G., Dresselhaus M.S. (1996) Edge state in graphene ribbons: Nanometer size effect and edge shape dependence, *Phys Rev B*, **54**, 17954–17961.

27. Fujita M., Wakabayashi K., Nakada K., Kusakabe K. (1996) Peculiar localized state at zigzag graphite edge, *J Phys Soc Jpn*, **65**, 1920–1923.

28. Wakabayashi K., Fujita M., Ajiki H., Sigrist M. (1999) Electronic and magnetic properties of nanographite ribbons, *Phys Rev B*, **59**, 8271–8282.

29. Ezawa M. (2006) Peculiar width dependence of the electronic properties of carbon nanoribbons, *Phys Rev B*, **73**,. 045432.

30. Brey L., Fertig H.A. (2006) Electronic states of graphene nanoribbons studied with the Dirac equation, *Phys Rev B*, **73**, 235411.

31. Sasaki K.I., Murakami S., Saito R. (2006) Gauge field for edge state in graphene, *J Phys Soc Jpn*, **75**, 074713.

32. Son Y.W., Cohen M.L., Louie S.G. (2006) Energy gaps in graphene nanoribbons, *Phys Rev Lett*, **97**, 216803.

33. Yang L., Park C.H., Son Y.W., Cohen M.L., Louie S.G. (2007) Quasiparticle energies and band gaps in graphene nanoribbons, *Phys Rev Lett*, **99**, 186801.

34. Son Y.W., Cohen M.L., Louie S.G. (2006) Half-metallic graphene nanoribbons, *Nature*, **444**, 347–349.

35. Katsnelson M.I., Novoselov K.S., Geim A.K. (2006) Chiral tunnelling and the Klein paradox in graphene, *Nat Phys*, **2**, 620–625.

36. Stander N., Huard B., Goldhaber-Gordon D. (2009) Evidence for Klein tunneling in graphene p–n junctions, *Phys Rev Lett*, **102**, 026807.

37. Young A.F., Kim P. (2009) Quantum interference and Klein tunnelling in graphene heterojunctions, *Nat Phys*, **5**, 222–226.

38. Steele G.A., Gotz G., Kouwenhoven L.P. (2009) *Nat. Nanotechnol.*, **4**, 363.

39. Dragoman D. (2009) Evidence against Klein paradox in graphene, *Phys Scripta*, **79**, 015003.

40. Suzuura H., Ando T. (2002) Crossover from symplectic to orthogonal class in a two-dimensional honeycomb lattice, *Phys Rev Lett*, **89**, 266603.

41. McCann E., Kechedzhi K., Fal'ko V.I., Suzuura H., Ando T., Altshuler B.L. (2006) Weak-localization magnetoresistance and valley symmetry in graphene, *Phys Rev Lett*, **97**, 146805.

42. Morozov S.V., Novoselov K.S., Katsnelson M.I., et al. (2006) Strong suppression of weak localization in graphene, *Phys Rev Lett*, **97**, 016801.

43. Morpurgo A.F., Guinea F. (2006) Intervalley scattering, long-range disorder, and effective time-reversal symmetry breaking in graphene, *Phys Rev Lett*, **97**, 196804.

44. Miao F., Wijeratne S., Zhang Y., Coskun U.C., Bao W., Lau C.N. (2007) Phase-coherent transport in graphene quantum billiards, *Science*, **317**, 1530–1533.

45. Danneau R., Wu F., Craciun M.F., et al. (2008) Shot noise in ballistic graphene, *Phys Rev Lett*, **100**, 196802.

46. Ziegler K. (2007) Minimal conductivity of graphene: Nonuniversal values from the kubo formula, *Phys Rev B*, **75**, 233407.

47. Adam S., Hwang E.H., Galitski V.M., Das Sarma S. (2007) A self-consistent theory for graphene transport, *Proc Natl Acad Sci USA,* **104**, 18392–18397.

48. Hwang E.H., Adam S.,Das Sarma S. (2007) Carrier transport in two-dimensional graphene layers, *Phys Rev Lett*, **98**, 186806.

49. Ando T. (2006) Screening effect and impurity scattering in monolayer graphene, *J Phys Soc Jpn*, **75**,074716.

50. Nomura K., MacDonald A.H. (2007) Quantum transport of massless Dirac fermions, *Phys Rev Lett*, **98**, 076602.

51. Cheianov V.V., Fal'ko V.I. (2006) Friedel oscillations, impurity scattering, and temperature dependence of resistivity in graphene, *Phys Rev Lett*, **97**, 226801.

52. V. F. Gantmakher and Lucia I. Man (2005) *Electrons and Disorder in Solids*, Pg.16–41, Clarendon Press, Oxford.

53. Imry Y. (2001) *Introduction to Mesoscopic Physics*, 2nd ed, Pg. 84–115, Oxford University Press, New York.

54. Yan X.Z., Ting C.S. (2008) Weak localization of Dirac fermions in graphene, *Phys Rev Lett*, **101**, 126801.

55. Wu X.S., Li X.B., Song Z.M., Berger C., de Heer W.A. (2007) Weak anti-localization in epitaxial graphene: Evidence for chiral electrons, *Phys Rev Lett*, **98**, 136801.

56. Tikhonenko F.V., Horsell D.W., Gorbachev R.V., Savchenko A.K. (2008) Weak localization in graphene flakes, *Phys Rev Lett*, **100**, 056802.

57. Gorbachev R.V., Tikhonenko F.V., Mayorov A.S., Horsell D.W., Savchenko A.K. (2007) Weak localization in bilayer graphene, *Phys Rev Lett*, **98**, 176805.

58. Pezzini S., Cobaleda C., Diez E., Bellani V. (2012) Quantum interference corrections to magnetoconductivity in graphene, *Phys Rev B*, **85**, 165451.

59. Tworzydlo J., Trauzettel B., Titov M., Rycerz A., Beenakker C.W.J. (2006) Sub-poissonian shot noise in graphene, *Phys Rev Lett*, **96**, 246802.

60. Katsnelson M.I. (2006) Zitterbewegung, chirality, and minimal conductivity in graphene, *Eur Phys J B*, **51**, 157–160.

61. Ludwig A.W.W., Fisher M.P.A., Shankar R., Grinstein G. (1994) Integer quantum Hall transition: An alternative approach and exact results, *Phys Rev B*, **50**, 7526–7552.

62. Ziegler K. (1998) Delocalization of 2D Dirac fermions: The role of a broken supersymmetry, *Phys Rev Lett*, **80**, 3113–3116.

63. Peres N.M.R., Guinea F., Neto A.H.C. (2006) Electronic properties of disordered two-dimensional carbon, *Phys Rev B*, **73**, 125411.

64. Cserti J. (2007) Minimal longitudinal dc conductivity of perfect bilayer graphene, *Phys Rev B*, **75**, 033405.

65. Ostrovsky P.M., Gornyi I.V., Mirlin A.D. (2006) Electron transport in disordered graphene, *Phys Rev B*, **74**, 235443.

66. Ryu S., Mudry C., Furusaki A., Ludwig A.W.W. (2007) Landauer conductance and twisted boundary conditions for Dirac fermions in two space dimensions, *Phys Rev B*, **75**, 205344.

67. Ziegler K. (2006) Robust transport properties in graphene, *Phys Rev Lett*, **97**, 266802.

68. Gusynin V.P., Sharapov S.G. (2005) Unconventional integer quantum Hall effect in graphene, *Phys Rev Lett*, **95**, 146801.

69. Martin J., Akerman N., Ulbricht G., et al. (2008) Observation of electron–hole puddles in graphene using a scanning single-electron transistor, *Nat Phys*, **4**, 144–148.

70. Zhang Y.B., Brar V.W., Girit C., Zettl A., (2009) Crommie M.F., Origin of spatial charge inhomogeneity in graphene, *Nature Physics*, **5**, 722–726.

71. Chen J.H., Jang C., Adam S., Fuhrer S., Williams E.D., Ishigami M. (2008) Charged-impurity scattering in graphene, *Nat Phys*, **4**, 377–381.

72. Tan Y.W., Zhang Y., Bolotin K., et al. (2007) Measurement of scattering rate and minimum conductivity in graphene, *Phys Rev Lett*, **99**, 246803.

73. Du X., Skachko I., Barker A., Andrei E.Y. (2008) Approaching ballistic transport in suspended graphene, *Nat Nanotechnol*, **3**, 491–495.

74. Bolotin K.I., Sikes K.J., Hone J., Stormer H.L., Kim P. (2008) Temperature-dependent transport in suspended graphene, *Phys Rev Lett*, **101**, 096802.

75. Stauber T., Peres N.M.R., Net A.H.C. (2008) Conductivity of suspended and non-suspended graphene at finite gate voltage, *Phys Rev B*, **78**, 085418.

76. Ni Z.H., Yu T., Luo Z.Q., et al. (2009) Probing charged impurities in suspended graphene using Raman spectroscopy, *ACS Nano*, **3**, 569–574.

77. Ponomarenko L.A., Yang R., Mohiuddin T.M., et al. (2009) Effect of high-k environment on charge carrier mobility in graphene, *Phys Rev Lett*, **102**, 206603.

78. Bolotin K.I., Sikes K.J., Jiang Z., et al. (2008) Ultrahigh electron mobility in suspended graphene, *Solid State Commun*, **146**, 351–355.

Chapter 3

Growth of Epitaxial Graphene on SiC

Xiaosong Wu

State Key Laboratory for Artificial Microstructure and Mesoscopic Physics,
Peking University, 5 Yiheyuan Rd, Haidian, Beijing 100871, China
xswu_at_pku.edu.cn

Applications of graphene in electronics demand high-mobility, high-uniformity and wafer-size materials. Epitaxial graphene (EG) grown on SiC provides a viable way to meet this requirement. Consequently, it has attracted great interest. After extensive studies, its structural and electronic properties are basically known. However, the growth mechanism, in particular, the growth kinetics, is less understood. Such understanding is crucial for improving the growth and technologically important. We review recent progresses made in the past few years for epitaxial graphene growth on silicon carbide, with emphasis on the growth mechanism and kinetics. By analysing common surface morphological features, the basic growth mechanism is inferred. Growth under different local pressure conditions indicates that silicon desorption is an essential process and it can be used to control the growth. The effect of the initial surface condition is discussed. Furthermore, it is shown that other processes, such as carbon diffusion and SiC recombination have important influences

---

Two-Dimensional Carbon: Fundamental Properties, Synthesis, Characterization, and Applications
Edited by Yihong Wu, Zexiang Shen, and Ting Yu
Copyright © 2014 Pan Stanford Publishing Pte. Ltd.
ISBN 978-981-4411-94-3 (Hardcover), 978-981-4411-95-0 (eBook)
www.panstanford.com

Growth of Epitaxial Graphene on SiC

on the growth and thus deserve more studies. Macroscopic island growth found on the C-face of SiC manifests another growth mode and its implication is discussed.

3.1 Introduction

3.1.1 Graphene

Owing to its exceptional properties, graphene have drawn enormous attention from researchers in various fields, shortly after its free-standing form was made available [62] and its potential electronic application was realized [5]. Among these properties many are the extreme. Graphene is a two-dimensional network of carbon atoms arranging in a honeycomb lattice. Each carbon atom covalently connects with three nearest neighbours by sp^2 bonding, one of the strongest chemical bonds in nature. The strong bonding makes graphene the strongest material ever measured, yet flexible because of its two-dimensionality. The measured Young's modulus of graphene is as high as 1 TPa [46]. The strong covalent bond, combined with the low atomic mass of carbon, gives rise to a very high speed of sound, which suggests a good ability to conduct heat. In addition, the two-dimensionality of graphene further enhances its thermal conductivity. A recently experiment show that its thermal conductivity is ~5000 W/mK [4], well surpassing the best bulk thermal conductor, diamond, whose thermal conductivity is about 2000 W/mK. Graphene is one atom thick, which gives it a large aspect ratio. Because it is thin, it is fairly transparent and absorbs only 2.3% of light [58]. Moreover, graphene is an extraordinary electrical conductor. It has the highest mobility at room temperature [9,57]. It is an ideal material for spintronics owing to its small spin-orbit interaction. It is the first material having demonstrated room temperature spin transport in a scale of micrometres [90]. Furthermore, because of its very peculiar band structure, electrons in graphene behave like Dirac fermions and obey chiral symmetry, which leads to many exotic effects, such as half integer quantum Hall effect [61,100], Klein tunnelling [43,99], weak anti-localization [97], Veselago lens for electrons [15] and fractional quantum Hall effect [8,23], etc. Thus, it provides an excellent playground for condensed matter physicists.

Currently, there are mainly four methods for obtaining graphene: Scotch tape method, chemical exfoliation, chemical vapour deposition (CVD) on metal surfaces and surface thermal decomposition of SiC. The Scotch type method is to mechanically exfoliate graphene flakes from graphite [62]. This method has been the most commonly used one due to the simplicity and the capability of producing high-quality samples. Although majority of studies on graphene employed this method, the inability of scaling up limits its use in future applications. Chemical exfoliation uses various chemical reactions to intercalate and exfoliate graphite [70]. It has the capacity of producing a large amount of material at low cost and may find its application in flexible electronics, supercapacitor, etc. CVD growth on metal surfaces produces uniform monolayer graphene in wafer size [3,49]. However, the film needs to be transferred to a semiconductor or insulator substrate. So far, the mobility of CVD grown graphene is up to 5000 $cm^2/V \cdot s$. Improvements on growth and transfer are needed to further increase the mobility. Graphene grown on SiC can have very high mobility. Wafer size growth has been realized [24]. Integrated circuitry has been demonstrated [50,51]. It shows great promise in future electronics applications.

3.1.2 Epitaxial Graphene on SiC

Graphene grown on SiC is considered as epitaxy, because its orientation either is strictly registered to the SiC substrate or displays certain preference, which will be explained below. So, it is commonly referred to as epitaxial graphene. Growth of EG on SiC can be traced back to the pioneer work of van Bommel et al. in 1975 [93]. Interest in the material had mainly come from the surface science community and little attention had been paid to the electronic properties of the material. The situation has changed since de Heer demonstrated that graphene grown on SiC has remarkable electrical properties, hence potential application in future electronics in 2004 [5].

SiC has over 200 polytypes. Most commonly used in EG growth are hexagonal SiC, i.e. 4H and 6H. Both can be seen as C-Si bilayers stacked in different arrangements. 4H-SiC is ABCB..., f o u r bilayers per unit cell, while 6H-SiC is ABCACB, 6 bilayers per unit cell. Based on the areal density of carbon atoms for graphene and a SiC bilayer, 3.14 SiC bilayers are needed to form a monolayer graphene. The stoichiometry requirement leads to suspicion that

6H-SiC, whose half unit cell consists of 3 bilayers, is more suitable for the growth. However, so far there is no evidence that there is a substantial difference between graphene grown on two polytypes. Consequently, we do not particularly underline the type of substrate in this review. On the other hand, we want to emphasize that SiC has two polar surfaces perpendicular to c-axis. One is SiC(0001) silicon-terminated face, called Si-face, while the other is SiC(0001) carbon-terminated face, called C-face. Significant differences in the growth and the structure of graphene exist for two polar surfaces. As a result, we divide the growth into two categories according to the polar surface, elucidated in Sections 3.2 and 3.3, respectively.

When the SiC substrate is heated to an elevated temperature, between 1150°C to 2000°C, depending on the growth condition, silicon desorption occurs and the liberated carbon reconstructs into graphene. Since the source of carbon comes from the substrate, in the multilayer case, new layers form underneath the existing ones [10,25,27,38,41,56]. The implication of this growth mode is that the growth is intrinsically self-limited. Once the first layer of graphene is grown, the following layers are hard to form because the first layer covers the SiC substrate and blocks the escape path for Si atoms. Particularly, the better crystal perfectness the first layer has, the better chance one would get a monolayer graphene.

EG on SiC has been extensively studied by variety of surface characterization tools. Its structural and electronic properties are understood. To briefly summarize, in the case of Si-face, on top of the substrate is a buffer layer, which is basically a defective graphene layer covalently bonded to the substrate [64]. It is also called the zeroth graphene layer because it does not have the band structure of graphene. The first layer graphene is on top of the buffer layer, behaving like a free-standing graphene. It is rotated by 30° with respect to the substrate. Multilayer graphene is AB stacked [35,47]. For C-face, there is no carbon rich buffer layer. Graphene is weakly bonded to the substrate. Multilayer graphene on this surface is rotationally disordered, meaning that graphene layers do not have a fixed orientation with respect to the substrate. However, they still exhibit a preferred orientation around 0° and 30°. Therefore, they are still considered epitaxy. Because of the rotation, which preserves the AB symmetry of graphene, each layer of multilayer graphene on the C-face behaves as a free-standing graphene. For detailed information on the structural and electronic properties of EG on SiC, we refer the reader to a few reviews [29,74,79,81]. In contrast, the formation of graphene is less understood. In this review, we will focus ourselves on recent progress, mainly since 2008, on understanding of the growth mechanism.

3.2 Growth on Si-Face

3.2.1 Growth Mechanism

A large portion of the studies in EG growth on SiC have been focused on the Si-face. This is because the growth rate on this polar surface is relatively slow. Thus, mono- or bilayer graphene can be readily made. Moreover, the orientation of graphene is registered to the SiC substrate by the so-called van der Waals epitaxy, indicating growth of single crystal graphene in a large size is possible.

Prior to the graphene growth, the SiC surface undergoes a serial of reconstructions, from 3×3 to 1×1, then to $\sqrt{3} \times \sqrt{3}$, as the temperature increases [31, 74]. The last stage before graphene forms is $(6\sqrt{3} \times 6\sqrt{3})R30°$, which exhibits three periodicities, $(6\sqrt{3} \times 6\sqrt{3})R30°$, (6×6) and (5×5). Its homogeneity was found to depend on the preparation condition [75,85]. The $(6\sqrt{3} \times 6\sqrt{3})R30°$ reconstruction is a C-rich buffer layer, having a graphene structure but with more defects [72,73,80]. Since it is covalently bonded to the substrate, it does not have a Dirac cone band structure as a quasi-free-standing graphene has. As the source of carbon is the decomposition of the SiC substrate, the further growth is believed to occur under the buffer layer. A new buffer layer forms underneath and the old buffer layer becomes the first graphene layer [10,38,56]. As a result, it is important to have a high-quality buffer layer in order to obtain high-quality EG.

The temperature for graphene growth on the Si-face is about 1400°C, but it varies significantly in different experiment setups, probably due to its sensitiveness to the growth conditions (see Table 1 in Ref. 52). Moreover, the surface morphology also varies. Consequently, before we discuss variety of experiment results obtained by different setups, it is worth pointing out that the dominant growth mechanism can be different in these experiments and discrepancies are expected.

The growth of graphene layers is not well staged. Coexistence of the buffer layer, monolayer graphene and multilayer graphene is commonly seen, in particular for Ultra High Vacuum (UHV) growth [33,65,75]. Variety of surface features, i.e. pits [28], fingers [7,39,84] and islands [91] have been observed. These surfaces features provide valuable information on the growth kinetics.

Hupalo et al. have studied the formation of island and finger features by annealing 6H-SiC samples at 1200°C under UHV for a short period of time [39]. They found that graphene islands formed first. With further annealing, islands merged into fingers. Eventually, continuous graphene formed. They argued that the decomposition of SiC began at the step edge and the growth pattern was related to the different erosion rate of SiC bilayers. By employing a rapid and shorter heating procedure, they were able to change the kinetics for the growth of the first graphene layer and obtain a better quality.

A continuum equation for step motion was later developed to explain the finger structure and the effect of temperature and Si background pressure [10]. It was proposed that the crystallization of free carbon atoms into the new buffer layer material releases heat, causing a positive feedback to decomposition of SiC steps, leading to a step-edge instability.

Similar finger structures have been observed for graphene grown at around 1500°C under high vacuum [7,84], but important differences can be seen in the morphological detail of the fingers. As seen in Fig. 3.1, fingers here are 0.2 nm below the original terrace and the area between fingers is further 0.8 nm lower than fingers. Based on the surface roughness measured by scanning tunnelling microscopy (STM) and atomic force microscopy (AFM) phase analysis, it was concluded that the terrace and the finger were the buffer layer, while the area between fingers was graphene. As the terrace was further eroded, the finger became islands and eventually disappeared. The finger acted as barriers preventing continuous graphene formation. Erosion rate difference for SiC bilayers was believed to be the origin of macroscopic step bunching. However, the source of the step-edge instability was not identified. It is also worth noting that in Hupalo's experiment, islands emerged first and then merged into fingers. Whereas these differences suggest that the dynamics in two growth conditions is not the same, it is clear that Si desorption mainly started from step edges.

In the same experiment, pits were also observed on terraces. This presents another growth mode, in which erosion of SiC starts somewhere on the terrace. By analysing the temperature dependence of the pit density, it was argued that pits originated from point defects, which were most likely carbon vacancies. Growth on terraces is confirm by another study, in which the surface morphology of UHV grown samples was investigated using low-energy electron microscopy (LEEM), STM and photoemission electron microscopy (PEEM) [34]. Hannon et al. have also studied the formation of pits and islands for 6H-SiC annealed in UHV [28]. Starting with triple SiC bilayer steps, the buffer layer formed below 1150°C by collecting carbon emitted by eroding the triple bilayer steps. What was surprising is that carbon did not form the buffer layer right on top of the eroded surface, instead it diffused a distance and reconstructed into the buffer layer on top of the un-eroded surface, forming islands. The reason for this was not given and it may be related to the annealing condition and/or the initial surface condition. However, the important implication is that carbon can diffuse for a long distance even at a temperature as low as 1060°C. Another observation in the experiment is that while the buffer layer protected steps from being eroded, other areas were further eroded, leading to formation of pits. It was then suggested that using a rapid, high-temperature annealing can increase the nucleation density, which could lead to a pit free growth.

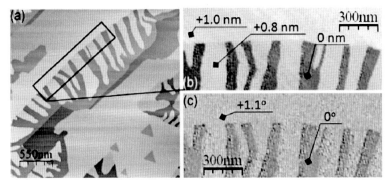

Figure 3.1 AFM images of finger structures on Si-face after graphitization at 1450°C for 10 min. (a) Topography. (b) A zoom-in topography of fingers showing three height levels. Fingers are 0.2 nm below the terrace from which they originate. (c) Corresponding phase image of (b) shows two regions, indicating the terrace and the finger is covered with the same material. Reprinted with permission from Ref. 7. Copyright 2009 American Physical Society.

3.2.2 Control the Growth

The UHV growth has an advantage of employing certain in situ characterization tools, such as low-energy electron diffraction (LEED), LEEM, photoemission spectroscopy (PES), STM, etc, which greatly help understand the growth mechanism. However, UHV grown graphene suffers from low quality, in particular, rough surfaces, non-uniform growth and low mobility, etc. Surprisingly, non-UHV conditions often lead to significantly improved film quality. de Heer's group at Georgia Tech has been pioneering in this field. The group has developed a radio frequency induction furnace method and has been able to produce EG of high mobility [5,6,19]. The technique is now referred to the confinement controlled sublimation (CCS) method [18], which will be explained in detail later. In the method, the silicon carbide substrate is confined in a graphite enclosure, whose purpose is to limit silicon from escaping. As a result, the silicon pressure remains high around the substrate and graphene growth proceeds close to thermodynamic equilibrium. Moreover, because the silicon desorption is suppressed, the growth temperature is substantially increased, hence the kinetics.

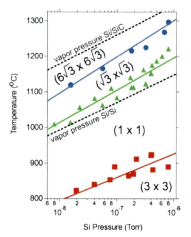

Figure 3.2 Pressure–temperature phase diagram for the 4H-SiC(0001) surface. Solid red squares, green triangles and blue circles mark the phase boundaries of $(3 \times 3) - (1 \times 1)$, $(1 \times 1) - (\sqrt{3} \times \sqrt{3})$, and $(\sqrt{3} \times \sqrt{3}) - (6 \times 3 \times 6\sqrt{3})$, respectively. Solid lines are guides to the eye. The lower and upper dashed lines are the vapour pressure of Si over Si and Si over SiC, respectively. Reprinted with permission from Ref. 91. Copyright 2009 American Physical Society.

Tromp and Hannon have quantitatively studied the effect of the background silicon pressure [91]. In their experiment, disilane was introduced into the growth chamber so as to increase the partial pressure of silicon and LEEM was used to monitor the evolution of surface structures in situ. Significant increases, up to 300°C, have been observed for the transition temperatures between different surface phases, including the growth temperature of graphene. The enhanced kinetics for transitions owing to a higher temperature improved the uniformity of the surface structures. Moreover, the pressure–temperature phase diagram for the surface closely followed the vapour pressure of Si over Si and Si over SiC, indicating that the phase transitions took place in a near equilibrium condition (see Fig. 3.2).

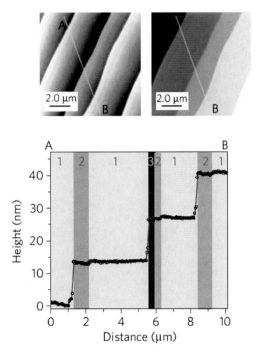

Figure 3.3 AFM images of EG grown on the Si-face of a 6H-SiC substrate in 900 mbar argon at 1650°C. The surface shows significant step bunching, with step height around 15 nm, see the line profile. Depressions of 4 and 8 Å are found at the step edges because of the nucleation of the second and third graphene layers. Reprinted with permission from Ref. 24. Copyright 2009 Nature Publishing Group.

44 | *Growth of Epitaxial Graphene on SiC*

In a different approach, SiC samples were annealed in an argon atmosphere [24,94,95]. The effect of the background argon atmosphere was to reflect desorbed silicon back onto the SiC surface, leading to suppression of silicon desorption. The growth temperature was consequently increased to 1650°C and even up to 2000°C. Seen in Fig. 3.3, uniform macro-step bunching formed and graphene films were predominately monolayer. Only at the step edges, two to three layers were found, which is in line with graphene growth starting from step edges. The results have been reproduced in a theoretical study [56]. Using kinetic Monte Carlo simulations and rate equations, the authors have identified two growth modes: climb-over and stripe coalescence. Depending on the nucleation energy and the propagation energy, different growth modes can dominate.

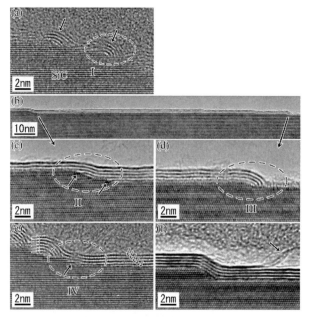

Figure 3.4 HRTEM (cross-sectional) images of graphene layers formed on the 6H-SiC(0001) surface. (a) Initial growth at step edges. (b) Wide range view of a few-layer graphene film. (c–f) Close-up images. (c) A new graphene layer nucleates under other layers at a step edge. (d) Multilayer graphene vertically intercept the lower terrace. (e) Coalescence of graphene layers occurring at a step edge. (f) Five layers continuously cover a step. Reprinted with permission from Ref. 59. Copyright 2010 Elsevier B.V.

The picture of step–edge nucleation is supported by a high resolution transmission electron microscopy (HRTEM) study on graphene on the Si-face, as seen in Fig. 3.4 [59]. Various graphene features were identified. It was clear that the graphene nucleation and coalescence took place at the step edges. Growth processes were proposed. Growth starts at step edges. It propagates towards the upper terrace, while it is pinned at the lower terrace. Coalescence happens when graphene on the upper terrace meets the one on the next terrace. Similar images have been obtained by another group [76]. Other experiments are consistent with this step-edge growth mode [71,76], too.

Given the role of step edges as the growth front, changing the step structure provides another means to improve the graphene quality [20,32,66, 95], in addition to tuning the silicon desorption rate. For UHV growth, pitting is commonly seen on the SiC surface. By systematically studying the UHV growth on substrates of a serials of miscut angles, 0.45°, 0.35°, 0.28°, 0.10°, 0.03° and 0.00°.

Dimitrakopoulos et al. have found that a large miscut angle led to a pit-free surface with denser steps [20]. Nevertheless, the mobility of graphene strongly depended on the step density and was hardly affected by the pit density.

Unlike the UHV growth, the non-UHV growth usually gives rise to flat terraces with significant step bunching. Virojanadara et al. have compared the EG growth in an argon atmosphere for SiC substrates with two different miscut angles, 0.03° and 0.25°, respectively [95]. It was revealed that a substrate with a smaller miscut angle exhibited wider terraces, smaller step heights and domains of multilayer graphene. On the other hand, a substrate with a larger miscut angle produced mainly one domain of monolayer graphene but narrower terraces and higher steps.

In another experiment, Oliveira et al. presented evidence that the step density had no influence on the structure quality of graphene. By employing different hydrogen etching conditions, initial surfaces with various step densities were prepared for graphene growth. It was found that significant surface modifications in a form of step bunching appeared after growth, while for the substrate with already significant step bunching prior to growth, no further bunching was observed. In both cases, the Raman D peak was very small, indicating similar graphene structural quality.

3.2.3 Beyond One Layer

Increasing experiment evidence supports the importance of the silicon kinetics for the graphene growth. Tuning the silicon desorption from the surface provides an efficient means to control the growth rate and increase the growth temperature. However, this may not be true once the first layer or even the buffer layer forms, because these layers likely act as a cap keeping silicon from desorption. In fact, pinning of the decomposition of SiC terraces by the buffer layer has been observed [28]. The capping effect adds a significant barrier for silicon desorption and may become the bottleneck for further growth. Kageshima et al. have studied the energetics of graphene formation on the Si-face [42]. They found that the formation of the first graphene layer requires overcoming a barrier 0.7 eV higher than that for the buffer layer. Consequently, changing the growth environment becomes ineffective at this stage. It is possible that the growth dynamics for the sequential layers is quite different from the one for the first (or buffer) layer.

To understand the growth mechanism after the buffer layer, one has to understand how silicon diffuses out of this layer. The buffer layer is more or less a graphene layer covalently bonded to the substrate. A first-principles study reveals that even helium atoms cannot penetrate a graphene film [48]. An experiment on hydrogen gas has confirmed that hydrogen molecules cannot penetrate it, too [86]. Thus, it is unlikely that silicon atoms can diffuse through the buffer layer and the graphene layers above it. Although it was suspected that Si atoms could go through graphene via point defects [41,88], a first-principles calculation suggests that the barrier for such diffusion remains very high [13,48]. Experimentally, Si $2p$ spectra obtained by micro photoemission spectroscopy indicate that no silicon exists either on top of graphene or in between layers, except under the buffer layer [94], consistent with no silicon penetration through graphene. Moreover, no Si–Si bond has been found. Therefore, the only path for escaping is to diffuse along the basal plane of the buffer layer and leak from domain boundaries.

The question is where these vents are. Few experiments actually have addressed this important question, although the deep pit seems a reasonable candidate. However, with improved growth method, high-quality graphene can be grown without deep pits and the topmost graphene layer is continuous across steps

[16,38,39,45]. One may conclude that the growth is self-limited for high-quality graphene. Tanaka et al. have studied the growth rate of EG as a function of temperature and time [88]. They observed that the growth was strongly suppressed once the first layer had formed. Moreover, the growth after the first layer was slower at high temperature than at low temperature. They argued that it was because higher temperature growth led to a better graphene film, hence a larger barrier for the sequential growth. EG on the C-face indeed shows large variations on the film thickness, indicating that the growth rate can be very different from spot to spot. We believe that it is because the quality and the domain size varies, which gives rise to different extents of self-limited growth.

Such self-limited growth suggests that the better quality of the graphene film, the harder the sequential layers can grow. It is known that Si-face EG is usually one to three layers, while C-face EG can easily grow up to tens of layers. In view of the self-limited growth, one may infer that the Si-face EG has a better quality than the C-face one, which, as we mention earlier, contradicts with experiments. The self-limited growth seems reasonable and consistent with many experiments. We suspect that the reason for the difference between two polar surfaces is not due to the different growth constrain. We propose that the difference lies in the buffer layer. On the Si-face, the buffer layer is strongly bonded to the substrate, imposing a significant barrier for silicon diffusion, leading to a suppression of graphene growth. On the contrary, there is no buffer layer on the C-face and the first layer graphene is weakly bonded to the substrate, making it much easier for silicon to escape.

3.2.4 More Than Just Si Desorption

When being annealed in argon or in a confinement, which is believed to suppress evaporation of silicon atoms, the growth temperature is significantly increased and EG with better quality is produced. It seems quite clear that the growth is mainly limited by silicon desorption [22]. However, Ohta et al. have proposed a different scenario. They have studied the growth of EG on the Si-face of SiC in argon and in UHV [63]. For samples annealed in argon, intriguing arrow features were found on a single bilayer terrace, as seen in Fig. 3.5. The arrow feature consists of a triangle arrowhead and a ribbon that connects with the head. Based on the heights of the surfaces, it was identified that the sides of the ribbon were narrow graphene stripes, while other areas of the arrow were the buffer layer. They believed that the graphene stripes acted as a sink to carbon emitted by erosion of a single bilayer step nearby. It also zipped up the step segment that it connects to so that the segment was protected from erosion. Consider a straight initial step. Once a short graphene stripe appears by fluctuations, the segment close to the stripe recesses at a higher speed, because carbon produced by erosion is quickly removed by the sink. The step is then further roughened. This instability leads to arrow features. So, the erosion of the SiC step and the growth of graphene is limited by the diffusion of carbon. However, if the step is a triple bilayer, a step flow type of growth can occur. This is because the carbon atom areal density for a triple bilayer is proximately equal to that for graphene. The formation of graphene does not require long path diffusion. When the step is neither a single bilayer nor a triple bilayer, complex growth patterns can appear, such as finger-like features.

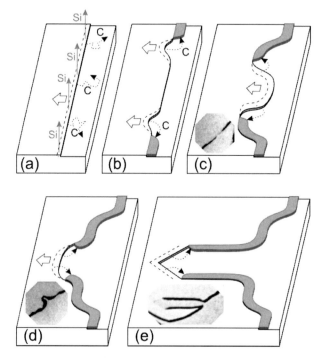

Figure 3.5 Schematic of the formation of an arrow feature on a single bilayer SiC terrace.

The insets show static LEEM images of SiC bilayer steps at similar stages of evolution. (a)

SiC starts to decompose at the bilayer step. (b) Graphene stripes, forming at the step edge, help decomposition of the bilayer nearby by acting as a carbon sink, which leads to fast erosion. (c) Two fast erosion fronts, accompanied by graphene stripes, approach each other. (d) Coalescence of two erosion fronts. (e) Graphene prevents erosion of the step behind it. Further proceeding of the erosion front leads to the arrow feature. Reprinted with permission from Ref. 63. Copyright 2010 American Physical Society.

From a naive point of view, the growth can be broken down to three sequential processes: decomposing of SiC, diffusion and desorption of silicon and diffusion and crystallization of carbon. This may be true for the UHV growth, where silicon leaves the surface and never comes back. However, for growth in a non-UHV condition, the evaporation of silicon takes place close to the equilibrium. As a result, the recombination of silicon and carbon atoms into SiC cannot be neglected. Although this process has rarely been discussed in the literature, it is implied by experiments. Starting with uniform terraces with a constant step height, the morphology of the SiC surface undergoes significant changes after growth in a non-UHV condition. For instance, uniform step bunching has been observed in the growth in argon [24]. The steps were more than 10 nm high (see Fig. 3.3). If the step bunching was only caused by decomposition of SiC, a simple calculation shows that the amount of carbon released can form quite a few layers of graphene covering the whole surface. However, the experiment has shown that monolayer areas dominated the surface and only two to three layers appeared at the step edges. The discrepancy indicates that step bunching is mainly caused by the decomposition and recombination of SiC. This dynamic equilibrium can not only shaping the SiC surface but also have profound effects on the formation of EG. Tuning the dynamics equilibrium could be a key to control the substrate morphology and obtain high-quality graphene.

3.3 Growth on C-Face

On the C-face, graphene can be produced with much better quality in terms of the carrier mobility and the defect density. The disadvantages are that the growth is faster and, hence, harder to control and the thickness variation is larger. Although multilayer graphene usually forms on the C-face, each individual layer behaves as a quasi-free-standing single layer because of an azimuthal disorder among layers. In some applications in which large electrical conductivity is desirable for graphene, multilayer has an advantage. Moreover, single layer C-face graphene can now be made owing to recent progress in the growth technique [96]. Therefore, C-face graphene could be a viable material for graphene electronics.

Similar to the Si-face, the C-face displays reconstructions prior to graphene growth. As the precursor of graphene, it is important to know how these reconstructions develop, and particularly how the last reconstruction evolves into the first layer graphene. It is known that there is a carbon buffer layer on the Si-face. However, there was little information about the existence of the interface layer on the C-face. Until 2007, Hass et al. used surface X-ray diffraction (SXRD) to carefully look at the interface between SiC and graphene [30]. Two models were proposed to explain the diffraction data. One is comprised of a carbon rich buffer layer, whose carbon density is approximately equal to that of four SiC bilayers. The other consists of two carbon layers and a silicon layer. The experiment could not distinguish two models. However, both models give a short distance of ~ 1.6 Å between the buffer layer and the first graphene layer, suggesting that the graphene layer is strongly bonded to the buffer layer. However, later experiments suggest a different structure. Emtsev et al. have studied the early stage of the growth of EG on the two polar surfaces of SiC [25]. With the angle-resolved photoelectron spectroscopy (ARPES), soft X-ray induced core-level spectroscopy and low-energy electron diffraction, they have found that even at a low carbon coverage, the spectral signature of a and n bands of graphene appeared, indicating absence of a C rich buffer layer. The ARPES data also indicated a weak coupling of the first layer graphene with the SiC (0001) surface. Other studies have not only confirmed the absence of a C-rich buffer layer and the weak coupling to the substrate, but also revealed that graphene on the C-face of 6H SiC grows on either the $(2 \times 2)_C$ or 3×3 reconstruction [36,85]. The interaction between the first graphene layer and either of two reconstructions is much weaker than the one on Si-face, although graphene appears to have slightly stronger interaction with the $(2 \times 2)_C$ reconstruction.

The structure of C-face multilayer graphene is relatively well understood. It is now clear that graphene layers exhibit stacking disorder, in contrast to the AB stack for the case of graphite and Si-face graphene. Graphene layers prefer a rotation of 30° or around 0° with respect to the SiC (2130) direction, instead of completely random orientations. Graphene grown by the CCS method exhibits stronger rotation preference than UHV grown graphene does [81]. Precisely because of the way these layers are stacked, each layer is electronically decoupled and behaves like an individual graphene layer [21,31,44,55,67,83]. However, a very recent study suggests that for the argon grown multilayer C-face graphene in the experiment, the distribution of rotation angles results from different graphene domains. Within each domain, graphene is AB or AA stacked [40]. For detailed information on the stacking and the electronic structure of C-face graphene, we refer the reader to a few reviews [29,81]. In the following, we will focus ourselves on the growth mechanism.

Figure 3.6 shows the evolution of the surface morphology for 4H-SiC substrates annealed in UHV condition [26]. Mainly beginning at the step edges, graphene grows progressively, while the SiC terrace is eroded. The SiC erosion front and the graphene growth front have irregular shapes. As a result, the original terraces gradually disappear. In some cases, SiC erosion and graphene growth start in the middle of the terrace, marked by circles in Fig. 3.6a,b. The dominant edge growth has also been observed for samples grown in a non-UHV condition [53,92].

A different growth mechanism was observed in a non-UHV condition. Camara et al. have studied the early stage of growth on the C-face of 6H-SiC under high vacuum [13]. Circular multilayer graphene islands formed at 1500°C (see Fig. 3.7). The centre of the island coincided with the dislocation of SiC. They argued that the extended defect acted as a chimney for venting silicon. At 1550°, the SiC surface was fully covered by graphene. The graphene film was not uniform in term of thickness, consisting of two types of areas. The area with more graphene layers was apparently reminiscence of graphene islands, while the area with less graphene layers filled in between islands. The latter had a rather smooth surface, indicating a second growth mechanism, which was not identified, but believed to be associated with a second type of defects. It was concluded that the low-pressure graphitization was not intrinsic, but related to the SiC defects. Later studies on the EG growth in argon have confirmed

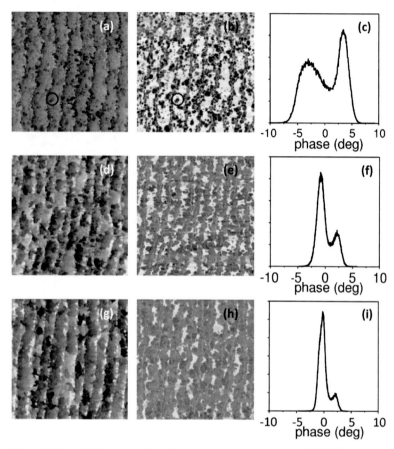

Figure 3.6 AFM images of graphene grown on the C-face of 4H-SiC. (a, d, g) topography, (b, e, h) phase, (c, f, i) histograms of phase values. The growth conditions are 10 (a, b, c), 20 (d, e, f) and 30 min (g, h, i) at 1150°C, respectively. Reprinted with permission from Ref. 26. Copyright 2011 American Institute of Physics.

that island nucleation and coalescence are a mechanism for the C-face graphene growth [37,89]. Using an electron channelling contrast imaging technique, defects were found at the centre of graphene islands acting as graphene nucleation centres and they were all threading screw dislocations. Graphene in the immediate vicinity of these dislocations was thicker and became increasingly thick as islands expanded. This accounts for the significant surface roughness commonly seen for C-face graphene.

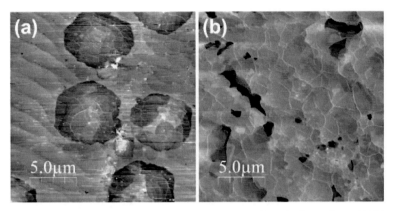

Figure 3.7 AFM topographic images of two samples graphitized at 1500°C and 1550°C. (a) Sample graphitized at 1500°C shows growth of circular graphene islands surrounded by the SiC steps. (b) Sample graphitized at 1550°C shows full coverage of graphene. Reprinted with permission from Ref. 13. Copyright 2008 American Institute of Physics.

One of the main differences between graphene grown on two polar surface of SiC is that C-face graphene displays pleats, also referred to puckers, giraffe stripes, wrinkles or ridges. The height of a pleat ranges from less than 1 nm to more than 10 nm, depending on the number of graphene layers. Thus, it can server as an indication of the graphene thickness. It can be a few tens of micrometres long and run across many SiC steps, which indicates the continuity of the graphene film. Because of a thermal expansion coefficient difference between the SiC substrate and graphene, the formation of pleats has been ascribed to a strain relaxation during cooling down from the growth temperature. Raman studies indeed found that graphene on the C-face experienced a compressive strain, consistent with the above explanation [77]. On the other hand, pleats were also found around the perimeter of graphene islands, which was believed to be unstrained. The observation led to a conclusion that a second mechanism could also be responsible for the formation of pleats. However, the assumption that graphene is unconstrained at the edge may not be true. There are HRTEM studies that show, in some cases, that graphene grow vertically into the SiC substrate [60]. Consequently, graphene edges are bonded to the substrate, thus pleats can still form at edges during cooling down.

As for the case on the Si-face, the growth on the C-face strongly depends on the pressure condition. Several methods have been employed to obtain a local silicon pressure, hence control the growth. Since 2003, de Heer's group has developed the CCS method, as seen in Fig. 3.8. In this method, the SiC substrate is placed in a graphite enclosure, which is heated to the growth temperature in an induction vacuum furnace. The enclosure prevents silicon from escaping, giving rise to a high local silicon pressure. Consequently, the desorption of silicon on the SiC surface takes place in a condition close to the thermodynamic equilibrium. The growth rate is then intimately related to the escaping rate of silicon from the enclosure, which can be controlled by the design of the enclosure, i.e. opening

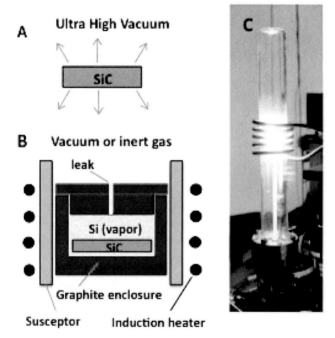

Figure 3.8   The Confinement Controlled Sublimation method. (A) UHV sublimation. Graphene growth is rapid and out of equilibrium. (B) Sublimation in a confinement. Silicon pressure is close to equilibrium vapour pressure. The growth rate is determined by the leak rate of silicon. (C) Photograph of the induction furnace. Reprinted with permission from Ref. 18. Copyright (2011) National Academy of Sciences, USA.

up a small hole. The CCS method has produced EG of exceptional quality on both polar surfaces of SiC. The mobility of CCS grown C-face graphene exhibits mobility up to 30,000 $cm^2/V \cdot s$ for the first layer and more than 250,000 $cm^2/V \cdot s$ for outermost layers in the case of multilayer graphene. The first direct observation of a perfect Dirac band structure of graphene by ARPES was achieved in this material [83]. Many important quantum transport phenomena of Dirac electrons have been demonstrated in CCS grown EG, such as quantum confinement of Dirac electrons [6], weak anti-localization [97], formation of Landau levels in a magnetic field [55,67,78], half integer quantum Hall effect [96] and oscillation of the thermoelectric effect [98], etc.

In a similar approach, Camara et al. have covered the SiC substrate with a graphite cap [11]. As a result, the growth temperature was significantly increased to over 1700°C. Extremely long monolayer graphene stripes were grown on the substrate. A defect was often found at the centre of a stripe, in agreement with previous defect-assisted growth. Still, in many cases, stripes without a visible defect were found, which either signals another growth mechanism or is simply because the type of defect is not observable by the technique employed. Using an 8° off-axis substrate, the quality of graphene was further improved and the quantum Hall effect was observed [12,14].

Growth under high-pressure argon atmosphere has also been studied on the C-face [37,40,60,89]. Formation of graphene islands in the early stage of the growth has been observed, as we already discussed above. Graphene films were relatively thick. Simultaneous growth of multilayers in islands was evident in HRTEM images [60]. It was argued that the number layers maintained constant during the growth of the island. Once the whole surface was covered by graphene, growth took place in the vertical direction, resulting in an increase of the number of layers. Since monolayer graphene has been successfully grown on the C-face by either the CCS method [96] or the graphite capping method [11,12,14], the multilayer growth mode is not universal and probably happens only in the particular growth condition. In sharp contrast to the success of argon growth on the Si-face, no significant improvement on the quality of C-face graphene has been achieved compared with the UHV growth. So far, the quantum Hall effect has not been realized in samples grown by this technique.

3.4 Conclusion

Islands, pits and fingers are often seen in graphene grown on the Si-face of SiC. The formation of these features implies the growth mechanism. It is believed that the decomposition of the SiC surface starts from the step edge and proceeds in a step flow fashion. Different erosion rate of SiC bilayers is responsible for step bunching. Deviation from the initial straight step results from a step-edge instability, which can be caused by a capping effect of the grown graphene, or a heating effect of carbon atoms crystallizing into graphene. The deviation eventually leads to formation of islands, pits and fingers. Another explanation for pits is decomposition of SiC starting from a defect. Nevertheless, it seems that step flow is the dominant growth mode.

Growth of graphene results from decomposition of SiC and desorption of silicon. Local silicon pressure suppresses the silicon desorption. So, the desorption can be maintained in a near equilibrium. Different techniques, i.e. CCS method and argon background pressure, were used to control the desorption and significant improvement to graphene quality were achieved.

In addition to the basic step flow growth and silicon equilibrium, other factors deserve attention. First, the growth of multilayer could be very different. Because of the capping effect of graphene that are already grown, the growth could be strongly suppressed. Because there is a buffer layer covalently bonded to the substrate for graphene on the Si-face, the capping effect is likely much stronger than on the C-face, which could be underlying mechanism for the difference observed in the growth on two polar surfaces. Second, the dynamic equilibrium of decomposing and recomposing of SiC constantly shapes the surface morphology. It is crucial to understand its effect on the growth.

Similar step flow growth, as well pit formation, has been observed on the C-face. However, the formation of macroscopic graphene islands in a non-UHV condition suggests a substantial difference in the growth mechanism. It also implies possibility of growth of a large single crystal graphene on this face. CCS method had great success in growth of high-quality graphene on the C-face.

Although the basic growth mechanism has been revealed, deep understanding of the growth under non-UHV conditions, the multilayer growth and the C-face growth is still lack, which hinders the progress on producing high mobility, large area, uniform graphene films on SiC.

Finally, it is worth briefly mentioning other methods of graphene growth on SiC. Growth of graphene has also been attempted on other surfaces of 4H-SiC and 3C-SiC. Sprinkle et al. have developed a scalable process and prepared graphene nanoribbons on high index surfaces of 4H-SiC [82]. Graphene has also been grown on 3C-SiC [2,17,68]. Carbon molecular beam epitaxy and CVD method have also been employed to grow graphene on SiC [1,54,69,87]. Studies on these methods not only provide alternatives for growing graphene on SiC but also may shed light on the growth mechanism for graphene on hexagonal SiC.

Acknowledgement

This work was supported by NSFC (project 11074007). We also acknowledge the International Science & Technology Cooperation Program of China Sino Swiss Science and Technology Cooperation Program (SSSTC, 2010DFA01810).

References

1. Al-Temimy, A., Riedl, C., and Starke, U. (2009). Low temperature growth of epitaxial graphene on SiC induced by carbon evaporation, *Appl. Phys. Lett.* **95**, 23, 231907.

2. Aristov, V. Y., Urbanik, G., Kummer, K., Vyalikh, D. V., Molodtsova, O. V., Preobrajenski, A. B., Zakharov, A. A., Hess, C., Hanke, T., Buchner, B., Vobornik, I., Fujii, J., Panaccione, G., Ossipyan, Y. A., and Knupfer, M. (2010). Graphene Synthesis on Cubic SiC/Si Wafers. Perspectives for Mass Production of Graphene-Based Electronic Devices, *Nano Lett.* **10**, 3, 992–995.

3. Bae, S., Kim, H., Lee, Y., Xu, X. F., Park, J. S., Zheng, Y., Balakrishnan, J., Lei, T., Kim, H. R., Song, Y. I., Kim, Y. J., Kim, K. S., Ozyilmaz, B., Ahn, J. H., Hong, B. H., and Iijima, S. (2010). Roll-to-roll production of 30-inch graphene films for transparent electrodes, *Nat. Nanotechnol.* **5**, 8, 574–578.

4. Balandin, A. A., Ghosh, S., Bao, W., Calizo, I., Teweldebrhan, D., Miao, F., and Lau, C. N. (2008). Superior Thermal Conductivity of Single-Layer Graphene, *Nano Lett.* **8**, 3, 902–907.

5. Berger, C., Song, Z. M., Li, T. B., Li, X. B., Ogbazghi, A. Y., Feng, R., Dai, Z. T., Marchenkov, A. N., Conrad, E. H., First, P. N., and de Heer, W. A. (2004). Ultrathinepitaxial graphite: 2D electron gas properties and a route toward graphene-based nanoelectronics, *J. Phys. Chem. B* **108**, 52, 19912–19916.

6. Berger, C., Song, Z. M., Li, X. B., Wu, X. S., Brown, N., Naud, C., Mayo, D., Li, T. B., Hass, J., Marchenkov, A. N., Conrad, E. H., First, P. N., and de Heer, W. A. (2006). Electronic confinement and coherence in patterned epitaxial graphene, *Science* **312**, 5777, 1191–1196.

7. Bolen, M. L., Harrison, S. E., Biedermann, L. B., and Capano, M. A. (2009). Graphene formation mechanisms on 4H-SiC(0001), *Phys. Rev. B* **80**, 11, 115433.

8. Bolotin, K. I., Ghahari, F., Shulman, M. D., Stormer, H. L., and Kim, P. (2009). Observation of the fractional quantum Hall effect in graphene, *Nature* **462**, 7270, 196–199.

9. Bolotin, K. I., Sikes, K. J., Hone, J., Stormer, H. L., and Kim, P. (2008). Temperature-dependent transport in suspended graphene, *Phys. Rev. Lett.* **101**, 9, 096802.

10. Borovikov, V., and Zangwill, A. (2009). Step-edge instability during epitaxial growth of graphene from SiC(0001), *Phys. Rev. B* **80**, 12, 121406.

11. Camara, N., Huntzinger, J.-R., Rius, G., Tiberj, A., Mestres, N., Perez-Murano, F., Godignon, P., and Camassel, J. (2009). Anisotropic growth of long isolated graphene ribbons on the C face of graphite-capped 6H -SiC, *Phys. Rev. B* **80**, 12, 125410.

12. Camara, N., Jouault, B., Caboni, A., Jabakhanji, B., Desrat, W., Pausas, E., Consejo, C., Mestres, N., Godignon, P., and Camassel, J. (2010a). Growth of monolayer graphene on 8° off-axis 4H-SiC (0001) substrates with application to quantum transport devices, *Appl. Phys. Lett.* **97**, 9, 093107.

13. Camara, N., Rius, G., Huntzinger, J.-R., Tiberj, A., Magaud, L., Mestres, N., Godignon, P., and Camassel, J. (2008). Early stage formation of graphene on the C face of 6H-SiC, *Appl. Phys. Lett.* **93**, 26, 263102.

14. Camara, N., Tiberj, A., Jouault, B., Caboni, A., Jabakhanji, B., Mestres, N., Godignon, P., and Camassel, J. (2010b). Current status of self-organized epitaxial graphene ribbons on the C face of 6H-SiC substrates, *J. Phys. D: Appl. Phys.* **43**, 37, 374011.

15. Cheianov, V. V., Fal'ko, V., and Altshuler, B. L. (2007). The Focusing of Electron Flow and a Veselago Lens in Graphene p–n Junctions, *Science* **315**, 5816, 1252–1255.

16. Choi, J., Lee, H., and Kim, S. (2010). Atomic-scale investigation of epitaxial graphene grown on 6H-SiC(0001) using scanning tunneling microscopy and spectroscopy, *J. Phys. Chem. C* **114**, 31, 13344–13348.

17. Coletti, C., Emtsev, K. V., Zakharov, A. A., Ouisse, T., Chaussende, D., and Starke, U. (2011). Large area quasi-free standing monolayer graphene on 3C-SiC(111), *Appl. Phys. Lett.* **99**, 8, 081904.

18. de Heer, W. A., Berger, C., Ruan, M., Sprinkle, M., Li, X., Hu, Y., Zhang, B., Hankinson, J., and Conrad, E. (2011). Large area and structured epitaxial graphene produced by confinement controlled sublimation of silicon carbide, *PNAS* **108**, 41, 16900–16905.

19. de Heer, W. A., Berger, C., Wu, X. S., First, P. N., Conrad, E. H., Li, X. B., Li, T. B., Sprinkle, M., Hass, J., Sadowski, M. L., Potemski, M., and Martinez, G. (2007). Epitaxial graphene, *Solid State Commun.* **143**, 1–2, 92–100.

20. Dimitrakopoulos, C., Grill, A., McArdle, T. J., Liu, Z. H., Wisnieff, R., and Antoniadis, D. A. (2011). Effect of SiC wafer miscut angle on the morphology and Hall mobility of epitaxially grown graphene, *Appl. Phys. Lett.* **98**, 22, 222105.

21. dos Santos, J. M. B. L., Peres, N. M. R., and Neto, A. H. C. (2007). Graphene Bilayer with a Twist: Electronic Structure, *Phys. Rev. Lett.* **99**, 25, 256802.

22. Drabinska, A., Grodecki, K., Strupinski, W., Bozek, R., Korona, K. P., Wysmolek, A., Stepniewski, R., and Baranowski, J. M. (2010). Growth kinetics of epitaxial grapheme on SiC substrates, *Phys. Rev. B* **81**, 24, 245410.

23. Du, X., Skachko, I., Duerr, F., Luican, A., and Andrei, E. Y. (2009). Fractional quantum Hall effect and insulating phase of Dirac electrons in graphene, *Nature* **462**, 7270, 192–195.

24. Emtsev, K. V., Bostwick, A., Horn, K., Jobst, J., Kellogg, G. L., Ley, L., McChesney, J. L., Ohta, T., Reshanov, S. A., Rohrl, J., Rotenberg, E., Schmid, A. K., Waldmann, D., Weber, H. B., and Seyller, T. (2009). Towards wafer-size graphene layers by atmospheric pressure graphitization of silicon carbide, *Nat. Mater.* **8**, 3, 203–207.

25. Emtsev, K. V., Speck, F., Seyller, T., Ley, L., and Riley, J. D. (2008). Interaction, growth, and ordering of epitaxial graphene on SiC0001 surfaces: A comparative photoelectron spectroscopy study, *Phys. Rev. B* **77**, 15, 155303.

26. Ferrer, F. J., Moreau, E., Vignaud, D., Deresmes, D., Godey, S., and Wallart, X. (2011). Initial stages of graphitization on SiC(0001), as studied by phase atomic force microscopy, *J. Appl. Phys.* **109**, 5, 054307.

27. Hannon, J. B., Copel, M., and Tromp, R. M. (2011). Direct Measurement of the Growth Mode of Graphene on SiC(0001) and SiC(000T), *Phys. Rev. Lett.* **107**, 166101.

28. Hannon, J. B., and Tromp, R. M. (2008). Pit formation during graphene synthesis on SiC(0001): In situ electron microscopy, *Phys. Rev. B* **77**, 24, 241404.

29. Hass, J., de Heer, W. A., and Conrad, E. H. (2008a). The growth and morphology of epitaxial multilayer graphene, *J. Phys.: Condens. Matter* **20**, 32, 323202.

30. Hass, J., Feng, R., Millan-Otoya, J. E., Li, X., Sprinkle, M., First, P. N., de Heer, W. A., Conrad, E. H., and Berger, C. (2007). Structural properties of the multilayer graphene/4H-SiC(0001) system as determined by surface X-ray diffraction, *Phys. Rev. B* **75**, 21, 214109.

31. Hass, J., Varchon, F., Millan-Otoya, J. E., Sprinkle, M., Sharma, N., de Heer, W. A., Berger, C., First, P. N., Magaud, L., and Conrad, E. H. (2008b). Why multilayer graphene on 4H-SiC(0001) behaves like a single sheet of graphene, *Phys. Rev. Lett.* **100**, 12, 125504.

32. Hattori, A. N., Okamoto, T., Sadakuni, S., Murata, J., Arima, K., Sano, Y., Hattori, K., Daimon, H., Endo, K., and Yamauchi, K. (2011). Formation of wide and atomically flat graphene layers on ultraprecision-figured 4H-SiC(0001) surfaces, *Surf. Sci.* **605**, 5–6, 597–605.

33. Hibino, H., Kageshima, H., Maeda, F., Nagase, M., Kobayashi, Y., and Yamaguchi, H. (2008). Microscopic thickness determination of thin graphite films formed on SiC from quantized oscillation in reflectivity of low-energy electrons, *Phys. Rev. B* **77**, 7, 075413.

34. Hibino, H., Kageshima, H., and Nagase, M. (2010). Epitaxial few-layer graphene: towards single crystal growth, *J. Phys. D: Appl. Phys.* **43**, 37, 374005.

35. Hibino, H., Mizuno, S., Kageshima, H., Nagase, M., and Yamaguchi, H. (2009). Stacking domains of epitaxial few-layer graphene on SiC(0001), *Phys. Rev. B* **80**, 8, 085406.

36. Hiebel, F., Mallet, P., Varchon, F., Magaud, L., and Veuillen, J.-Y. (2008). Graphene-substrate interaction on 6H-SiC(0001): A scanning tunneling microscopy study, *Phys. Rev. B* **78**, 153412.

37. Hite, J. K., Twigg, M. E., Tedesco, J. L., Friedman, A. L., Myers-Ward, R. L., Eddy, C. R., and Gaskill, D. K. (2011). Epitaxial graphene nucleation on C-face silicon carbide, *Nano Lett.* **11**, 3, 1190–1194.

38. Huang, H., Chen, W., Chen, S., and Wee, A. T. S. (2008). Bottom-up growth of epitaxial graphene on 6H-SiC(0001), *ACS Nano* **2**, 12, 2513–2518.

39. Hupalo, M., Conrad, E. H., and Tringides, M. C. (2009). Growth mechanism for epitaxial graphene on vicinal 6H-SiC(0001) surfaces: A scanning tunneling microscopy study, *Phys. Rev. B* **80**, 4, 041401.

40. Johansson, L. I., Watcharinyanon, S., Zakharov, A. A., Iakimov, T., Yakimova, R., and Virojanadara, C. (2011). Stacking of adjacent graphene layers grown on C-face SiC, *Phys. Rev. B* **84**, 12, 125405.

41. Kageshima, H., Hibino, H., Nagase, M., and Yamaguchi, H. (2009). Theoretical study of epitaxial graphene growth on SiC(0001) surfaces, *Appl. Phys. Exp.* **2**, 6, 065502.

42. Kageshima, H., Hibino, H., Yamaguchi, H., and Nagase, M. (2011). Theoretical study on epitaxial graphene growth by si sublimation from SiC(0001) surface, *Jpn. J. Appl. Phys.* **50**, 9, 095601.

43. Katsnelson, M. I., Novoselov, K. S., and Geim, A. K. (2006). Chiral tunnelling and the Klein paradox in graphene, *Nat. Phys.* **2**, 620.

44. Latil, S., Meunier, V., and Henrard, L. (2007). Massless fermions in multilayer graphitic systems with misoriented layers: Ab initio calculations and experimental fingerprints, *Phys. Rev. B* **76**, 201402.

45. Lauffer, P., Emtsev, K. V., Graupner, R., Seyller, T., Ley, L., Reshanov, S. A., and Weber, H. B. (2008). Atomic and electronic structure of few-layer graphene on SiC(0001) studied with scanning tunneling microscopy and spectroscopy, *Phys. Rev. B* **77**, 15, 155426.

46. Lee, C., Wei, X., Kysar, J. W., and Hone, J. (2008). Measurement of the elastic properties and intrinsic strength of monolayer graphene, *Science* **321**, 5887, 385–388.

47. Lee, K., Kim, S., Points, M. S., Beechem, T. E., Ohta, T., and Tutuc, E. (2011). Magnetotransport properties of quasi-free-standing epitaxial graphene bilayer on SiC: evidence for Bernal stacking, *Nano Lett.* **11**, 9, 3624–3628.

48. Leenaerts, O. (2008). Graphene: A perfect nanoballoon, *Appl. Phys. Lett.* **93**, 19, 193107.

49. Li, X., Cai, W., An, J., Kim, S., Nah, J., Yang, D., Piner, R., Velamakanni, A., Jung, I., Tutuc, E., Banerjee, S. K., Colombo, L., and Ruoff, R. S. (2009). Large-area synthesis of high-quality and uniform graphene films on copper foils, *Science* **324**, 1312.

50. Lin, Y.-M., Dimitrakopoulos, C., Jenkins, K. A., Farmer, D. B., Chiu, H.-Y., Grill, A., and Avouris, P. (2010). 100-GHz Transistors from wafer-scale epitaxial graphene, *Science* **327**, 5966, 662.

51. Lin, Y. M., Valdes-Garcia, A., Han, S. J., Farmer, D. B., Meric, I., Sun, Y. N., Wu, Y. Q., Dimitrakopoulos, C., Grill, A., Avouris, P., and Jenkins, K.

A. (2011). Wafer-scale graphene integrated circuit, *Science* **332**, 6035, 1294–1297.

52. Lu, W. J., Boeckl, J. J., and Mitchel, W. C. (2010). A critical review of growth of low-dimensional carbon nanostructures on SiC (0001): impact of growth environment, *J. Phys. D: Appl. Phys.* **43**, 37, 374004.

53. Luxmi, Fisher, P. J., Srivastava, N., Feenstra, R. M., Sun, Y. G., Kedzierski, J., Healey, P., and Gu, G. (2009). Morphology of graphene on SiC(0001) surfaces, *Appl. Phys. Lett.* **95**, 7, 073101.

54. Michon, A., Vezian, S., Ouerghi, A., Zielinski, M., Chassagne, T., and Portail, M. (2010). Direct growth of few-layer graphene on 6H-SiC and 3C-SiC/Si via propane chemical vapor deposition, *Appl. Phys. Lett.* **97**, 17, 171909.

55. Miller, D. L., Kubista, K. D., Rutter, G. M., Ruan, M., de Heer, W. A., First, P. N., and Stroscio, J. A. (2009). Observing the quantization of zero mass carriers in graphene, *Science* **324**, 5929, 924–927.

56. Ming, F., and Zangwill, A. (2011). Model for the epitaxial growth of graphene on 6H-SiC(0001), *Phys. Rev. B* **84**, 115459.

57. Morozov, S. V., Novoselov, K. S., Katsnelson, M. I., Schedin, F., Elias, D. C., Jaszczak, J. A., and Geim, A. K. (2008). Giant intrinsic carrier mobilities in graphene and its bilayer, *Phys. Rev. Lett.* **100**, 1, 016602.

58. Nair, R. R., Blake, P., Grigorenko, A. N., Novoselov, K. S., Booth, T. J., Stauber, T., Peres, N. M. R., and Geim, A. K. (2008). Fine structure constant defines visual transparency of graphene, *Science* **320**, 5881, 1308.

59. Norimatsu, W., and Kusunoki, M. (2010). Formation process of graphene on SiC (0001), *Physica E* **42**, 4, 691–694.

60. Norimatsu, W., Takada, J., and Kusunoki, M. (2011). Formation mechanism of graphene layers on SiC (0001) in a high-pressure argon atmosphere, *Phys. Rev. B* **84**, 3, 035424.

61. Novoselov, K. S., Geim, A. K., Morozov, S. V., Jiang, D., Katsnelson, M. I., Grigorieva, I. V., Dubonos, S. V., and Firsov, A. A. (2005). Two-dimensional gas of massless Dirac fermions in graphene, *Nature* **438**, 7065, 197–200.

62. Novoselov, K. S., Geim, A. K., Morozov, S. V., Jiang, D., Zhang, Y., Dubonos, S. V., Grigorieva, I. V., and Firsov, A. A. (2004). Electric field effect in atomically thin carbon films, *Science* **306**, 5296, 666–669.

63. Ohta, T., Bartelt, N. C., Nie, S., Thiirmer, K., and Kellogg, G. L. (2010). Role of carbon surface diffusion on the growth of epitaxial graphene on SiC, *Phys. Rev. B* **81**, 12, 121411.

64. Ohta, T., Bostwick, A., McChesney, J. L., Seyller, T., Horn, K., and Rotenberg, E. (2007). Interlayer interaction and electronic screening in multilayer graphene investigated with angle-resolved photoemission spectroscopy, *Phys. Rev. Lett.* **98**, 20, 206802.

65. Ohta, T., El Gabaly, F., Bostwick, A., McChesney, J. L., Emtsev, K. V., Schmid, A. K., Seyller, T., Horn, K., and Rotenberg, E. (2008). Morphology of graphene thin film growth on SiC(0001), *New J. Phys.* **10**, 023034.

66. Oliveira, M. H., Schumann, T., Ramsteiner, M., Lopes, J. M. J., and Riechert, H. (2011). Influence of the silicon carbide surface morphology on the epitaxial graphene formation, *Appl. Phys. Lett.* **99**, 11, 111901.

67. Orlita, M., Faugeras, C., Plochocka, P., Neugebauer, P., Martinez, G., Maude, D. K., Barra, A.-L., Sprinkle, M., Berger, C., de Heer, W. A., and Potemski, M. (2008). Approaching the Dirac point in high-mobility multilayer epitaxial graphene, *Phys. Rev. Lett.* **101**, 26, 267601.

68. Ouerghi, A., Kahouli, A., Lucot, D., Portail, M., Travers, L., Gierak, J., Penuelas, J., Jegou, P., Shukla, A., Chassagne, T., and Zielinski, M. (2010). Epitaxial graphene on cubic SiC(111)/Si(111) substrate, *Appl. Phys. Lett.* **96**, 19, 191910.

69. Park, J., Mitchel, W. C., Grazulis, L., Smith, H. E., Eyink, K. G., Boeckl, J. J., Tomich, D. H., Pacley, S. D., and Hoelscher, J. E. (2010). Epitaxial graphene growth by carbon molecular beam epitaxy (CMBE), *Adv. Mater.* **22**, 37, 4140–4145.

70. Park, S., and Ruoff, R. S. (2009). Chemical methods for the production of graphenes, *Nat. Nanotechnol.* **4**, 4, 217–224.

71. Poon, S. W., Chen, W., Tok, E. S., and Wee, A. T. S. (2008). Probing epitaxial growth of graphene on silicon carbide by metal decoration, *Appl. Phys. Lett.* **92**, 10, 104102.

72. Qi, Y., Rhim, S. H., Sun, G. F., Weinert, M., and Li, L. (2010). Epitaxial graphene on SiC(0001): more than just honeycombs, *Phys. Rev. Lett.* **105**, 8, 085502.

73. Riedl, C., Coletti, C., Iwasaki, T., Zakharov, A. A., and Starke, U. (2009). Quasi-free-standing epitaxial graphene on SiC obtained by hydrogen intercalation, *Phys. Rev. Lett.* **103**, 24, 246804.

74. Riedl, C., Coletti, C., and Starke, U. (2010). Structural and electronic properties of epitaxial graphene on SiC(0001): A review of growth, characterization, transfer doping and hydrogen intercalation, *J. Phys. D: Appl. Phys.* **43**, 37, 374009.

75. Riedl, C., Starke, U., Bernhardt, J., Franke, M., and Heinz, K. (2007). Structural properties of the graphene-SiC(0001) interface as a key

for the preparation of homogeneous large-terrace graphene surfaces, *Phys. Rev. B* **76**, 24, 245406.

76. Robinson, J., Weng, X. J., Trumbull, K., Cavalero, R., Wetherington, M., Frantz, E., LaBella, M., Hughes, Z., Fanton, M., and Snyder, D. (2010). Nucleation of epitaxial graphene on SiC(0001), *ACS Nano* **4**, 1, 153–158.

77. Rohrl, J., Hundhausen, M., Emtsev, K. V., Seyller, T., Graupner, R., and Ley, L. (2008). Raman spectra of epitaxial graphene on SiC(0001), *Appl. Phys. Lett.* **92**, 20, 201918.

78. Sadowski, M. L., Martinez, G., Potemski, M., Berger, C., and de Heer, W. A. (2006). Landau level spectroscopy of ultrathin graphite layers, *Phys. Rev. Lett.* **97**, 26, 266405.

79. Seyller, T., Bostwick, A., Emtsev, K. V., Horn, K., Ley, L., McChesney, J. L., Ohta, T., Riley, J. D., Rotenberg, E., and Speck, F. (2008). Epitaxial graphene: a new material, *Physica Status Solidi (b)* **245**, 7, 1436–1446.

80. Speck, F. (2011). The quasi-free-standing nature of graphene on H-saturated SiC(0001), *Appl. Phys. Lett.* **99**, 12, 122106.

81. Sprinkle, M., Hicks, J., Tejeda, A., Taleb-Ibrahimi, A., Le Fevre, P., Bertran, F., Tinkey, H., Clark, M. C., Soukiassian, P., Martinotti, D., Hass, J., and Conrad, E. H. (2010a). Multilayer epitaxial graphene grown on the SiC (0001) surface; structure and electronic properties, *J. Phys. D: Appl. Phys.* **43**, 37, 374006.

82. Sprinkle, M., Ruan, M., Hu, Y., Hankinson, J., Rubio-Roy, M., Zhang, B., Wu, X., Berger, C., and de Heer, W. A. (2010b). Scalable templated growth of grapheme nanoribbons on SiC, *Nat. Nanotechnol.* **5**, 10, 727–731.

83. Sprinkle, M., Siegel, D., Hu, Y., Hicks, J., Tejeda, A., Taleb-Ibrahimi, A., Fevre, P. L., Bertran, F., Vizzini, S., Enriquez, H., Chiang, S., Soukiassian, P., Berger, C., de Heer, W. A., Lanzara, A., and Conrad, E. H. (2009). First direct observation of a nearly ideal graphene band structure, *Phys. Rev. Lett.* **103**, 22, 226803.

84. Sprinkle, M. W. (2010). Epitaxial graphene on silicon carbide: Low-vacuum growth, characterization, and device fabrication, PhD thesis, Georgia Institute of Technology.

85. Starke, U., and Riedl, C. (2009). Epitaxial graphene on SiC(0001) and SiC(0001): From surface reconstructions to carbon electronics, *J. Phys.: Condens. Matter* **21**, 13, 134016.

86. Stolyarova, E., Stolyarov, D., Bolotin, K., Ryu, S., Liu, L., Rim, K. T., Klima, M., Hybertsen, M., Pogorelsky, I., Pavlishin, I., Kusche, K., Hone, J., Kim, P., Stormer, H. L., Yakimenko, V., and Flynn, G. (2009). Observation of

graphene bubbles and effective mass transport under graphene films, *Nano Lett.* **9**, 1, 332–337.

87. Strupinski, W., Grodecki, K., Wysmolek, A., Stepniewski, R., Szkopek, T., Gaskell, P. E., Gruneis, A., Haberer, D., Bozek, R., Krupka, J., and Baranowski, J. M. (2011). Graphene epitaxy by chemical vapor deposition on SiC, *Nano Lett.* **11**, 4, 17861791.

88. Tanaka, S., Morita, K., and Hibino, H. (2010). Anisotropic layer-by-layer growth of graphene on vicinal SiC(0001) surfaces, *Phys. Rev. B* **81**, 4, 041406.

89. Tedesco, J. L., Jernigan, G. G., Culbertson, J. C., Hite, J. K., Yang, Y., Daniels, K. M., Myers-Ward, R. L., C. R. Eddy, J., Robinson, J. A., Trumbull, K. A., Wetherington, M. T., Campbell, P. M., and Gaskill, D. K. (2010). Morphology characterization of argon-mediated epitaxial graphene on C-face SiC, *Appl. Phys. Lett.* **96**, 22, 222103.

90. Tombros, N., Jozsa, C., Popinciuc, M., Jonkman, H. T., and van Wees, B. J. (2007). Electronic spin transport and spin precession in single graphene layers at room temperature, *Nature* **448**, 7153, 571–574.

91. Tromp, R. M., and Hannon, J. B. (2009). Thermodynamics and kinetics of graphene growth on SiC(0001), *Phys. Rev. Lett.* **102**, 10, 106104.

92. Ushio, S., Yoshii, A., Tamai, N., Ohtani, N., and Kaneko, T. (2011). Wide-range temperature dependence of epitaxial graphene growth on 4H-SiC (0001): A study of ridge structures formation dynamics associated with temperature, *J. Cryst. Growth* **318**, 1, 590–594.

93. Van Bommel, A., Crombeen, J., and Van Tooren, A. (1975). LEED and Auger electron observations of the SiC(0001) surface, *Surf. Sci.* **48**, 2, 463–472.

94. Virojanadara, C., Syvajarvi, M., Yakimova, R., Johansson, L. I., Zakharov, A. A., and Balasubramanian, T. (2008). Homogeneous large-area graphene layer growth on 6H-SiC(0001), *Phys. Rev. B* **78**, 24, 245403.

95. Virojanadara, C., Yakimova, R., Osiecki, J. R., Syvajarvi, M., Uhrberg, R. I. G., Johansson, L. I., and Zakharov, A. A. (2009). Substrate orientation: A way towards higher quality monolayer graphene growth on 6H-SiC(0001), *Surf. Sci.* **603**, 15, L87–L90.

96. Wu, X. S., Hu, Y., Ruan, M., Madiomanana, N. K., Hankinson, J., Sprinkle, M., Berger, C., and de Heer, W. A. (2009). Half integer quantum Hall effect in high mobility single layer epitaxial graphene, *Appl. Phys. Lett.* **95**, 22, 223108.

97. Wu, X. S., Li, X. B., Song, Z. M., Berger, C., and de Heer, W. A. (2007). Weak anti-localization in epitaxial graphene: Evidence for chiral electrons, *Phys. Rev. Lett.* **98**, 136801.

98. Wu, Z., Zhai, F., Peeters, F. M., Xu, H. Q., and Chang, K. (2011). Valley-dependent Brewster angles and Goos–Hanchen effect in strained graphene, *Phys. Rev. Lett.* **106**, 17, 176802.

99. Young, A. F., and Kim, P. (2009). Quantum interference and Klein tunnelling in graphene heterojunctions, *Nat Phys* **5**, 3, 222–226.

100. Zhang, Y. B., Tan, Y. W., Stormer, H. L., and Kim, P. (2005). Experimental observation of the quantum Hall effect and Berry's phase in graphene, *Nature* **438**, 7065, 201204.

Chapter 4

Chemical Vapor Deposition of Large-Area Graphene on Metallic Substrates

Wei Wu[a] and Qingkai Yu[b]

[a]*Department of Electrical and Computer Engineering, Center for Advanced Materials, University of Houston, Houston, TX 77204, USA*
[b]*Ingram School of Engineering and Materials Science, Engineering and Commercialization Program, Texas State University, San Marcos, TX 78666, USA*
wwu12@mail.uh.edu, qingkai.yu@txstate.edu

4.1 Introduction

Graphene is the name given to a single atomic layer of sp^2-bonded carbon atoms densely packed in a two-dimensional honeycomb lattice, and is the basic building block of all graphitic forms [1]. Theoretically, graphene has been studied since 1947 [2], but was experimentally isolated for the first time only recently in 2004 [3]. As a rising star in solid state physics, chemistry, and materials science, graphene has been attracting significant attention due to its extraordinary properties, such as anomalous quantum Hall effect [1, 4–6], extremely high mobility [1, 7], high elasticity [8], and

---

Two-Dimensional Carbon: Fundamental Properties, Synthesis, Characterization, and Applications
Edited by Yihong Wu, Zexiang Shen, and Ting Yu
Copyright © 2014 Pan Stanford Publishing Pte. Ltd.
ISBN 978-981-4411-94-3 (Hardcover), 978-981-4411-95-0 (eBook)
www.panstanford.com

optical transparency [9]. Such unique properties make graphene a potential candidate for various applications, e.g., transistors [10, 11], transparent electrodes [12–15], and chemical/bio sensors [16–18]. To further envision graphene technology, it is critical to synthesize large-scale, high-quality graphene films. This chapter provides an overview of graphene synthesis methods, with the particular focus on chemical vapor deposition (CVD) synthesis of large-area graphene on Ni and Cu substrates. In addition, the recent progresses in the synthesis of single-crystal graphene films/grains by CVD are also briefly reviewed.

4.2 Graphene Synthesis Methods

Since the first mechanical isolation of graphene from graphite crystal in 2004 [3], intense efforts have been made to develop technologies for large-area graphene synthesis, including reduction of graphite oxide, thermal decomposition of SiC, and CVD of hydrocarbons on transition metals. In particular, graphene synthesized by CVD on Ni and Cu substrates has shown great promise owing to its large scale, high quality, and transferability to arbitrary substrates for various applications. On the other hand, epitaxial graphene on SiC is more convenient to be used for microelectronics.

4.2.1 Mechanical Exfoliation

The first isolated graphene was obtained by mechanical exfoliation of graphite in 2004 at the University of Manchester [3]. An adhesive tape (Scotch tape) is used to repeatedly stick and peel off graphite crystals into increasingly thinner pieces, which are then transferred on an oxidized Si substrate (300 nm SiO_2) simply by a gentle press of the tape. The transferred graphene adds a tiny optical path to the Fabry–Perrot cavity created by the 300 nm SiO_2 layer on Si. An optical contrast between graphene and the substrate is maximized at about 12% under a light with wavelength 550 nm, where the sensitivity of the human eye is optimal. Graphene becomes visible and can be easily identified under an ordinary optical microscope. Single-layer graphene deposited on 300 nm SiO_2 appears light purple, and the color turns into darker purple for few-layer graphene samples.

This mechanical exfoliation approach is quite simple, and so far gives the best graphene samples in terms of structural and electronic quality, primarily because of the high-quality nature of the starting graphite crystal source, e.g., highly oriented pyrolytic graphite (HOPG). Graphene flakes with size up to millimeters can be obtained through this method. However, the approach has disadvantages of low efficiency, and is impractical for large-scale applications.

4.2.2 Graphite Oxide Reduction

Chemical reduction of aqueous dispersions of graphite oxide (GO) can produce large quantities of graphene flakes. Currently, reducing agents, such as hydrazine [19, 20], $NaBH_4$ [21], HI [22], and Fe [23] are used to reduce GO. Advantages of the GO reduction method are its low cost and straightforward scalability. The starting material is simple graphite, and the process can be performed in suspension. However, quality of graphene obtained by this route is much lower than that of mechanically exfoliated graphene. For example, mobility in reduced GO is typically ranging from 0.001 to 10 $cm^2V^{-1}s^{-1}$, depending on the reduction conditions and film thickness [24], but value up to 365 $cm^2V^{-1}s^{-1}$ has been reported for well-reduced few-layer films [25]. These values are several orders of magnitude lower than those of graphene prepared by mechanical exfoliation [1]. One reason is that the reduction of GO is incomplete, and the resulting material contains both graphene and graphene oxide, with a typical C/O atomic ratio of 5–12 [23]. Additionally, a high density of defects can also be induced during the formation of GO colloidal suspension and the following reduction step.

4.2.3 Thermal Decomposition of SiC

Another method of obtaining graphene is through thermal treatment of silicon carbide (SiC) at high temperatures (>1100 °C) under vacuum conditions. The graphitization consists in the sublimation of Si atoms and reorganization of the remaining C. This growth method enables large-scale production of graphene with good mobilities (1000–10000 $cm^2V^{-1}s^{-1}$) over entire surface of SiC wafers, and processing of graphene field-effect devices on insulating SiC substrates. Lin et al. reported the fabrication of arrays of top-gated graphene field-effect transistors (FETs) on a 2 in SiC wafer, and the transistors exhibit modest mobility between 900 and 1520 $cm^2V^{-1}s^{-1}$ [10]. But it may be limited to devices on SiC only, since the transfer of graphene from SiC onto other substrates such as SiO_2/Si or glass has rarely been reported [26]. In addition, the need of high growth temperature can also be a limitation.

4.2.4 CVD of Hydrocarbons on Metals

Graphene has been grown on a variety of transition metals such as Pt [27], Ru [28], Ir [29], Ni [13, 30, 31], and Cu [12, 18, 32] by CVD of hydrocarbons. The growth can be a surface adsorption process or a carbon segregation process by cooling depending on the carbon solubility in the metal substrate and the growth conditions. Recently, results have shown that CVD onto relatively inexpensive polycrystalline Ni [13] and Cu [12, 33] substrates is a highly promising approach for producing high-quality graphene in a large scale. In addition, the transfer of graphene grown on Ni and Cu onto arbitrary substrates can be readily achieved since the wet etching of Ni and Cu is feasible and straightforward. Few-layer graphene films grown on thin Ni films and transferred on SiO_2 substrates exhibit mobility greater than 3700 $cm^2V^{-1}s^{-1}$ and half-integer quantum Hall effect [13]. Single-layer graphene films of up to 30 in grown on Cu have been obtained by Bae et al. using low pressure CVD, and the transferred films show mobilities up to 7350 $cm^2V^{-1}s^{-1}$ at low temperature [12].

4.3 CVD Graphene on Nickel

Transition metals with low carbon affinity but still having the ability to stabilize carbon on their surfaces are favorable catalysts for graphitic carbon formation. Among various transition metals, Ni has been widely used as a catalyst for producing high-quality graphite [34, 35], diamond films [36], nanotubes [37], and most recently graphene [13, 30, 31].

The CVD synthesis of planar thick nano-graphene films (up to 35 layers of graphene) on Ni substrates (2×2 cm^2 sheets) was first reported in 2006 [38]. A natural hydrocarbon source, camphor

(0.1–0.5 g), is used as the precursor, which is first evaporated at 180 °C and then pyrolyzed at 700–850 °C with Ar as a carrier gas. After deposition, the sample is naturally cooled down. This study opens up the possibility of large-scale synthesis of graphene films by CVD based methods, although several issues remain to be solved, including controlling the number of graphene layers and transferring the graphene films onto other substrates for device fabrication. Shortly thereafter, Obraztsov et al. synthesized few-layer graphene films of nanometer thickness (1.5 ± 0.5 nm) by low pressure CVD of CH_4 on polycrystalline Ni sheets (25 × 25 mm^2, 0.5 mm in thickness) at 950 °C activated by a DC discharge [39]. The films have many atomically smooth micrometer-size areas separated by wrinkles or ridges with a typical height of 10–50 nm (Fig. 4.1). The formation of wrinkles is believed to be closely related to the difference in the in-plane thermal expansion behaviors between graphene and Ni (thermal expansion coefficient $\alpha_{graphene} = -6 \times 10^{-6}$ K^{-1} at 27 °C, $\alpha_{Ni} = 12.89–21.0 \times 10^{-6}$ K^{-1} between 27 and 1000 °C) [39, 40]. Mechanical stress on graphene is induced during the cooling process. The stress is then released through the formation of graphene wrinkles. Swells are also observed in regions where poor adhesion between the film and the substrate is poor (Fig. 4.1b, indicated by arrow).

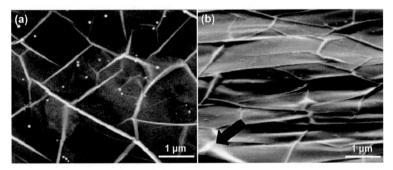

Figure 4.1 SEM images of graphene films grown on Ni substrates (a) Top view. (b) 75° side view. An example of swell is marked by an arrow in image (b). Reprinted from [39], Copyright 2007, with permission from Elsevier.

The transfer of CVD graphene on Ni was demonstrated by Yu et al. in 2008 [31]. High-quality single- to few-layer graphene films (3–4 layers) are synthesized on polycrystalline Ni substrates (5 × 5 mm^2,

0.5 mm in thickness) by ambient-pressure CVD method at 1000 °C with CH_4 as the carbon source, and the as-produced films are successfully transferred onto other substrates (glass plates) by wet etching of the Ni using diluted HNO_3 solution while still maintaining their high quality as confirmed by Raman spectroscopy. A thin layer of silicone is used as a medium during the transfer process. It was proposed that CVD growth of graphene on Ni occurs by a carbon segregation process due to the high solubility of carbon in Ni at elevated temperatures. At high temperature (1000 °C), hydrocarbon gas (CH_4) decomposes at the catalyst Ni surface, and a large amount of carbon is dissolved in Ni. Graphene forms when the sample is cooled down and carbon segregates on the Ni surface. Different sample cooling rates cause different carbon segregation behaviors, strongly affecting the thickness and the quality of the graphene films (Fig. 4.2). With an extremely fast cooling rate (quench effect), carbon atoms lose the mobility before they can diffuse any further. A wide range of medium cooling rate (~10 °C s^{-1} in this case) allows a finite amount of carbon to segregate on the Ni surface, forming graphene films. An extremely slow cooling rate results in a deeper diffusion of carbon atoms, so no segregation is found on the metal surface. This study demonstrates the significance of controlling sample cooling rates in graphene segregation on Ni substrates, and sheds light on the growth mechanism of CVD synthesis of graphene on Ni surface.

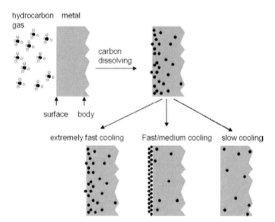

Figure 4.2 Schematic representation of carbon segregation process on Ni surface. Reprinted with permission from [31]. Copyright 2008, American Institute of Physics.

Later, large-scale CVD synthesis of few-layer graphene films on polycrystalline Ni and transfer of the films to arbitrary substrates were further demonstrated by the Kong's group [30]. Electron beam (e-beam) evaporated Ni films (500 nm in thickness) on SiO_2/Si wafer (1–2 cm^2) were used as substrates and annealed at 900–1000 °C in Ar and H_2 atmosphere for 10–20 min before CVD synthesis to generate single-crystal Ni grains of size up to 20 μm. The synthesis of graphene was carried out at 900 or 1000 °C by introducing CH_4 for 5–10 min at ambient pressure. An optical image of as-produced graphene films on Ni is shown in Fig. 4.3a. The transfer of the graphene films to other substrates was realized by a poly(methyl methacrylate) (PMMA)-assisted wet-transfer method. A thin layer of PMMA was coated on the graphene/Ni sample as a supporting material for graphene transfer. The underlying Ni was etched off in HCl solution (~3% vol.), resulting in a free-standing PMMA/graphene membrane that could be readily placed on any target substrate. The PMMA layer was finally removed by acetone and DI water. The transferred graphene films (on SiO_2/Si substrate) were found to be continuous over the entire surface and exhibited a large fraction of single- and bilayer graphene regions with up to 20 μm in lateral size (black arrows in Fig. 4.3b). The color variation in the optical image indicates the massive non-uniformity in the film thickness (1–12 layers of graphene, as verified by atomic force microscopy (AFM) measurement). It was also observed that the morphologies of the graphene films (Fig. 4.3b) were correlated to the microstructure of the Ni film (Fig. 4.3a). The regions of single- and bilayer graphene were grown on the large Ni grains with almost the same lateral size (black arrows in Fig. 4.3a), while most of the multi-layer graphene regions were grown on the grain boundaries of the Ni film. In fact, the segregation of carbon occurs preferentially and heterogeneously at the grain boundaries of polycrystalline Ni substrates so that the thickness of as-segregated graphene films at the grain boundaries is significantly larger than within the Ni grains. It was proposed that further work was needed on engineering the polycrystalline Ni substrates in order to improve the quality of the graphene films. Moreover, graphene films were also transferred on glass substrate for the measurements of optical transmittance and sheet resistance. The films showed sheet resistance of 770–1000 Ω sq^{-1} with ~90% transmittance in 500–1000 nm wavelength range, suggesting promising applications as transparent conductive electrodes. Similar results were also obtained by Arco et al. [41], showing the synthesis of few-layer graphene films by ambient pressure CVD of CH_4 on e-beam evaporated Ni films (100 nm thick) at 800 °C and the transfer of the graphene films to glass substrates for their characterization as transparent electrodes and to SiO_2/Si for device fabrication (graphene based FETs). Their typical graphene samples were relatively thicker, transmitting ~80% of light in the visible wavelength range, and the samples had very high sheet resistance of ~68 kΩ sq^{-1}, which indicates large amount of defects in the transferred graphene films. The high defects could be formed during CVD synthesis and/or the film transfer process.

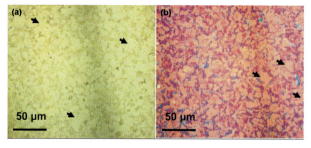

Figure 4.3 Optical images of (a) Ni film after CVD process with graphene films on the surface. (b) The same graphene films transferred to SiO_2/Si substrate showing continuity and high density of single- and bilayer graphene regions with size of 1–20 μm (indicated by black arrows). These regions are grown on the large Ni grains with almost the same lateral sizes (identified by black arrows in the image a). Reprinted (adapted) with permission from [30]. Copyright (2009) American Chemical Society.

At about the same time, an independent study was reported by Kim et al. on CVD synthesis and transfer of high-quality graphene films in a large scale, and the applications of the films in flexible, stretchable, foldable transparent electronics were demonstrated [13]. The graphene films were grown on 300 nm Ni film (e-beam evaporated on SiO_2/Si substrate) using gas mixtures (CH_4, H_2, Ar) at 1000 °C, followed by a fast cooling (~10 °C s^{-1}) to room temperature in Ar atmosphere. Such cooling rate was found to be critical in the formation of few-layer graphene films on Ni substrate [31], and it was proposed here that this fast cooling rate is also important for an efficient transfer of the films. Two methods were developed in this study to transfer the graphene films. As-produced graphene samples here typically had a structure of graphene/Ni/SiO$_2$/Si, and used an aqueous FeCl$_3$ (1 M) solution to remove the Ni layer, so that the graphene film separated from the substrate and floated on the solution surface (Fig. 4.4b), ready to be transferred to other substrates. By using FeCl$_3$ solution instead of a strong acid such as HNO$_3$, the etching of the Ni layers is a slow and mild process without causing any gaseous products or precipitates, while the acid etchant often produces hydrogen bubbles and damages the graphene films. A dry-transfer process was also developed using a soft substrate such as polydimethylsiloxane (PDMS) stamp. In this dry-transfer process, a PDMS stamp was first attached to the graphene sample and the Ni layer was etched away by FeCL$_3$ solution, leaving the adhered graphene film on the PDMS substrate (Fig. 4.4c). Graphene film was then transferred simply by conformal contact with any target substrates (Fig. 4.4d, e). Figure 4.5a shows an optical image of graphene film transferred on 300 nm SiO$_2$/ Si substrate, and the corresponding Raman mapping is shown in Fig. 4.5b. Raman spectroscopy is a powerful yet relatively simple method to identify the number of graphene layers and the presence of defects in graphene [42–44]. The most intense Raman features of graphene are the G band at ~1580 cm^{-1} and the 2D band at ~2700 cm^{-1}. The G band is associated with the doubly degenerate phonon mode (E$_{2g}$ symmetry) at the center of the Brillouin zone and the 2D band corresponds to the second-order of zone-boundary phonons. Another feature, the so-called disorder-induced D band, at ~1350 cm^{-1}, is completely absent for high-quality graphene (such as graphene obtained by mechanical exfoliation) but can be observed when samples contain high density of defects or impurities. The Raman fingerprint for single-, bi, and few-layer graphene is largely connected to the 2D band's shape, position, and intensity relative to the G band. The 2D band for single-layer graphene exhibits a symmetrical Lorentzian peak with a full width at half maximum of ~24 cm^{-1}, and the intensity ratio of 2D band to G band (I_{2D}/I_G) is larger than 2. The value of I_{2D}/I_G decreases as the number of graphene layers increases, with 1–2 for bilayer graphene and less than 1 for 3 or more layers. As indicated in Fig. 4.5c (the typical Raman spectra obtained from the marked spots in Fig. 4.5a, b), the film mostly consists of less than a few layers of graphene, and the D band from all the spectra is in the noise level, indicating the overall good quality of the graphene film. The transferred graphene films on the SiO$_2$/Si substrate show electron mobility larger than 3700 cm^{2}V^{-1}s^{-1} at low temperature, and exhibit the half-integer quantum Hall effect [4, 5], indicating that the quality of the films is comparable to that of mechanically exfoliated graphene [3]. Moreover, the graphene films transferred on quartz plate show very low sheet resistance of 280 Ω sq^{-1} with ~80% optical transmittance. In addition to the good electrical and optical properties, the graphene films also show excellent mechanical stability. Graphene films transferred to ~100 μm polyethylene terephthalate (PET) substrate coated with ~200 μm PDMS layer resist up to ~18.7% strain, which obviously outperforms the conventional indium tin oxide (ITO)-based flexible electronics [45].

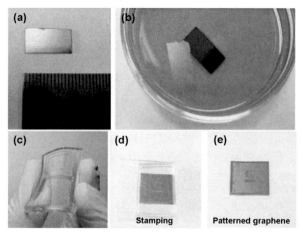

Figure 4.4 (a) As-produced graphene sample with structure of graphene/Ni (300 nm)/SiO$_2$ (300 nm)/Si. (b) Graphene film floating on 1M FeCl$_3$ aqueous solution after the nickel layer is etched off. (c) Graphene film adhered on the PDMS stamp showing transparency and flexibility. (d) Pressing the PDMS stamp against a SiO$_2$ substrate. (e) Peeling back the stamp and leaving graphene film on the SiO$_2$ substrate. Reprinted with permission from Macmillan Publishers Ltd: Nature [13], copyright 2009.

The CVD graphene on polycrystalline Ni catalyst typically features non-uniform film thickness of single- and few-layer regions over the entire substrate surface. The lack of control of the number of graphene layers can be largely attributed to the fact that the segregation of carbon upon sample cooling occurs preferentially

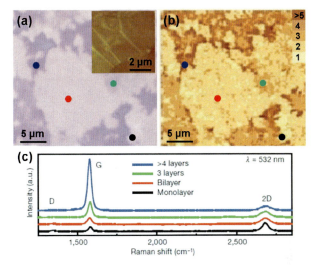

Figure 4.5 (a) Optical image of the graphene film transferred on 300 nm SiO$_2$/Si. The inset is the AFM image showing typical graphene wrinkles. (b) Raman mapping of the same area in image a. (c) Raman spectra measured from the marked spots in image a and b. Laser excitation wavelength is 532 nm. Reprinted with permission from Macmillan Publishers Ltd: Nature [13], copyright 2009.

at the grain boundaries of the polycrystalline Ni substrates. Thus, the thickness and quality of graphene films differ significantly along the surface of Ni. Recently, Iwasaki et al. used epitaxial Ni (111) thin film (150 nm) on MgO (111) as the substrate for graphene growth and obtained large-scale single-crystal single-layer graphene films as determined by low-energy electron diffraction, angle-resolved ultraviolet photoelectron spectroscopy, and X-ray photoelectron spectroscopy [46]. The graphene growth was performed at a pressure of 1.0×10^{-6} mbar for 5 min with C$_3$H$_6$ as the carbon source, and the growth temperature was set at 600–680 °C (chosen to prevent carbon atoms from diffusing into the Ni film). It was found that the most important factor for the growth of such single-crystal graphene was the preparation of the single-crystal Ni (111) films. After growth, the film was also transferred onto SiO$_2$/Si substrate using polycarbonate (PC) as the supporting material and 1 M FeCl$_3$ solution as the etchant. Figure 4.6a shows a photograph of the transferred graphene film on SiO$_2$/Si. The film is continuous with a millimeter-scale single domain structure. Optical image of the sample (Fig. 4.6b) shows uniform contrast indicating no variation in film thickness and no grain boundaries over the whole area. While, as shown in Fig. 4.3, graphene grown on polycrystalline Ni film by high-temperature segregation exhibits massive non-uniformity in the film thickness consisting of single-, bi, and few-layer graphene regions [30]. It was noticed that there were some holes in the graphene film which could be caused by defects of the Ni (111) film or by damages during the transfer process. Raman spectrum (Fig. 4.6c) shows the two primary features of graphene, a G band at ~1587 cm^{-1}, and a 2D band at ~2700 cm^{-1}. The I_{2D}/I_G intensity ratio of ~3.3 and the single Lorentz lineshape for the 2D band indicate that the film is a single-layer graphene. The D band (~1356 cm^{-1}) corresponding to the defect level in graphene is of comparable intensity as reported for graphene grown on polycrystalline Ni substrates at high growth temperatures [30].

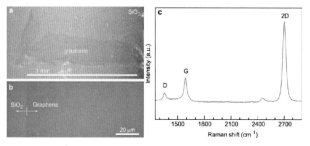

Figure 4.6 (a) Photograph of transferred graphene film on SiO$_2$/Si substrate. (b) Optical image of the same sample showing the boundary between graphene and SiO$_2$. (c) Raman spectrum of the graphene film. Excitation laser wavelength is 488 nm. Reprinted (adapted) with permission from [46]. Copyright (2011) American Chemical Society.

4.4 CVD Graphene on Copper

Unlike Ni having a very high carbon solubility (>0.1 atomic %), Cu shows a much lower solubility of carbon (<0.001 atomic %) and it does not form any metal-carbide compounds. The differences in the affinity towards carbon between Ni and Cu can be attributed to the different electron configurations of these two metals. Cu has a symmetrical distribution of electrons in the filled 3d shell [Ar] $3d^{10}$ $4s^1$ which is the most stable configuration (along with the half filling $3d^5$), while, the asymmetrical distribution of electrons in the 3d shell of Ni [Ar] $3d^8 4s^2$ makes Ni more active with carbon. Taking advantage of the low activity with carbon, Cu has been considered as a special catalyst for the growth of several carbon allotropes such as diamond [47, 48] and carbon nanotubes [49, 50]. The use of Cu for graphene growth was reported very recently [32, 51]. Before that, the growth of graphite on Cu was unintentionally achieved by Lee et al. in 1991 [52]. Diamond was initially expected to grow on Cu surfaces using carbon-ion-implantation-outdiffusion method. Specifically, Cu (100), (110), (111), and (210) single crystals are implanted by carbon ions to doses up to 10^{18} ions cm^{-2} at 200 keV in the temperature range of 850–1000 °C, and then held at the high temperatures leading to the outdiffusion of implanted carbon to the surface where 40–90 nm thick graphitic films are formed. But it was found that the as-produced films on Cu surfaces by the implantation-outdiffusion mechanism were actually (0001)-oriented graphite instead of diamond. In a later study of diamond nucleation and growth on single-crystal Cu by Ong et al. [53], they found that the graphite films formed on Cu surfaces by the implantation-outdiffusion technique can, in fact, greatly enhance diamond crystallite nucleation.

The CVD growth of graphene on Cu is in principle simple and straightforward due to the extremely low solubility of carbon in Cu. It occurs by a surface adsorption process. At high temperatures, Cu catalytically decomposes hydrocarbon gases and carbon diffuses on the Cu surface to form graphene with minimal diffusion into the Cu substrate. Once the Cu surface is fully covered by a layer of graphene, graphene growth terminates, because of the absence of a catalyst (exposed Cu surface) to further decompose the hydrocarbons. Thus, the growth process is somehow self-limiting, and enables that the formed graphene film has uniform thickness (typically single-layer graphene) all over the substrate surface. The initial [32] and subsequent follow-on studies [12, 18, 54–60] have already demonstrated the great potential of using Cu as a catalyst to produce large-scale, high-quality graphene films, transferrable to arbitrary substrates for graphene based electronic and optoelectronic applications.

In 2009, Ruoff's group managed to grow large-scale (of the order of centimeters) single-layer graphene films (>95% area) on 25-

μm thick Cu foils by low pressure CVD of CH_4 at 1000 °C [32]. By studying the graphene samples on Cu grown for different times, it was found that the graphene growth on Cu involves the nucleation and growth of individual graphene grains which upon time coalesce together into a large and continuous graphene film. The self-limiting growth of graphene on Cu was demonstrated by the fact that the graphene films still show very similar features (>95% single-layer) even with a much longer growth time of 60 min. It was claimed that once a continuous graphene film formed on the catalytic Cu surface, the growth of extra graphene layer is inhibited because of the absence of Cu to catalytically decompose the carbon precursor gas. The transfer of graphene was achieved by etching the Cu foils in an aqueous solution of iron nitrate (0.05 g ml^{-1}) with PMMA or PDMS as the supporting material. The etching process usually takes over night. It should be noted that since the two sides of Cu surfaces are exposed to CH_4, graphene is grown on both sides of the Cu foil. Although two films can be exfoliated during the etching process, in general, only one film is transferred for simplifying the film processing. Before the etching of Cu foils, graphene on the side of Cu without PMMA/PDMS cover usually will be removed by a simple polishing process [14]. Alternatively, the etching of the uncovered graphene can also be performed using a gentle O_2 plasma treatment [61–63]. The graphene based dual-gated FETs fabricated on SiO_2/Si substrates show high electron mobilites of up to 4050 cm^{2}V^{-1}s^{-1} at room temperature, suggesting good quality of the films. For the transfer of graphene by the PMMA-assisted method, it was found by Li et al. that the transfer process can cause graphene films to form cracks [14]. When graphene protected with a PMMA layer on the top (PMMA/graphene) is transferred on some other substrate (e.g., SiO_2/Si),small gaps appear between the graphene and the SiO_2/Si substrate surface, i.e., the graphene does not make full contact with the substrate. In this case, the graphene film will easily crack when the PMMA layer is dissolved away. An improved transfer process was then developed by introducing a second PMMA coating after the PMMA/graphene is placed on the SiO_2/Si substrate. The new liquid PMMA drops partially or fully dissolve the precoated PMMA layer, mechanically relaxing the underlying graphene and leading to a better contact with the substrate. Single-layer graphene transferred on glass by this improved PMMA-assisted technique shows a sheet resistance of 2.1 kΩ sq^{-1} with ~97.7% transmittance, while for the four layers of such graphene (transferred four times, layer by layer transfer), the sheet resistance is as low as 350 Ω sq^{-1} with transmittance of ~90%. The high electrical conductivity and high optical transmittance of the CVD-grown graphene films using Cu as catalyst make them attractive for transparent electrode applications.

The production of single-layer graphene can be in large scale using the developed CVD method on Cu substrates. The size of graphene is only limited by the size of the used substrates and the CVD system [12, 18]. Hong's group adapted the Cu-based CVD technique to a scalable, industrial manufacturing process by roll-to-roll production of 30 in predominantly single-layer graphene films [12]. The graphene is grown on a roll of Cu foil with dimensions of 30 in in diagonal length in an 8 in wide tubular quartz reactor. The growth temperature and growth pressure are 1000 °C and 460 mtorr, respectively, in the mixtures of CH_4 and H_2. Before graphene growth, the Cu foil is annealed in H_2 atmosphere at 1000 °C for 30 min. The annealed Cu foil shows smoother surface, and the Cu grain size increases up to ~100 μm from a few micrometers. Annealing is a heat treatment which can be used to soften metals and alloys, relieve internal stresses, reduce defects, and refine the microstructures. The process includes heating (above the recrystallization temperature), holding, and then cooling. During annealing, deformed and stressed crystals (e.g., Cu) are transformed into unstressed crystals by recovery, recrystallization, and grain growth. It has been found that such annealing process, prior to graphene deposition, is critical for high-quality graphene growth [32, 54, 64]. The graphene films are transferred on target substrates by a cost- and time-effective roll-to-roll transfer technique as illustrated in Fig. 4.7. Multi-layer graphene films (e.g., four layers) can be prepared simply by repeating the transfer steps on the same substrate. One layer of such graphene film (transferred once on quartz substrate) shows single-layer feature of ~97.4% optical transmittance, and has a very low sheet resistance of ~125 Ω sq^{-1}. Electrical transport measurements of the single-layer graphene film show an effective Hall mobility of 7350 cm^{2}V^{-1}s^{-1} at low temperatures, close to that of mechanically exfoliated graphene samples from HOPG, and the half-integer quantum Hall effect, suggestive of high quality of the film. By stacking four layers of graphene (transferred four times) and applying chemical doping (p-doping with HNO3 in this study), the sheet resistance is measured to be as low as ~30 Ω sq^{-1} but still with ~90% transmittance. The graphene-based touch panels are found to resist up to 6% strain, and it was claimed that this is not limited by the graphene itself, but by the printed silver electrodes. A good industrial standard ITO film can transmit ~90% of light while have a sheet resistance of less than 100 Ω sq^{-1}. But ITO can easily wear out when used in flexible electronics (easily breaks under ~2–3% strain) [45]. Obviously, the CVD-grown graphene films obtained by Hong and co-workers can be a promising candidate replacing ITO as transparent electrodes for electronics and optoelectronics.

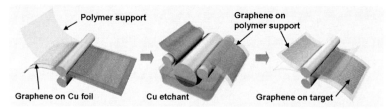

Figure 4.7 Schematic representation of the roll-to-roll production of graphene films grown on Cu foils. The graphene film grown on a Cu foil is attached to a polymer support by rolling. An etching bath then removes the Cu foil, and graphene is transferred from the polymer support onto a target substrate. The process can be repeated to get stacked multi-layer graphene films. Wet-chemical doping can be carried out in a bath stage similar to that used for etching. Reprinted with permission from Macmillan Publishers Ltd: Nature Nanotechnology [12], copyright 2010.

CVD graphene on Cu has been proposed to grow by a surface adsorption process because of the very limited carbon solubility in Cu. The growth is considered to be self-limiting, where the produced films are single-layer graphene with a very small percentage (less than 5%) of multi-layer domains [32]. Evidence for growth mechanism of CVD graphene on Cu at low pressure had been provided by Li et al. using carbon isotope labeling of the CH$_4$ precursor gas in conjunction with Raman mapping [65]. By taking advantage of the separation of the ^{12}C and ^{13}C Raman modes, they sequentially introduced ^{12}CH$_4$ and ^{13}CH$_4$ at growth stage and monitored the spatial distribution of graphene domains during the growth process. Raman results clearly show that graphene growth on Cu is dominated by surface adsorption, where the spatial distribution of ^{12}C and ^{13}C follows the introduce sequence of the two types of precursors. The growth is self-limiting since the deposition of the first layer of graphene leads to the passivation of the catalytic Cu surface so that the multi-layer growth is severely hindered. In contrast, in the case of CVD growth of graphene on Ni, growth is dominated by a carbon segregation process. Raman analysis reflects that the introduced isotopic carbon first diffused into the Ni substrate at high temperatures, mixes and then segregates at the surface of Ni when the sample is cooled down forming few-layer graphene films with a uniform mixture of ^{12}C and ^{13}C. However, recent studies have demonstrated that the previously observed self-limiting growth of graphene on Cu by low pressure CVD does not apply under certain conditions [55, 59, 66–69]. Lee et al. reported low pressure CVD growth of wafer-scale bilayer graphene films on Cu [55]. They speculate that the key points for the bilayer graphene growth are the depletion of hydrogen, high vacuum, and, most importantly, slower cooling process ($\sim$18 °C min^{-1}) compared to previous single-layer graphene synthesis (40–300 °C min^{-1}) [32]. Sample cooling rate has been found to be critical in determining the uniformity and thickness of CVD graphene films on Ni substrates since the segregation of carbon occurs heterogeneously during the cooling process [31]. But when graphene is grown on Cu, cooling rate has not been expected to affect the quality of graphene films in terms of thickness and uniformity, considering Cu's much lower affinity towards carbon and its surface-catalyzed growth process [59, 65]. Growth of multi-layer graphene on Cu has also been frequently observed under ambient pressure CVD conditions [59, 67]. Bhaviripudi et al. argued that the reason ambient pressure favors the growth of non-uniform multi-layer graphene is the lowered mass transport rate of active carbon species through the boundary layer (due to steady state gas flow) [68]. A variation in the thickness of the boundary layer can result in the thickness non-uniformity of as-grown graphene films. The reduction of carbon precursor supply to the ppm levels is needed to deposit uniform single-layer graphene on Cu at such high ambient pressure. On the other hand, at lower growth pressures, mass transport through the boundary layer is no longer the rate limiting step but the surface reaction is rate limiting. As long as the temperature across the growth substrate is maintained uniform, the thickness of the graphene will be uniform.

4.5 Single-Crystal Graphene

The large-scale synthetic CVD graphene films produced on Ni/Cu substrates so far are typically polycrystalline, consisting of numerous crystal grains separated by grain boundaries. In the case of using Cu as catalyst more suitable for the synthesis of graphene films with uniform thickness, a typical CVD process starts with the nucleation of individual graphene grains randomly distributed across the Cu surface. These grains continue to grow with time and eventually merge together to form the continuous polycrystalline films. Recent results have shown that the individual graphene grains before the formation of continuous films can be in different shapes and crystal structures depending on CVD conditions [56, 57, 67, 70, 71]. Yu et al. reported the growth of single-crystal single-layer graphene grains in a hexagonal shape on Cu by ambient pressure CVD [71], and found that the grains showed no definite epitaxial relationship with the underlying Cu substrate and can be grown continuously across the Cu grain boundaries without any apparent shape distortion (Fig. 4.8a), indicating weak graphene–Cu interactions. The result is consistent with recent STM studies of CVD graphene on polycrystalline Cu and theoretical calculations of graphene on single-crystal Cu (111) [72–75]. The growth of hexagonally shaped few-layer graphene grains by CVD on Cu at ambient pressure has also been reported by Robertson et al. [67], who characterized these grains as single crystals with AB Bernal stacking without the intrinsic rotational stacking faults (Fig. 4.8b). On the other hand, CVD graphene on Cu grown at low pressure (0.1–1 Torr) and ultrahigh vacuum before the formation of continuous films can be flower-shaped single-crystal single-layer grains and four-lobed polycrystalline single-layer islands, respectively [32, 57, 70]. According to the thermodynamics-based Wulff construction [76, 77], the ultimate equilibrium shape for crystals dictated by surface energy minimization must be a polyhedron. This growth shape is thermodynamically stable. However, the flower/lobe shape of the graphene grains is obviously contrary to the Wulff construction, and should not be determined by minimizing surface energy, instead, it could be a consequence of growth kinetics. By monitoring the profile of a lobe as a function of time using low-energy electron microscopy, Wofford et al. proposed a growth model for the shape evolution of the four-lobed graphene islands that is dominated by edge kinetics with an angularly dependent growth velocity [70]. This growth kinetics determines the steady-state shape for the lobes than can be determined from the thermodynamics-based Wulff construction.

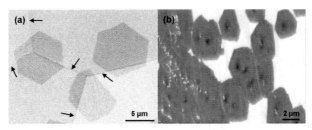

Figure 4.8 SEM images of (a) hexagonally shaped single-layer graphene grains grown continuously across Cu grain boundaries. Cu grain boundaries are highlighted by black arrows [71]. (b) Hexagonally shaped few-layer graphene grains on Cu substrates. Cu surface without graphene is relatively brighter (images a and b). Few-layer graphene domains are relatively darker (image b). Reprinted (adapted) with permission from [67]. Copyright (2011) American Chemical Society.

Upon coalescence of the individual graphene grains which may have different orientations into continuous films, graphene grain boundaries are formed. Grain boundaries (which, by definition, are defective) in graphene films have been found to consist of a series of pentagons, heptagons, and distorted hexagons [78], and severely affect the quality of graphene, e.g., scattering charge carriers [79], and weakening mechanical strength [80]. The polycrystalline nature of large-scale CVD graphene can be a problem for graphene based devices, since, so far, no effective method can avoid grain boundaries in the fabricated graphene devices, especially in the case of device arrays and circuits. It is therefore necessary to synthesize either high-quality graphene films with large grain size (few or even no grain boundaries), or individual single-crystal graphene grains in a controllable arrangement. Recently, Li et al. developed a two-step low-pressure CVD process to synthesize single-layer graphene films with larger grain size up to hundreds of square micrometers, compared with grain size of only tens of square micrometers in graphene films obtained by the previous one-step CVD [32, 56]. The idea of the two-step growth process is based on the observations that high temperature and low CH_4 partial pressure prefer to generate a low density of graphene nucleation, while high CH_4 partial pressure is preferred for the growth of continuous graphene films. Electrical transport measurements performed on back-gated FETs show that graphene films with large grain size have higher mobilities (up to 16000 cm^{2}V^{-1}s^{-1}) than those with small grains (mobilities in the range of 800–7000 cm^{2}V^{-1}s^{-1}), highly suggesting the detrimental effect of grain boundaries in graphene on electronic transport. In a follow-up study from the same group, single-crystal graphene grains with size up to 0.5 mm were grown on the inside of Cu foil enclosures by low-pressure CVD of CH$_4$ at 1035 °C. The CVD conditions were similar to those previously reported [56, 65], but had slightly lower CH$_4$ partial pressures in this case. Fig. 4.9a shows a Cu foil enclosure used as the growth substrate. The large size graphene single crystals were found to grow on the inside of the Cu enclosure (Fig. 4.9b), while, the graphene growth on the outside showed similar results as that reported previously [56], smaller grains in a high density. The differences in graphene growth on the inside and outside of the Cu enclosure was attributed to the much lower partial pressure of CH$_4$ and the "static" environment inside the enclosure which enable a sparse nucleation of graphene and the growth of graphene grains in a large size. The mobility for the graphene films with such large single-crystal grains (0.5 mm) is found to be larger than 4000 cm^{2}V^{-1}s^{-1}. This study offers an approach for producing large size single-crystal graphene grains, but continuing to increase the grain size would be difficult since the density of graphene nucleation on Cu has largely been determined by the CVD parameters and the Cu surface conditions. In addition, the lack of control in grain distribution may also limit its further applications.

Very recently, Yu et al. developed a method using pre-patterned growth seeds (small graphene flakes) to control CVD graphene nucleation on Cu, and synthesized arrays of single-crystal graphene grains on Cu at the pre-determined sites [71]. The seed crystals are patterned from a pre-grown CVD multi-layer graphene film on Cu by e-beam lithography, and serve as graphene nucleation centers during a CVD re-growth process. Figure 4.10a shows an array of the patterned seeds on Cu foil. After re-growth, the graphene grains are well arranged as shown in Fig 4.10b. This seeded growth method smartly bypasses the challenge of synthesis of wafer-scale single-crystal graphene films and offers a highly potential way to fabricate graphene based devices free of graphene grain boundaries and with better performances. But more work is still needed to determine the effects of different types and properties of seeds on the growth of graphene grains, and to further improve the yield of single-crystal grains.

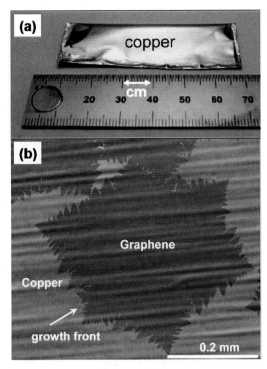

Figure 4.9 (a) Cu foil enclosure used as the growth substrate. (b) Graphene single crystals grown on the inside of the Cu enclosure. Reprinted (adapted) with permission from [57]. Copyright (2011) American Chemical Society.

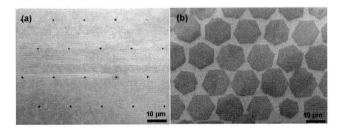

Figure 4.10 SEM images of (a) An array of growth seeds (seen as dots) patterned from a pre-grown CVD multi-layer graphene film on Cu by e-beam lithography. (b) Typical graphene grain array grown from the seeds [71].

4.6 Summary

Large-scale and transferable graphene films grown on transition metals (such as Ni and Cu) by CVD of hydrocarbons hold great promise for future graphene based nanotechnology. Although significant advances have been made in optimizing CVD parameters for the synthesis of high-quality graphene and utilizing graphene in electronics and optoelectronics, great challenges still exist before graphene can fully realize its promise in future nanotechnology applications. For example, it has been proposed that the CVD growth of graphene on metals can be a surface adsorption process (e.g., on Cu) or a carbon segregation process upon cooling (e.g., on Ni) depending on the carbon solubility in the metals. However, to further improve the quality of such CVD graphene films (e.g., uniform thickness, large grain size, low defects), a better understanding and optimization of the growth process is still required. In addition, direct synthesis of graphene films with controllable number of layers would be a key breakthrough leading to new possibilities of graphene based electronic and photonic devices. Another challenging issue is the development of a more reliable and also manufacturable transfer technique aiming to maintain the quality of graphene films after being transferred on arbitrary substrates. Effort is also needed in controlling doping of graphene during the CVD growth and the following transfer processes. It can be expected that by overcomingthese challenges, CVD graphene films could revolutionize the future of nanotechnology.

References

1. Geim A.K., Novoselov K.S. (2007) The rise of graphene, *Nat Mater*, **6**, 183–191.

2. Wallace P.R. (1947) The band theory of graphite, *Phys Rev*, **71**, 622–634.

3. Novoselov K.S., Geim A.K., Morozov S.V., et al. (2004) Electric field effect in atomically thin carbon films, *Science*, **306**, 666–669.

4. Novoselov K.S., Geim A.K., Morozov S.V., et al. (2005) Two-dimensional gas of massless Dirac fermions in graphene, *Nature*, **438**, 197–200.

5. Zhang Y.B., Tan Y.W., Stormer H.L., Kim, P. (2005) Experimental observation of the quantum Hall effect and Berry's phase in graphene, *Nature*, **438**, 201–204.

6. Ozyilmaz B., Jarillo-Herrero P., Efetov D., Abanin D.A., Levitov L.S., Kim P. (2007) Electronic transport and quantum Hall effect in bipolar graphene p-n-p junctions, *Phys Rev Lett*, **99**, 166804.

7. Bolotin K.I., Sikes K.J., Jiang Z., et al. (2008) Ultrahigh electron mobility in suspended graphene, *Solid State Commun*, **146**, 351–355.

8. Lee C., Wei X.D., Kysar J.W., Hone J. (2008) Measurement of the elastic properties and intrinsic strength of monolayer graphene, *Science*, **321**, 385–388.

9. Nair R.R., Blake P., Grigorenko A.N., et al. (2008) Fine structure constant defines visual transparency of graphene, *Science*, **320**, 1308.

10. Lin Y.M., Dimitrakopoulos C., Jenkins K.A., et al. (2010) 100-GHz transistors from wafer-scale epitaxial graphene, *Science*, **327**, 662.

11. Lin Y.M., Jenkins K.A., Valdes-Garcia A., Small J.P., Farmer D.B., Avouris P. (2009) Operation of graphene transistors at gigahertz frequencies, *Nano Lett*, **9**, 422–426.

12. Bae S., Kim H., Lee Y., et al. (2010) Roll-to-roll production of 30-inch graphene films for transparent electrodes, *Nat Nanotechnol*, **5**, 574–578.

13. Kim K.S., Zhao Y., Jang H., et al. (2009) Large-scale pattern growth of graphene films for stretchable transparent electrodes, *Nature*, **457**, 706–710.

14. Li X.S., Zhu Y.W., Cai W.W., et al. (2009) Transfer of large-area graphene films for high-performance transparent conductive electrodes, *Nano Lett*, **9**, 4359–4363.

15. Park H., Rowehl J.A., Kim K.K., Bulovic V., Kong J. (2010) Doped graphene electrodes for organic solar cells, *Nanotechnology*, **21**, 505204.

16. Mohanty N., Berry V. (2008) Graphene-based single-bacterium resolution biodevice and DNA transistor: Interfacing graphene derivatives with nanoscale and microscale biocomponents, *Nano Lett*, **8**, 4469–4476.

17. Ohno Y., Maehashi K., Yamashiro Y., Matsumoto K. (2009) Electrolyte-gated graphene field-effect transistors for detecting pH protein adsorption, *Nano Lett*, **9**, 3318–3322.

18. Wu W., Liu Z.H., Jauregui L.A., et al. (2010) Wafer-scale synthesis of graphene by chemical vapor deposition and its application in hydrogen sensing, *Sens Actuators B: Chem*, **150**, 296–300.

19. Stankovich S., Dikin D.A., Piner R.D., et al. (2007) Synthesis of graphene-based nanosheets via chemical reduction of exfoliated graphite oxide, *Carbon*, **45**, 1558–1565.

20. Tung V.C., Allen M.J., Yang Y., Kaner R.B. (2009) High-throughput solution processing of large-scale graphene, *Nat Nanotechnol*, **4**, 25–29.

21. Shin H.J., Kim K.K., Benayad A., et al. (2009) Efficient reduction of graphite oxide by sodium borohydrilde and its effect on electrical conductance, *Adv Funct Mater*, **19**, 1987–1992.

22. Pei S.F., Zhao J.P., Du J.H., Ren W.C., Cheng H.M. (2010) Direct reduction of graphene oxide films into highly conductive and flexible graphene films by hydrohalic acids, *Carbon*, **48**, 4466–4474.

23. Fan Z.J., Wang K., Yan J., et al. (2011) Facile synthesis of graphene nanosheets via Fe reduction of exfoliated graphite oxide, *ACS Nano*, **5**, 191–198.

24. Eda G., Fanchini G., Chhowalla M. (2008) Large-area ultrathin films of reduced graphene oxide as a transparent and flexible electronic material, *Nat Nanotechnol*, **3**, 270–274.

25. Wang S., Ang P.K., Wang Z.Q., Tang A.L.L., Thong J.T.L., Loh K.P. (2010) High mobility, printable, and solution-processed graphene electronics, *Nano Lett*, **10**, 92–98.

26. Unarunotai S., Murata Y., Chialvo C.E., et al. (2009) Transfer of graphene layers grown on SiC wafers to other substrates and their integration into field effect transistors, *Appl Phys Lett*, **95**, 202101.

27. Land T.A., Michely T., Behm R.J., Hemminger J.C., Comsa G. (1992) STM investigation of single layer graphite structures produced on Pt(111) by hydrocarbon decomposition, *Surf Sci*, **264**, 261–270.

28. Sutter P.W., Flege J.I., Sutter E.A. (2008) Epitaxial graphene on ruthenium, *Nat Mater*, **7**, 406–411.

29. Coraux J., N'Diaye A.T., Busse C., Michely T. (2008) Structural coherency of graphene on Ir(111), *Nano Lett*, **8**, 565–570.

30. Reina A., Jia X.T., Ho J., et al. (2009) Large-area, few-layer graphene films on arbitrary substrates by chemical vapor deposition, *Nano Lett*, **9**, 30–35.

31. Yu Q.K., Lian J., Siriponglert S., Li H., Chen Y. P., Pei S.S. (2008) Graphene segregated on Ni surfaces and transferred to insulators, *Appl Phys Lett*, **93**(3), 113103.

32. Li X.S., Cai W.W., An J.H., et al. (2009) Large-area synthesis of high-quality and uniform graphene films on copper foils, *Science*, **324**, 1312–1314.

33. Cao H.L., Yu Q.K., Jauregui L.A., et al. (2010) Electronic transport in chemical vapor deposited graphene synthesized on Cu: Quantum Hall effect and weak localization, *Appl Phys Lett*, **96**, 122106.

34. Banerjee B.C., Walker P.L., Hirt T.J. (1961) Pyrolytic carbon formation from carbon suboxide, *Nature*, **192**, 450–451.

35. Karu A.E., Beer M. (1966) Pyrolytic formation of highly crystalline graphite films, *J Appl Phys*, **37**, 2179–2181.

36. Belton D.N., Schmieg S.J. (1989) Loss of epitaxy during diamond film growth on ordered Ni(100), *J Appl Phys*, **66**, 4223–4229.

37. Helveg S., Lopez-Cartes C., Sehested J., et al. (2004) Atomic-scale imaging of carbon nanofibre growth, *Nature*, **427**, 426–429.

38. Somani P.R., Somani S.P., Umeno M. (2006) Planer nano-graphenes from camphor by CVD, *Chem Phys Lett*, **430**, 56–59.

39. Obraztsov A.N., Obraztsova E.A., Tyurnina A.V., Zolotukhin A.A. (2007) Chemical vapor deposition of thin graphite films of nanometer thickness, *Carbon*, **45**, 2017–2021.

40. Bao W.Z., Miao F., Chen Z., et al. (2009) Controlled ripple texturing of suspended graphene and ultrathin graphite membranes, *Nat Nanotechnol*, **4**, 562–566.

41. De Arco L.G., Zhang Y., Kumar A., Zhou C.W. (2009) Synthesis, transfer, and devices of single- and few-layer graphene by chemical vapor deposition, *IEEE Trans Nanotechnol*, **8**, 135–138.

42. Ferrari A.C. (2007) Raman spectroscopy of graphene and graphite: Disorder, electron–phonon coupling, doping and nonadiabatic effects, *Solid State Commun*, **143**, 47–57.

43. Ferrari A.C., Meyer J.C., Scardaci V., et al. (2006) Raman spectrum of graphene and graphene layers, *Phys Rev Lett*, **97**, 187401.

44. Malard L.M., Pimenta M.A., Dresselhaus G., Dresselhaus, M.S. (2009) Raman spectroscopy in graphene, *Phys Rep—Rev Sect Phys Lett*, **473**, 51–87.

45. Cairns D.R., Witte R.P., Sparacin D.K., et al. (2000) Strain-dependent electrical resistance of tin-doped indium oxide on polymer substrates, *Appl Phys Lett*, **76**, 1425–1427.

46. Iwasaki T., Park H.J., Konuma M., Lee D.S., Smet J.H., Starke U. (2011) Long-range ordered single-crystal graphene on high-quality heteroepitaxial Ni thin films grown on MgO(111), *Nano Lett*, **11**, 79–84.

47. Arnault J.C., Demuynck L., Constant L., Speisser C., Le Normand F. (1999) *Surf Eng: Sci Technol I*, 343–354.

48. Constant L., Speisser C., LeNormand F. (1997) HFCVD diamond growth on Cu(111). Evidence for carbon phase transformations by in situ AES and XPS, *Surf Sci*, **387**, 28–43.

49. Tao X.Y., Zhang X.B., Cheng J.P., Wang Y.W., Liu F., Luo Z.Q. (2005) Synthesis of novel multi-branched carbon nanotubes with alkali-element modified Cu/MgO catalyst, *Chem Phys Lett*, **409**, 89–92.

50. Zhou W.W., Han Z.Y., Wang J.Y.,et al. (2006) Copper catalyzing growth of single-walled carbon nanotubes on substrates, *Nano Lett*, **6**, 2987–2990.

51. Yu Q.K., Lian J., Siriponglert S., Li H., Chen Y.P., Pei S.S. (2008) Graphene synthesis by surface segregation on Ni and Cu, *arXiv:0804.1778v1*.

52. Lee S.T., Chen S., Braunstein G., Feng X., Bello I., Lau W.M. (1991) Heteroepitaxy of carbon on copper by high-temperature ion-implantation, *Appl Phys Lett*, **59**, 785–787.

53. Ong T.P., Xiong F.L., Chang R.P.H., White C.W. (1992) Mechanism for diamond nucleation and growth on single crystal copper surfaces implanted with carbon, *Appl Phys Lett*, **60**, 2083–2085.

54. Lee Y., Bae S., Jang H., et al. (2010) Wafer-scale synthesis and transfer of graphene films, *Nano Lett*, **10**, 490–493.

55. Lee Y.H., Lee J.H. (2010) Scalable growth of free-standing graphene wafers with copper(Cu) catalyst on SiO_2/Si substrate: Thermal conductivity of the wafers, *Appl Phys Lett*, **96**, 083101.

56. Li X.S., Magnuson C.W., Venugopal A., et al. (2010) Graphene films with large domain size by a two-step chemical vapor deposition process, *Nano Lett*, **10**, 4328–4334.

57. Li X.S., Magnuson C.W., Venugopal A., et al. (2011) Large-area graphene single crystals grown by low-pressure chemical vapor deposition of methane on copper, *J Am Chem Soc*, **133**, 2816–2819.

58. Sun Z.Z., Yan Z., Yao J., Beitler E., Zhu Y., Tour J.M. (2010) Growth of graphene from solid carbon sources, *Nature*, **468**, 549–552.

59. Yao Y.G., Li Z., Lin Z.Y., Moon K.S., Agar J., Wong C.P. (2011) Controlled growth of multilayer, few-layer, and single-layer graphene on metal substrates, *J Phys Chem C*, **115**, 5232–5238.

60. Gao L.B., Ren W.C., Zhao J.P., Ma L.P., Chen Z.P., Cheng H.M. (2010) Efficient growth of high-quality graphene films on Cu foils by ambient pressure chemical vapor deposition, *Appl Phys Lett*, **97**, 183109.

61. Aleman B., Regan W., Aloni S., et al. (2010) Transfer-free batch fabrication of large-area suspended graphene membranes, *ACS Nano*, **4**, 4762–4768.

62. Huh S., Park J., Kim K.S., Hong B.H., Bin Kim, S. (2011) Selective n-type doping of graphene by photo-patterned gold nanoparticles, *ACS Nano*, **5**, 3639–3644.

63. van der Zande A.M., Barton R.A., Alden J.S., et al. (2010) Large-scale arrays of single-layer graphene resonators, *Nano Lett*, **10**, 4869–4873.

64. Cai W.W., Zhu Y.W., Li X.S., Piner R.D., Ruoff R.S. (2009) Large area few-layer graphene/graphite films as transparent thin conducting electrodes, *Appl Phys Lett*, **95**, 123115.

65. Li X.S., Cai W.W., Colombo L., Ruoff, R.S. (2009) Evolution of graphene growth on Ni and Cu by carbon isotope labeling, *Nano Lett*, **9**, 4268–4272.

66. Lee S., Lee K., Zhong Z.H. (2010) Wafer scale homogeneous bilayer graphene films by chemical vapor deposition, *Nano Lett*, **10**, 4702–4707.

67. Robertson A.W., Warner J.H. (2011) Hexagonal single crystal domains of few-layer graphene on copper foils, *Nano Lett*, **11**, 1182–1189.

68. Bhaviripudi S., Jia X.T., Dresselhaus M.S., Kong J. (2010) Role of kinetic factors in chemical vapor deposition synthesis of uniform large area graphene using copper catalyst, *Nano Lett*, **10**, 4128–4133.

69. Srivastava A., Galande C., Ci L., et al. (2010). Novel liquid precursor-based facile synthesis of large-area continuous, single, and few-layer graphene films, *Chem Mater*, **22**, 3457–3461.

70. Wofford J.M., Nie S., McCarty K.F., Bartelt N.C., Dubon O.D. (2010) Graphene islands on Cu foils: The interplay between shape, orientation, and defects, *Nano Lett*, **10**, 4890–4896.

71. Yu Q.K., Jauregui L.A., Wu W., et al. (2011) Control and characterization of individual grains and grain boundaries in graphene grown by chemical vapor deposition, *Nat Mater*, **10**, 443–449.

72. Cho J., Gao L., Tian J.F., et al. (2011) Atomic-scale investigation of graphene grown on Cu foil and the effects of thermal annealing, *ACS Nano*, **5**, 3607–3613.

73. Khomyakov P.A., Giovannetti G., Rusu P.C., Brocks G., van den Brink J., Kelly P.J. (2009) First-principles study of the interaction and charge transfer between graphene and metals, *Phys Rev B*, **79**, 195425.

74. Rasool H.I., Song E.B., Allen M.J., et al. (2011) Continuity of graphene on polycrystalline copper, *Nano Lett*, **11**, 251–256.

75. Vanin M., Mortensen J.J., Kelkkanen A.K., Garcia-Lastra J.M., Thygesen K.S., Jacobsen K.W. (2010) Graphene on metals: A van der Waals density functional study, *Phys Rev B*, **81**, 081408.

76. Du D.X., Srolovitz D.J. (2006) Crystal morphology evolution in film growth: A general approach, *J Cryst Growth*, **296**, 86–96.

77. Herring C. (1951) Some theorems on the free energies of crystal surfaces, *Phys Rev*, **82**, 87.

78. Kim K.K.K., Lee Z., Regan W., Kisielowski C., Crommie M.F., Zettl A. (2011) Grain boundary mapping in polycrystalline graphene, *ACS Nano*, **5**, 2142–2146.

79. Gao L., Guest J.R., Guisinger N.P. (2010) Epitaxial graphene on Cu(111), *Nano Lett*, **10**, 3512–3516.

80. Huang P.Y., Ruiz-Vargas C.S., van der Zande A.M., et al. (2011) Grains and grain boundaries in single-layer graphene atomic patchwork quilts, *Nature*, **469**, 389–392.

Chapter 5

Growth and Electrical Characterization of Carbon Nanowalls

Yihong Wu

Department of Electrical and Computer Engineering, National University of Singapore, 4 Engineering Drive 3, Singapore 117576, Singapore
elewuyh@nus.edu.sg

5.1 Introduction

The last few chapters discussed the growth of graphene sheets on a foreign substrate. In these cases, the graphene sheets are mostly "lying down" on the substrate surface; therefore, interactions between graphene and the substrate are unavoidable. Although these graphene sheets are useful for certain types of applications such as electronic devices, they are not suitable for applications which rely on the large surface areas or sharp edges of graphene such as energy storage, field emission, thermal diffuser, etc. For these applications, vertically aligned and free-standing graphene sheets are more desirable. Considering the fact that 0D and 1D nanocarbon could be routinely synthesized using different types of techniques, there should not be a surprise to the fact that 2D carbon was found

---

Two-Dimensional Carbon: Fundamental Properties, Synthesis, Characterization, and Applications
Edited by Yihong Wu, Zexiang Shen, and Ting Yu
Copyright © 2014 Pan Stanford Publishing Pte. Ltd.
ISBN 978-981-4411-94-3 (Hardcover), 978-981-4411-95-0 (eBook)
www.panstanford.com

to co-exist with 0D and 1D structures during the preparation of the latter using laser ablation and arc discharge [1, 2]. Ando et al. [3] found a large amount of petal-like graphite sheets outside the flame region of the arc discharge as well as on the graphite wall surrounding the anode and cathode. These nanosheets were highly curved and interlaced with one another, forming some sponge-like structures. Wu et al. succeeded in the growth of vertically aligned 2D carbon nanostructures on various types of substrates, dubbed carbon nanowalls (CNWs), using microwave plasma-enhanced vapor deposition (MPECVD) [4–7]. In addition to MEPCVD [8–15], similar types of 2D carbon nanostructures have also been successfully grown using other techniques such as RF plasma enhanced chemical vapor deposition [16–25] and hot filament chemical vapor deposition (HFCVD) [26–29]. The unique surface morphology makes CNWs distinguish themselves from graphene sheets grown or laid down on a flat substrate [30]. A monograph on carbon nanowalls has been recently written by Hiramatsu and Hori [31]. In this chapter, we will discuss the growth and characterization of the carbon nanowalls by focusing mainly on our own work. For the characterization part, we will focus mainly on the electrical and magnetic properties, and the structural properties will be covered by Tachibana in Chapter 6.

5.2 Synthesis of Carbon Nanowalls by MPECVD

5.2.1 Growth Apparatus and Surface Morphology

Figure 5.1 shows the MPECVD system used by Wu et al. [4, 6, 7, 32] to grow carbon nanowalls. It consists of a vertical quartz tube located inside a microwave cavity with a 500 W microwave source. Inside the quartz tube are two parallel plate electrodes, placed 2 cm away from each other in the longitudinal direction of the tube, for applying a DC bias during the growth. The gases used are mixtures of CH_4 and H_2. This specific system does not come with an independent substrate heater, and the substrate temperature is controlled by the microwave power to be 650–700°C. Apart from the temperature, other important parameters which affect the growth of carbon nanostructures are H_2/CH_4 flow rate ratio and electrical field. The latter consists of both a global DC field, adjustable by the applied DC bias, and a localized AC field due to the plasma itself. The local electrical field plays an important role in determining the orientation of the individual pieces of carbon nanowalls [5].

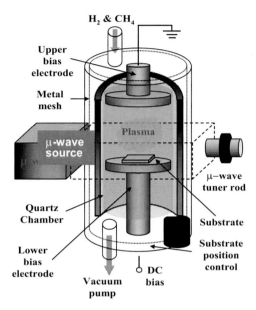

Figure 5.1 Schematic representation of the MPECVD system used by Wu et al. [33] to grow carbon nanowalls.

Wu et al. carried out a systematic study on the growth of carbon nanowalls using different H_2/CH_4 flow rate ratios [7]. It was found that there was hardly any growth with a H_2/CH_4 flow rate ratio >50, for a duration of ~5 min. When the H_2/CH_4 flow rate ratio was reduced to 30, some columnar structure of amorphous carbon were formed. Further decrease in the gas flow rate ratio led to the formation of a mixture of carbon fibers/tubes and 2D nanographite sheets. A pure form of carbon nanowalls forms when the gas flow rate ratio is in the range of 4–8. Too low a gas flow rate ratio would again lead to the formation of amorphous carbon or no growth at all due to the etching effect of hydrogen.

Figure 5.2a shows a typical SEM image of the carbon nanowalls grown under optimum conditions [7]. The distribution of the nanowalls was found to be remarkably uniform over the whole substrate surface area that was typically 1 cm × 1 cm. Figure 5.2b shows some of the nanowalls peeled off from the substrate, lying down on top of the nanowall samples. As the nanowalls are only a few nanometers thick, they appear almost transparent under SEM. The nanowalls grow very fast during the first 1–2 min and nearly stop growing after they reach a height of about 2 µm, though the exact height depends on the growth conditions. Width of the wall is in the range of 0.1–2 µm; it increases with decreasing the nanowall density. The thickness of the nanowalls is typically in the range of one to several nanometers (Fig. 5.2c), although sheets as thin as two monolayers were also observed by high-resolution transmission electron microscope (HRTEM) (Fig. 5.2d). As can be seen from Fig. 5.2c, the CNW consists of graphene sheets with different heights, and also the interlayer spacing appears to be larger than that of pristine graphite. These two features make it possible to study the electrical transport properties of point-contact between graphene and metal, which will be discussed shortly. Given the limited number of TEM experiments, it would not be a surprise that single layer graphene may also exist in the nanowalls. In fact, similar technique has been used to synthesize free-standing single layer graphene sheets [34]. Detailed TEM and Raman spectroscopy studies suggest that the CNWs consist of small graphitic domains embedded in a matrix of disordered carbon [6, 35–37], with the domain size depending on the growth techniques and conditions. More details are discussed by Tachibana in Chapter 6.

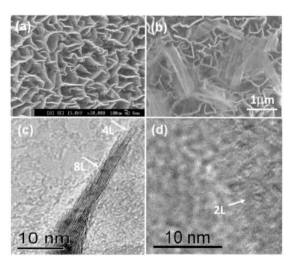

Figure 5.2   SEM (a and b) and HRTEM (c and d) images of carbon nanowalls grown at a H_2/CH_4 flow rate ratio of 4. Scale bars (a) 100 nm and (b) 1 µm. (a) was taken at a tilt angle of 25°.

5.2.2 Growth Mechanism

The carbon nanowalls or nanosheets can be grown on any type of substrate without a catalyst as long as the substrate can sustain the growth temperature which is typically below 800°C [7, 17]. Unlike the carbon nanotube case, the growth mechanism of 2D carbon by CVD is still not well-understood. Zhu et al. proposed a model for the growth of 2D carbon nanosheets using rf-PECVD [38]. In their experiments, the nanosheets were deposited on a variety of substrates in an inductively coupled plasma from a gas mixture of CH_4 (5–100%)/H_2, at a total gas pressure of 20–400 mTorr, substrate temperature of 600–950°C, and RF power of 400–1200 W. According to this model, the nanosheet initially grows parallel to the substrate up to a thickness of 1–15 nm before the onset of the vertical growth. The formation of parallel layer has been confirmed by surface x-ray scattering measurements [39]. The latter is presumably caused by the building up of upward curling force at the grain boundaries of nanographite domains. Once the nanosheet is oriented in the vertical direction, it grows much faster in the direction parallel to the sheet due to the very high surface mobility of incoming C atoms or CH_x radicals and polarization of the graphitic layers induced by the local electric field in the sheath layer. The fast diffusion of carbon-bearing species and etching by hydrogen radicals strongly suppress the growth in thickness direction. Once they reach the edges, however, the carbon-bearing species form bonds with the edge atoms, leading to the growth in height direction. Similar mechanism has also been confirmed by Kondo et al. in the growth of carbon nanowalls using radical injection plasma-enhanced chemical vapor deposition [40]. Using combined technique of atomic force microscopy and Raman spectroscopy, Yoshimura et al. [41] found that nanodiamond particles play an important role in the initial stage of CNW growth on Si substrate by dc plasma-enhanced chemical vapor deposition. It was found that nanodiamond particles with highly defective graphene layers were initially formed over the substrate, which promoted the growth of nanographite grains. The density of these nanographite grains increases with increasing deposition time, and they eventually coalesce to form a continuous graphite film. The growth proceeds further to form vertically aligned CNWs on the graphite film.

In order to obtain direct evidence of this two-stage growth model, Wu et al. performed Auger element mapping on carbon nanowalls grown on Cu substrates, which could be peeled off easily due to the weak bonding between carbon and Cu [42]. After peeling-off, a thin layer of carbon was often found to be present on the Cu substrate near the unpeeled region (see the darker region in Fig. 5.3a). Auger element mapping has been carried out for three different regions of a sample: (i) region with CNWs, (ii) exposed Cu substrate covered by a thin layer of graphene sheets, and (iii) completely exposed Cu substrate. The Auger spectra taken from regions (i) and (ii) resemble closely the Auger spectra of few layer graphene reported by Xu et al. [43], whereas those of region (iii) are dominated by peaks associated with copper. Based on the Auger spectra, element mappings are obtained for carbon and copper and the results are shown in Fig. 5.3b and c, respectively. As can be seen in Fig. 5.3b and c, the darker region appearing in Fig. 5.3a near the CNWs is actually the copper substrate covered by a thin layer of carbon. Depending on the thickness of this layer, it can be either completely or partially peeled off from the substrate with the nanowalls, as shown in Fig. 5.3d. The Auger results are in good agreement with the growth mechanism reported in literature, i.e., a thin layer of flat graphite sheets is formed on the substrate at the initial stage, followed by the growth of vertical nanosheets [38, 40]. As we will discuss shortly, the existence of both flat and vertical graphene sheets provides a convenient way to study the contact orientation dependence of electrical transport properties across the metal–graphene junctions.

5.3 Electrical Transport Properties

The ideal graphene is a semimetal because of its zero bandgap and vanished density of states at the Fermi level. However, the situation changes for nanometer sized graphene sheets due to the edge and surface effect. Theoretical studies have suggested that paramagnetic, ferromagnetic, antiferromagnetic, and superconducting phases may appear along or coexist with one another in nanometer sized graphite ribbons or graphene sheets [44–48], depending on the nature of electron–electron interactions and atomic configurations at some boundaries or defects. Of particular interest is the theoretical prediction of Fermi surface instability in nanographite ribbons due to edge states [49] and superconducting instability at high temperature in graphene sheets due to topological disorders [44]. Based on

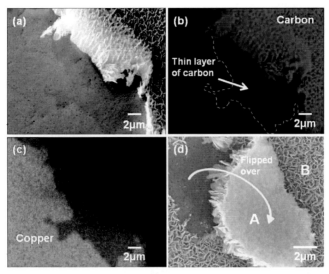

Figure 5.3 (a) SEM image of CNWs grown on Cu substrate. A portion of CNW has been peeled off by tweezers. The darker region (ii) near the unpeeled portion is covered by a thin layer of carbon. (b) Auger mapping of carbon using peaks associated with graphene sheets of the sample whose SEM image is shown in (a). Graphene sheets were found to exist in both regions (i) and (ii). (c) Copper mapping of the same sample. (d) SEM image of CNW sample (B) with a piece of flipped over CNW (A). Auger mapping confirms that region A consists of flat graphene sheets. Portions A and B were used to form side- and edge-contacts, respectively [42].

this background, and considering the unique surface morphology of carbon nanowalls, we have carried out electrical transport studies of CNWs using different types of electrode materials and configurations which have been summarized in Fig. 5.4. In the first type of measurements (see Fig. 5.4a), Au electrodes are deposited directly from the top surface of the nanowalls [7]. The spacing and length of the electrodes are chosen such that a large number of CNWs are present between the two inner electrodes from which voltage is measured. The purpose of this type of measurement is to study the electrical transport across a large number of CNW junctions. This is of great interest, in particular at low temperatures, when possible superconducting instability develops at low temperatures. The second type of measurements (Fig. 5.4b) is designed to study the electrical transport across a single piece or few pieces of CNWs [50]. In this case, both normal metal and superconducting electrodes are used so as to find out if there is any proximity effect on the CNWs. The last series of experiments was performed to study the effect of contact orientation on the electrical transport properties of 2D carbon [42]. Due to the length constraint, here we only describe the results obtained in configurations (a) and (c) of Fig. 5.4.

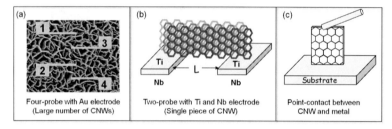

Figure 5.4 Different electrical contacts used to study the electrical transport properties of CNWs. (a) Au electrodes for studying the contacts between CNWs, (b) normal metal and superconducting electrodes for studying the electrical transport of single or few pieces of CNW, and (c) point-contact for studying the contact orientation dependence of electrical transport between metal and 2D carbon. All these measurements, to the best of our knowledge, have not been reported by other researchers.

5.3.1 Electrical Transport Across the CNW Junctions

We first discuss the results obtained by the electrode configuration shown in Fig. 5.4a. The Au electrodes were deposited by evaporation using a shadow mask fabricated by laser cutting. The number of pieces of CNWs between the voltage electrodes (2 and 3) along the current path was estimated to be around 7000. The DC electrical measurements were carried out using a superconducting quantum interference device (SQUID, Quantum Design MPMSXL) with variable temperature (2 K to 400 K) and magnetic field (0 to 7 T). The magnetic field could be applied either parallel or perpendicular to the substrate surface by setting the sample in different directions with respect to the field direction. During the transport measurement, a DC current of 1 mA was supplied from a Keithley current source from the two outer Au electrodes and the electrical potential drop across the two inner electrodes was measured using a Keithley 2182 nanovoltmeter.

The width of the current path was about 60 μm, leading to an estimated average current of 4 μA flowing through each junction of the 2D nanocarbons. Figure 5.5 shows the temperature dependence of resistance in the temperature range from 2 K to 10 K at applied magnetic fields between 0 Oe and 400 Oe with an interval of 100 Oe. The inset shows the temperature dependence of resistance in the temperature range from 2 K to 300 K. As can be seen from the inset, resistance decreases with decreasing the temperature from 300 K to about 106 K, and then shows an upturn at lower temperatures. The temperature (T) dependence of resistance (R) can be fitted very well using the following equation [51]:

$$R = 0.0018 \ T + 2.5 \ \exp\left(\frac{-127}{T}\right) + 10.8 \ \exp\left(\frac{15.1}{T+34.5}\right), \quad (5.1)$$

where the first term represents the metallic contribution, the second term is due to the quasi-1D characteristic of the network structure, and the last term accounts for the hopping/tunnelling resistance of the junctions. The units of R and T are in Ω and K, respectively. The junction resistance dominates the total resistance in the entire temperature range, in particular, at low temperatures. Hopping is thermally activated at high temperature, but at low temperature tunnelling is dominant. Also shown in the inset of Fig. 5.5 is the rate of change of resistance with temperature (dR/dT). We now turn to the results below 10 K. At zero applied field, the resistance continues to increase with decreasing the temperature, reaching the first local maximum at about 4.2 K; after a local minimum is reached at about 3.6 K, it increases again until the temperature decreases to 3 K below which the resistance decreases monotonically until reaching 2 K, which is the lowest temperature of the SQUID setup used in this experiment.

The characteristic of the resistance–temperature curve agrees well with the superconducting phase of a Josephson junction array (JJA) [52] or a weakly linked granular superconductor [53]. When a magnetic field is applied in the direction perpendicular to the substrate surface, i.e., almost along the surface of the carbon nanosheets, both local maxima of the curve shift to lower temperatures. The resistance, in general, also increases with the magnetic field. It is interesting to note that the resistance is not affected by the applied field at temperature higher than 7 K. The result indicates that the superconducting instability may have been developed below 7 K in the 2D carbon nanosheets. The slow decrease of resistance with temperature is due to the existence of junctions between the nanowalls. In other words, the superconducting phase is mainly contained inside the nanowalls because the junctions are still quite resistive.

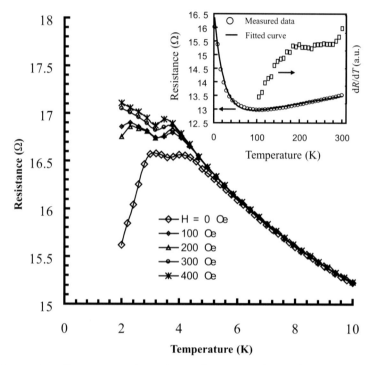

Figure 5.5 Temperature-dependence of resistance (R–T) of CNWs without (diamond) and with an applied field of 400 Oe (triangle), measured using the electrode configuration shown in Fig. 5.4a. Inset shows the R–T at zero field from 2 K to 300 K. Also shown in the inset is the rate of change of resistance with respect to temperature.

As the junctions between the CNWs are not chemically bonded, they can be treated as weak links electronically [54]. The non-chemical bonding nature of the junctions is apparent from Fig. 5.2b where CNWs peeled off from the substrate are isolated graphite sheets. This has conveniently led to the formation of disordered JJAs in a self-assembled fashion if the CNWs are indeed superconducting. One of the characteristics of JJAs is the oscillation of resistance with the magnetic field. To verify if there is any oscillation in the magnetoresistance, we performed the measurements by applying the magnetic field in both vertical (field perpendicular to the substrate) and traverse directions (field parallel to the substrate). In the former case, we observed clear oscillations from 2 K to 7 K, which dies out abruptly above this temperature [7]. The amplitude of oscillation increases by more than 3 orders when the temperature is decreased from 6 K to 2 K. The oscillations are, in general, quasi-periodic; however, the periodicity improves with temperature. Fourier spectrum of the magnetoresistance curve at 4.31 K shows three peaks corresponding to different periodicities. Although oscillations in magnetoresistance also occur in various types of multiply connected conductors of normal metals, the oscillations observed in this study are presumably caused by the flux quantization effect in JJAs, as supported by the temperature and field dependence of resistance data shown in Fig. 5.5 and also the strong temperature-dependence of the oscillation amplitude. The average period of the main oscillation is 495 Oe; this corresponds to an enclosed region with an area of 0.04 μm^2 if the oscillation has a period of one quantum flux (Φ_0). If we take into account the tilt of $19°$ of the nanosheets with respect to the field direction (we will come back to this point shortly), the area is about 0.042 μm^2. This value is roughly the area of the smallest region bounded by the carbon nanosheets. The fluctuation in the oscillation periods caused by the variations in the areas scales as $\Delta H = -(\Delta S/<S>)<H>$, where $<S>$ is the average area and $<H>$ is the corresponding period. A clear oscillation should be observable if ΔS is smaller than $<S>/2$ which may explain why the positional randomness did not destroy the oscillation in this case. On the other hand, the other two shorter periods occurring almost exactly at $\Phi_0/2$ and $\Phi_0/3$ might be caused by the so-called frustration in JJAs [52].

In addition to the measurement with the field being applied along the wall direction, we also observed clear but weaker oscillations up to 9 K when the magnetic field was applied along the substrate surface (Fig. 5.6). The periodicity was well developed from 6 K to 9 K, but the oscillation disappeared below 4 K. The occurrence of oscillation with a traverse applied field is not contradictory to theory because the carbon nanosheets are not perfectly aligned in the vertical direction, as can be seen from the SEM image shown in Fig. 5.2a which was taken at a tilt angle of $15°$. The oscillation period for the case when measuring with a traverse applied field is 1438 Oe; this in combination with the period observed in a perpendicular field gives a tilt angle of $\theta = \tan^{-1}(495/1437) = 19°$, agreeing well with the morphology observed experimentally.

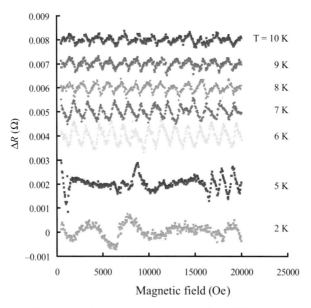

Figure 5.6   Magnetoresistance curves measured at different temperatures with the magnetic field applied parallel to the substrate surface.

Before we end this section, we have to point out that the occurrence of any oscillation in magnetoresistance depends strongly on the sample structure. In fact, the oscillation has been observed in only a limited number of samples. The oscillation in some samples disappears after the sample is taken out from the cryostat and put in again for repeated measurements. This is presumably caused by the fact that the nature of the junction might change during this process due to thermal expansion/contraction.

5.3.2 Electrical Transport in Carbon–Metal Point-Contact

Perfect graphene is an ideal 2D crystal. A question naturally arises here: how do electrons travel back and forth between a 3D metal and 2D carbon? The question is valid because in a perfect 2D crystal electrons are not supposed to have any motion or momentum in the direction which is perpendicular to the 2D plane. Alternatively, will it make any difference if the electrons enter graphene from the edges and side surfaces? To answer this question experimentally, one must be able to make reliable electrical contacts with both the edges and the side surfaces of graphene. However, compared to the side-contact, which is routinely employed for electrical transport measurement of graphene or graphene electronic devices, it remains a great challenge to form a pure edge-contact with graphene without touching its surface due to its ultra-small thickness. The unique topography of carbon nanowalls provides a convenient way to realize edge-contact with the help of a position-controllable nanoprobe.

Figure 5.7a shows the schematic representation of an edge-contact between CNW and a tungsten (W) probe. In order to form a reliable contact, our measurements were performed using an Omicron ultrahigh vacuum (UHV) system with a base pressure in the range of $3–8 \times 10^{-11}$ Torr. Equipped in the UHV system are a scanning electron microscope (SEM) and four independently controllable nanoprobes with auto-approaching capability, which allow position-specific measurements with good reproducibility. Figure 5.7b shows an example of an edge-contact formed by a tungsten probe and a piece of CNW. The size of the contact is determined mainly by the thickness of the CNW which is about one to several nanometers at the edge and is adjustable through monitoring the zero-bias contact resistance. In order to compare the results with graphene sheets, similar edge-contact experiments have also been performed on a second type of 2D carbon which was obtained in situ through mechanical exfoliation of highly ordered pyrolytic graphite (HOPG) by using a large-size probe which itself also forms a low-resistance contact with HOPG during the subsequent electrical measurements (see Fig. 5.7c). As can be seen in this figure, an edge-contact can be readily formed between a second W probe and the edge of an exfoliated 2D carbon sheets. The precise positioning of probe allows the formation of contacts between the probe and different points of the edge. Again, the actual contact size can be adjusted manually through monitoring the contact resistance. Alternatively, an edge-contact can also be formed by probing directly the edge of a small piece of HOPG flake placed with an off-angle from the flat surface of a substrate holder. All the electrical measurements were performed using a standard lock-in technique at room temperature. During the measurements, one of the probes was used to form a low-resistance contact and the other was adjusted manually to have a different contact resistance. The automatic approaching function helps to make reliable and reproducible contact without damaging the sample and the probe, unless the substrate is an insulator in which case the first probe has to be sacrificed by being pressed manually on the sample.

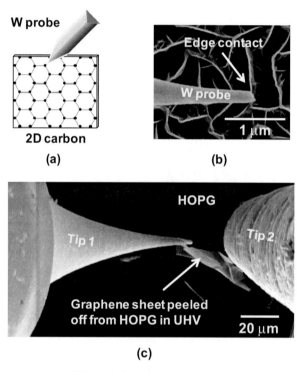

Figure 5.7 Setup of differential conductance measurement. (a) Illustration of edge-contact between W probe and 2D carbon; (b) SEM image showing experimental realization of edge-contact between W probe and edge of CNW; (c) SEM image showing experimental realization of edge-contact between W probe and graphene sheets obtained by in situ exfoliation of HOPG inside the UHV chamber [42].

Figure 5.8a shows the dI/dV curves as a function of bias voltage obtained by first forming an ohmic contact with CNWs grown on

SiO_2/Si using one of the probes and then varying the sample–probe distance of the second probe. In order to avoid causing damage to the nanowalls, the second probe was first placed on the nanowalls through the auto-approaching function and then the sample-probe distance was adjusted manually by monitoring the zero-bias resistance (ZBR) using a small AC current. The dI/dV versus bias voltage curves were then recorded by sweeping the DC bias current within a predefined range and measuring dI/dV using the lock-in technique. The different curves shown in Fig. 5.8a are corresponding to contacts with different ZBR values ranging from 4.6 to 26.1 kΩ. For the sake of clarity, all the curves other than the one with highest ZBR (the bottom curve) are shifted upward in the figure. The symbols are the measurement results and solid-lines are linear fittings to the experimental data. When both probes were in ohmic contact with the nanowalls, the resistance measured was in the range of 200–600 Ω, for a probe distance of 1–20 μm. In this case, the dI/dV curve is almost independent of the bias voltage. Therefore, the measured conductance shown in Fig. 5.8a is dominantly due to the second CNW-probe contact. Although the CNWs are few layer graphene sheets, the electrical contact was presumably formed with the layer (or layers) of highest height. Under the ballistic transport approximation, the conductance of one transverse mode of a graphene point-contact is $4e^2/h$, corresponding to a resistance of 6.45 kΩ. Therefore, during the measurements, the ZBR has been varied in the range of ~3–30 kΩ. Although there is no clear definition of the point-contact and tunneling regimes, the point-contact regime can be considered as being in the range where ZBR is close to the resistance quanta of one conduction channel. When the ZBR reaches a value which is several times that of the resistance quanta, it is more appropriate to treat the contact as being in the tunneling regime. As can be seen from Fig. 5.8a, the differential conductance is linear to the bias voltage for all the ZBR values that have been measured. The experiments have been repeated on different CNW samples and also at many different locations for a sample, which exhibited excellent reproducibility.

After a series of measurements were completed on bare CNWs, the sample was coated in situ with a thin layer of Fe in the preparation chamber and then transferred back to the measurement chamber for performing the same series of electrical measurements without breaking the vacuum. The coating was uniform and had a nominal thickness of about 30 nm. The dI/dV curves for the Fe-coated sample are shown in Fig. 5.8b. In a sharp contrast to the case of bare CNWs, the dI/dV curves exhibit a well-defined parabolic shape. Compared with the bare CNW sample, it is generally more difficult to form a stable contact between the probe and Fe-coated CNWs. Therefore, after several runs of measurements, the probe has to be lifted for re-establishing a new contact. This resulted in dI/dV curves with different rates of change with respect to the bias voltage. Nevertheless, all the experimental data (symbols) are fitted well using the relation dI/d$V \propto V^2$, with V being the probe–sample bias voltage. This kind of parabolic dI/dV curve is normally obtained in metal–insulator–metal tunnel junction [55]. The good agreement between theoretical and experimental curves for the Fe-coated sample confirms that the linear curves shown in Fig. 5.9a are due to the CNW–W edge-contact.

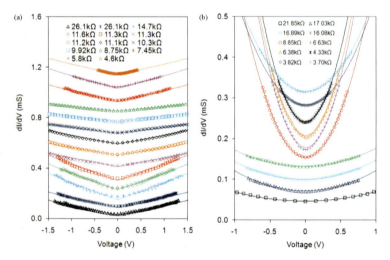

Figure 5.8 (a) dI/dV curves as a function of bias voltage for CNW edge-contacts, at different ZBR values (4.6–26.1 kΩ); (b) dI/dV curves as a function of bias voltage for Fe-coated CNW contacted at the edge by W probe, at different values of ZBR (3.7–21.9 kΩ). The Fe-coating was performed in situ in the same UHV system. For clarity, all the curves other than the one with highest ZBR have been shifted upward. The ZBR values (resistance at the lowest point of the dI/dV curve) for different contacts have been listed on the top of the figure in kΩ [42].

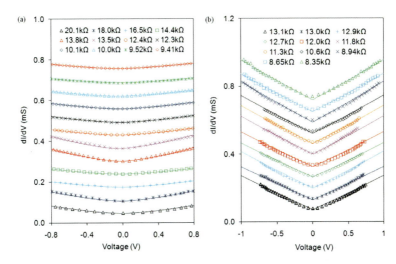

Figure 5.9 (a) dI/dV curves as a function of bias voltage for the back surface of CNWs (region A of Fig. 5.4d), at different ZBR values (9.41–20.1 kΩ), which corresponds to a side-contact. The CNWs were peeled off and flipped over in situ in the same UHV system; (b) dI/dV curves as a function of bias voltage for CNW edge-contacts (region B of Fig. 5.4d), at different ZBR values (8.35–13.1 kΩ) [42].

The same measurements were then performed on CNWs grown on a Cu substrate, which allows partial peeling-off of CNW from the substrate. One should note that the conducting substrate does not affect the measurement results because the resistance measured mainly comes from the contact with a larger resistance. Figure 5.9a and b shows the dI/dV curves obtained from the back surface (after peeling-off from the substrate) and front end of CNWs, corresponding to regions A and B of Fig. 5.3d, respectively. The experimental data (symbols) are fitted well to the relation dI/d$V \propto V^{3/2}$ for (a) and dI/d$V \propto V$ for (b), respectively. Again, for the sake of clarity, all the curves other than the one with the highest ZBR (the bottom curve) have been shifted upward in the figure. As the measurements on regions A and B of Fig. 5.3d correspond to side- and edge-contact, respectively, the results demonstrate that the dI/dV curves are indeed dependent on the relative orientation between the probe (or more accurately, current direction) and base plane of the carbon lattice.

In order to further confirm the results shown in Figs. 5.8 and 5.9, we repeated the same series of experiments on HOPG and exfoliated graphene sheets. In order to reduce the influence of surface contaminants, the top layer of HOPG was peeled off in situ using a probe. Compared to CNWs, it was found that it is generally more difficult to form reliable side-contact with thick HOPG plates due to its flatness and hardness. However, once a stable contact is formed, the measurement results are reproducible at different locations on the HOPG surface. On the other hand, it is relatively easy to form a stable contact from the edges for both thick flakes and few-layer graphene sheets. In the former case, although the flake is thick, contact is presumably only formed at point (or points) with highest protrusions. Therefore, the actual contact only occurs at the edge of the graphene sheets. On the other hand, in the latter case, it is determined dominantly by the thickness of the graphene sheets. The dI/dV curves obtained from both the side- and the edge-contacts are shown in Fig. 5.10a and b, respectively. Again, the dI/dV curves in Fig. 5.10a are fitted well by the relation $dI/dV \propto V^{3/2}$ for all ZBR values. On the other hand, the dI/dV curves in Fig. 5.10b follow closely a linear-dependence on the bias voltage, as is the case in Figs. 5.8a and 5.9b. In addition to the measurements carried out for the contacts with ZBRs in the range of 4.7–13.1 kΩ, we also measured the dI/dV of edge-contacts with a ZBR in the range of 256–476 Ω. The dI/dV still exhibits a linear-dependence with the bias voltage. This shows clearly that the bias-dependence of dI/dV is mainly determined by the intrinsic property of graphene rather than the W probe. In order to compare the results with Fe-coated CNWs, we performed similar experiments on Fe-coated HOPG. As with the case of Fe-coated CNWs, the Fe coating has a nominal thickness of 30 nm. The experiments were repeated on two samples with separately coated Fe layer and on different experiment runs. In both cases, the dI/dV-V curves exhibited a well-defined parabolic shape. We repeated the same measurements on CNWs and graphite irradiated by a focused ion beam with different doses (not shown here). dI/dV curves similar to those of amorphous carbon reported in literature were obtained at high doses. This confirms again that the measured dI/dV curves are indeed the true reflection of the density of states of the sample.

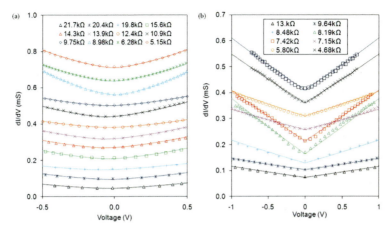

Figure 5.10 (a) dI/dV curves as a function of bias voltage for HOPG surface-contact, at different ZBR values (5.15–21.7 kΩ). The top layer of HOPG was peeled off in situ using one of the probes to avoid the influence of surface contamination; (b) dI/dV curves as a function of bias voltage for HOPG edge-contact, at different values of ZBR (4.71–13.1 kΩ) [42].

The aforementioned results can be accounted for theoretically by taking into account the band structure and density of states of graphene [42]. Intuitively, it can be understood as follows. Differential conductance is the measure of how easily the electrons travel back and forth between the metal tip and graphene. At equilibrium state, electrons inside the metals can move in all the directions, whereas electrons inside graphene can only travel in the graphene plane. If the periodicity in lateral direction is maintained, inside the metal–graphene gap only electrons with non-zero k_z component will contribute to the current or conductance. Here, k_z is the wavevector in vertical direction. For side-contacts, there is no k_z component for the wavevector inside the graphene; this "wavevector mismatch" suppresses the conductance at low bias, as compared to the edge-contact. For a more quantitative discussion, the reader may refer to ref. [42].

The results obtained in the point-contact study show that the characteristics of metal–graphene contacts are determined not only by the type of the metal but also the contact orientation. This is unique to graphene because of its 2D dimensionality which is not seen in other types of materials.

5.4 Magnetic Properties

Perfect carbon is not expected to exhibit ferromagnetism. Although "ferromagnetism-like" behavior has been observed in various types of carbon-based materials; their origin is still not well understood [56]. The proposed possible explanations include under-coordinated atoms [57], adatoms [58], carbon vacancies [59], uneven hydrogen terminators [48], etc. Vacancy or vacancy–hydrogen complex induced magnetism is supported by the experiment of proton-irradiation triggered magnetism in graphite [60]. Putting the controversy aside, it would be of interest to find out if there is any magnetism in carbon nanowalls because they have all the "ingredients" for observing magnetism in carbon: small size, point and extended defects, and edges.

In order to have a sufficient magnetic moment for performing an accurate measurement, the carbon nanowalls are peeled off from six pieces of small samples (total surface area of about 8 cm^2), and wrapped with a short segment of a drinking straw which is normally used to hold samples for SQUID measurement. In order to prevent the movement of the carbon samples, the straw is heated and pressed to flat before it is inserted into another straw holder for the magnetic measurement. Based on the average height, width, and thickness of 2 µm, 0.5 µm, and 5 nm, respectively, and an average nanowall density of 4 pieces per µm^2, the total volume of the nanowalls alone is about 2×10^{-5} cm^3. The total edge length, counting both the top and the bottom side, is approximately 3.2×10^9 µm. If we assume that all the edges have a zigzag shape, this corresponds to 2×10^{14} carbon atoms.

A series of measurements have been carried out to study the magnetic properties of the nanowall samples which include the measurement of hysteresis curves (M–H loops) at different temperatures, zero-field cooling and field cooling measurements at different magnetic fields, and temperature dependences of magnetization at different applied fields. Figure 5.11a shows the typical M–H loops at 2 K and 400 K, respectively, after a temperature-independent diamagnetism contribution was subtracted off from the original data. Clear hysteresis with a coercivity of 150 Oe was obtained at 400 K. Although ferromagnetic ordering may occur at the edges or extended defects of individual nanowalls, there is no exchange coupling among different edges or defects; therefore, the pile of nanowalls can be considered as a random ferromagnet which distinguishes itself from a bulk magnet. Figure 5.11b shows the temperature dependence of thermoremanent magnetic moment which was obtained by first cooling the samples with a 1 T field from 400 K to 2 K and then measuring the magnetic moment without field while the sample is warmed up from 2 K to 400 K. The non-Brillouin shape might be caused by factors like distribution of T_c among the edges and defects, formation of magnetic polarons near the edges or defects , though further studies are required to understand the true mechanism. Kawabata et al. [61] have also observed similar type of temperature-dependence of magnetization in pyrolytic carbon grown by chemical vapor deposition and pointed out that highly oriented structure and a large number of unpaired electrons are the key to observing ferromagnetism. This agrees well with the features of carbon nanowalls. As our samples were grown using pure hydrogen and methane gases without the use of any catalysts, the magnetism is very unlikely to be caused by ferromagnetic impurities. The saturation moment of 4×10^{-5} emu at 2 K corresponds to $4.3-10^{15}$ μB. This value is about one order of magnitude higher than the estimated value if we assume that each atom at the edge contributes 1.25 μB [62]. The discrepancy between experimental and estimated theoretical values can be understood as being caused by the fact that the extended defects on the nanowall surfaces dominate the magnetic properties of our samples. This is in good agreement with the observation of a very strong D band in the Raman spectra of carbon nanowalls [63].

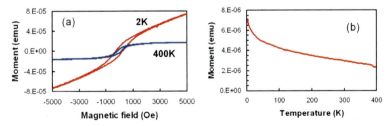

Figure 5.11 (a) *M–H* curves of carbon nanowalls at 5 K and 400 K, respectively, in which the diamagnetic contribution has been subtracted off from measured data; (b) Temperature dependence of thermoremanent magnetic moment.

5.5 Summary

In this chapter, after an introduction to the growth of carbon nanowalls, we have focused on their electrical transport studies. The carbon nanowalls are unique in several aspects as compared to graphene sheets discussed in Chapters 3 and 4: (1) the CNWs are standing up on the substrate surface; therefore, their top edges are easily accessible; (2) they are from the interaction with substrates; (3) they can be synthesized with ease in large quantity. All these features make it possible to study the intrinsic properties of 2D carbon without too much concern about the influence of surrounding materials or process-induced fluctuations. In particular, the studies of point-contact and carbon–carbon junctions are unique and so far similar experiments have not been reported for flat graphene sheets.

References

1. Ebbesen T.W., Ajayan P.M. (1992) Large-scale synthesis of carbon nanotubes, *Nature*, **358**, 220–222.

2. Iijima S., Wakabayashi T., Achiba Y. (1996) Structures of carbon soot prepared by laser ablation, *J Phys Chem*, **100**, 5839–5843.

3. Ando Y., Zhao X., Ohkohchi M. (1997) Production of petal-like graphite sheets by hydrogen arc discharge, *Carbon*, **35**, 153–158.

4. Wu Y.H., Chong T.C. (16–20 April 2001) *Carbon nano-flakes grown by microwave cvd*, paper presented at MRS Spring Meeting, San Francisco, p. W8.3.

5. Wu Y.H. (2002) Effects of localized electric field on the growth of carbon nanowalls, *Nano Lett*, **2**, 355–359.

6. Wu Y.H., Qiao P.W., Chong T.C., Shen Z.X. (2002) Carbon nanowalls grown by microwave plasma enhanced chemical vapor deposition, *Adv Mater*, **14**, 64–67.

7. Wu Y.H., Yang B.J., Zong B.Y., Sun H., Shen Z.X., Feng Y.P. (2004) Carbon nanowalls and related materials, *J Mater Chem*, **14**, 469–477.

8. Tanaka K., Yoshimura M., Okamoto A., Ueda K. (2005) Growth of carbon nanowalls on a SiO_2 substrate by microwave plasma-enhanced chemical vapor deposition, *Jpn J Appl Phys, Part 1* **44**, 2074–2076.

9. Chuang A.T.H., Boskovic B.O., Robertson J. (2006) Freestanding carbon nanowalls by microwave plasma-enhanced chemical vapour deposition, *Diamond Relat Mater*, **15**, 1103.

10. Le Brizoual L., Belmahi M., Chatei H., Assouar M.B., Bougdira J. (2007) Transmission electron microscopy study of carbon nanostructures grown by MPACVDm in CH_4/CO_2 gas mixture, *Diamond Relat Mater*, **16**, 1244–1249.

11. Zeng L.Y., Lei D., Wang W.B., et al. (2008) Preparation of carbon nanosheets deposited on carbon nanotubes by microwave plasma-enhanced chemical vapor deposition method, *Appl Surf Sci*, **254**, 1700–1704.

12. Vizireanu S., Nistor L., Haupt M., Katzenmaier V., Oehr C., Dinescu G. (2008) Carbon nanowalls growth by radiofrequency plasma-beam-enhanced chemical vapor deposition, *Plasma Processes and Polym*, **5**, 263–268.

13. Malesevic A., Vitchev R., Schouteden K., et al. (2008) Synthesis of few-layer graphene via microwave plasma-enhanced chemical vapour deposition, *Nanotechnology*, **19**.

14. Yuan G.D., Zhang W.J., Yang Y., et al. (2009) Graphene sheets via microwave chemical vapor deposition, *Chem Phys Lett*, **467**, 361–364.

15. Chuang A.T.H., Robertson J., Boskovic B.O., Koziol K.K.K. (2007) Three-dimensional carbon nanowall structures, *Appl Phys Lett*, **90**, 123107.

16. Wang J.J., Zhu M.Y., Outlaw R.A., Zhao X., Manos D.M., Holloway B.C. (2004) Synthesis of carbon nanosheets by inductively coupled radio-frequency plasma enhanced chemical vapor deposition, *Carbon*, **42**, 2867–2872.

17. Wang J.J., Zhu M.Y., Outlaw R.A., et al. (2004) Free-standing subnanometer graphite sheets, *Appl Phys Lett*, **85**, 1265–1267.

18. Hiramatsu M., Shiji K., Amano H., Hori M. (2004) Fabrication of vertically aligned carbon nanowalls using capacitively coupled plasma-enhanced chemical vapor deposition assisted by hydrogen radical injection, *Appl Phys Lett*, **84**, 4708–4710.

19. Nishimura K., Jiang N., Hiraki A. (2003) Growth and characterization of carbon nanowalls, *IEICE Trans Electron*, **E86C**, 821–824.

20. Lin C.R., Su C.H., Chang C.Y., Hung C.H., Huang Y.F. (2006) Synthesis of nanosized flake carbons by RF-chemical vapor method, *Surf Coat Technol*, **200**, 3190–3193.

21. Sato G., Morio T., Kato T., Hatakeyama R. (2006) Fast growth of carbon nanowalls from pure methane using helicon plasma-enhanced chemical vapor deposition, *Jpn J Appl Phys, Part 1*, **45**, 5210–5212.

22. Malesevic A., Vizireanu S., Kemps R., Vanhulsel A., Van Haesendonck C., Dinescu G. (2007) Combined growth of carbon nanotubes and carbon

nanowalls by plasma-enhanced chemical vapor deposition, *Carbon*, **45**, 2932–2937.

23. Kondo S., Hori M., Yamakawa K., Den S., Kano H., Hiramatsu M. (2008) Highly reliable growth process of carbon nanowalls using radical injection plasma-enhanced chemical vapor deposition, *J Vac Sci Technol B*, **26**, 1294–1300.

24. Zhu M.Y., Wang J.J., Outlaw R.A., Hou K., Manos D.M., Holloway B.C. (2007) Synthesis of carbon nanosheets and carbon nanotubes by radio frequency plasma enhanced chemical vapor deposition, *Diamond Relat Mater*, **16**, 196–201.

25. Kurita S., Yoshimura A., Kawamoto H., et al. (2005) Raman spectra of carbon nanowalls grown by plasma-enhanced chemical vapor deposition, *J Appl Phys*, **97**, 104320.

26. Shang N.G., Au F.C.K., Meng X.M., Lee C.S., Bello I., Lee S.T. (2002) Uniform carbon nanoflake films and their field emissions, *Chem Phys Lett*, **358**, 187–191.

27. Dikonimos T., Giorgi L., Giorgi R., Lisi N., Salernitano E., Rossi R. (2007) DC plasma enhanced growth of oriented carbon nanowall films by HFCVD, *Diamond Relat Mater*, **16**, 1240–1243.

28. Giorgi L., Makris T.D., Giorgi R., Lisi N., Salernitano E. (2007) Electrochemical properties of carbon nanowalls synthesized by hf-cvd, *SensActuators, B: Chem*, **126**, 144–152.

29. Shemabukuro S., Hatakeyama Y., Takeuchi M., Itoh T., Nonomura, S. (2008) Preparation of carbon nanowall by hot-wire chemical vapor deposition and effects of substrate heating temperature and filament temperature, *Jpn J Appl Phys*, **47**, 8635–8640.

30. Suarez-Martinez I., Grobert N., Ewels C.P. (2012) Nomenclature of sp(2) carbon nanoforms, *Carbon*, **50**, 741–747.

31. Hiramatsu M., Hori M., SpringerLink (Online service) (2010) Carbon nanowalls synthesis and emerging applications, Springer, Wien, pp. Online resource (x, 161 p.).

32. Wu Y.H., Yang B.J. (2002) Effects of localized electric field on the growth of carbon nanowalls, *Nano Lett*, **2**, 355–359.

33. Wu Y., Qiao P., Chong T., Shen Z. (2002). Carbon nanowalls grown by microwave plasma enhanced chemical vapor deposition, *Adv Mater*, **14**, 64–67.

34. Dato A., Radmilovic V., Lee Z., Phillips J., Frenklach M. (2008) Substrate-free gas-phase synthesis of graphene sheets, *Nano Lett*, **8**, 2012–2016.

35. Kobayashi K., Tanimura M., Nakai H., et al. (2007) Nanographite domains in carbon nanowalls, *J Appl Phys*, **101**, 094306.

36. Kurita S., Yoshimura A., Kawamoto H., et al. (2005) Raman spectra of carbon nanowalls grown by plasma-enhanced chemical vapor deposition, *J Appl Phys*, **97**.

37. Ni Z.H., Fan H.M., Feng Y.P., Shen Z.X., Yang B.J., Wu Y.H. (2006) Raman spectroscopic investigation of carbon nanowalls, *J Chem Phys*, **124**, 204703.

38. Zhu M.Y., Wang J.J., Holloway B.C., et al. (2007) A mechanism for carbon nanosheet formation, *Carbon*, **45**, 2229–2234.

39. French B.L., Wang J.J., Zhu M.Y., Holloway B.C. (2005) Structural characterization of carbon nanosheets via x-ray scattering, *J Appl Phys*, **97**, 114317.

40. Kondo S., Kawai S., Takeuchi W., et al. (2009) Initial growth process of carbon nanowalls synthesized by radical injection plasma-enhanced chemical vapor deposition, *J Appl Phys*, **106**, 094302.

41. Yoshimura A., Yoshimura H., Shin S.C., Kobayashi K., Tanimura M., Tachibana M. (2012) Atomic force microscopy and Raman spectroscopy study of the early stages of carbon nanowall growth by dc plasma-enhanced chemical vapor deposition, *Carbon*, **50**, 2698-2702.

42. Wu Y.H., Wang Y., Wang J.Y., et al. (2012) Electrical transport across metal/two-dimensional carbon junctions: Edge versus side contacts, *AIP Adv*, **2**, 012132.

43. Xu M.S., Fujita D., Gao J.H., Hanagata N. (2010) Auger electron spectroscopy: A rational method for determining thickness of graphene films, *ACS Nano*, **4**, 2937–2945.

44. Wakabayashi K., Fujita M., Ajiki H., Sigrist M. (1999) Electronic and magnetic properties of nanographite ribbons, *Phys Rev B*, **59**, 8271–8282.

45. Oshiyama A., Okada S., Saito S. (2002) Prediction of electronic properties of carbon-based nanostructures, *Physica B-Condens Matter*, **323**, 21–29.

46. Wakabayashi K., Harigaya K. (2003) Magnetic structure of nano-graphite mobius ribbon, *J Phys Soc Jpn*, **72**, 998–1001.

47. Gonzalez J., Guinea F., Vozmediano M.A.H. (2001) Electron–electron interactions in graphene sheets, *Phys Rev B*, **63**, 134421.

48. Kusakabe K., Maruyama M. (2003) Magnetic nanographite, *Phys Rev B*, **67**, 092406.

49. González J., Guinea F., Vozmediano M.A.H. (2001) Electron–electron interactions in graphene sheets, *Phys Rev B - Condens Matter Mater Phys*, **63**, 1344211–1344218.

50. Wu Y.H., Wang H.M., Choong C. (2011) Growth of two-dimensional carbon nanostructures and their electrical transport properties at low tempertaure, *Jpn J Appl Phys*, **50**, 01AF02.

51. Shiraishi M., Ata M. (2002) Conduction mechanisms in single-walled carbon nanotubes, *Synth Met*, **128**, 235–239.

52. Newrock R.S., Lobb C.J., Geigenmuller U., Octavio M. (2000) The two-dimensional Physics of josephson junction arrays, in *Solid State Physics: Advances in Research and Applications*, vol. **54**, pp. 263–512.

53. Muroi M., Street R. (1993) Percolative superconducting transition in y1-xprxba2cu3oy, *Physica C*, **216**, 345–364.

54. Wu Y.H., Yang B.J., Han G.C., et al. (2002) Fabrication of a class of nanostructured materials using carbon nanowalls as the templates, *Adv Funct Mater*, **12**, 489–494.

55. Wolf E.L. (1989) *Principles of Electron Tunneling Spectroscopy*, Oxford University Press, New York.

56. Makarova T., Palacio Parada F. (2006) *Carbon-Based Magnetism*, Elsevier, Amsterdam, The Netherlands; San Diego, CA, pp. xi, 564 p.

57. Kim Y. H., Choi J., Chang K.J., Tománek D. (2003) Defective fullerenes and nanotubes as molecular magnets: An ab initio study, *Phys Rev B*, **68**, 125420.

58. Coey M., Sanvito S. (2004) The magnetism of carbon, *Phys World*, **17**, 33–37.

59. Lehtinen P.O., Foster A.S., Ma Y., Krasheninnikov A.V., Nieminen R.M. (2004) Irradiation-induced magnetism in graphite: A density functional study, *Phys Rev Lett*, **93**, 187202.

60. Esquinazi P., Spemann D., Höhne R., Setzer A., Han K.H., Butz T. (2003) Induced magnetic ordering by proton irradiation in graphite, *Phys Rev Lett*, **91**, 227201.

61. Kawabata K., Mizutani M., Fukuda M., Mizogami S. (1989) Ferromagnetism of pyrolytic carbon under low-temperature growth by the cvd method, *Synth Met*, **33**, 399–402.

62. Esquiazi P., Spemann D., Schindler K., et al. (2006) Proton irradiation effects and magnetic order in carbon structures, *Thin Solid Films*, **505**, 85–89.

63. Ni Z., Fan H., Feng Y., Shen Z., Yang B., Wu Y. (2006) Raman spectroscopic investigation of carbon nanowalls, *J Chem Phys*, **124**, 204703.

Chapter 6

Structural Characterization of Carbon Nanowalls and Their Potential Applications in Energy Devices

Masaru Tachibana

Department of Nanosystem Science, Yokohama City University,
22-2 Seto, Kanazawa-ku, Yokohama 236-0027, Japan
tachiban@yokohama-cu.ac.jp

Carbon nanowalls (CNWs) are basically two-dimensional graphite sheets that are typically oriented vertically on a substrate. Each CNW originates from the stacking of several graphene sheets. A more detailed structure of each CNW exhibits domain structure that consists of nanographite domains that are several tens of nanometers in size. Such unique morphology and structure of CNWs have attracted much attention for various potential applications such as electronic devices and energy devices. In this chapter, the structural characterization of CNWs by Raman spectroscopy and transmission electron microscopy has been discussed. In addition, potential applications of CNWs as negative electrodes in lithium ion batteries and as catalytic supports in fuel cells have been reviewed in light of their unique structure.

Two-Dimensional Carbon: Fundamental Properties, Synthesis, Characterization, and Applications
Edited by Yihong Wu, Zexiang Shen, and Ting Yu
Copyright © 2014 Pan Stanford Publishing Pte. Ltd.
ISBN 978-981-4411-94-3 (Hardcover), 978-981-4411-95-0 (eBook)
www.panstanford.com

6.1 Introduction

Carbon nanostructures such as fullerenes and carbon nanotubes possess unique structure and dimensionality, leading to new mechanical, chemical, and electronic properties. Recently two-dimensional carbon nanostructures such as graphene sheets and carbon nanowalls (CNWs) have attracted much attention for various potential applications.

Graphene is a monatomic layer of carbon atoms arranged on a honeycomb lattice, and an ideal two-dimensional carbon material. Graphene possesses many extraordinary properties. Graphene was reported to exhibit many superior properties such as high ballistic electron mobility (>200,000 $cm^2V^{-1}s^{-1}$ for particular samples) [1], high thermal conductivity (5000 W $m^{-1}K^{-1}$) [2], Young's modulus (approximately 1150 GPa), fracture strength (125 GPa) [3], and a high specific surface area (approximately 2600 m^2g^{-1}) [4]. These properties are very attractive for various applications such as electronic and energy devices.

CNWs were first reported by Wu et al. in 2002 [5]. They observed wall-like carbon nanostructures perpendicular to a substrate, and named them CNWs. Since this report, a lot of groups have carried out the synthesis of CNWs mainly by plasma-enhanced chemical vapor deposition (PECVD) [6–8] and their characterization mainly by scanning electron microscopy (SEM), transmission electron microscopy (TEM) [9,10], and Raman spectroscopy [8,11]. TEM studies revealed that CNWs are basically two-dimensional graphite sheets with an average thickness of several nanometers, which are typically oriented vertically on a substrate. Each CNW originates from the stacking of several graphene sheets. This suggests that CNWs might exhibit similar properties as graphene. A more detailed analysis by TEM showed that each CNW has a domain structure which consists of nanographite domains which are several tens of nanometers in size [10]. Such unique morphology and structure of CNWs have stimulated not only fundamental studies on transport properties [12–15] and hydrogen absorption [16–18] but also various applications to electron field emitters [19–23], negative electrodes in lithium ion batteries [24,25], and catalyst supports in fuel cells [26–28].

In this chapter, the structural characterization of CNWs by Raman spectroscopy and TEM has been discussed. In addition, the potential applications of CNWs as negative electrodes in lithium ion batteries and as catalyst supports in fuel cells have also been reviewed in the light of their unique structure.

6.2 Synthesis of CNWS

CNWs have been synthesized on various substrates such as Si, quartz, and Cu by microwave PECVD [5,13,29], rf-PECVD [6,7,30,31], dc-PECVD [24,32], and hot-filament CVD [33,34]. Note that no catalyst is required for the growth of CNWs. Figure 6.1 shows a schematic representation of dc-PECVD apparatus used by Tachibana's group. The apparatus is equipped with an electron gun for generating the plasma and two coils to generate magnetic field which can form sheet-like plasma. The gases used are mixtures of CH_4, H_2, and Ar. The plasma reactor is evacuated down to about 0.5 Pa, and the substrate is subsequently heated to temperatures ranging from 500 to 800 °C. The typical flow rates of CH_4, H_2, and Ar are 10, 10, and 80 sccm (sccm denotes cubic centimeter per minute at STP), respectively. The typical dc plasma power is 3.5 kW. During the growth of CNWs, the dc plasma reactor is kept at about 1.3 Pa.

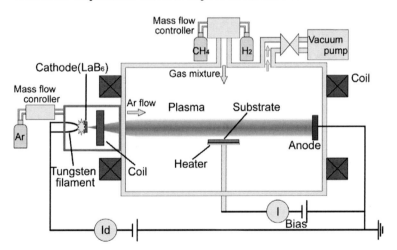

Figure 6.1 Schematic illustration of dc-PECVD apparatus used by Tachibana's group.

Typical SEM images of the top and tilted views of CNWs grown vertically on the quartz substrate by dc-PECVD are shown in Fig. 6.2a and b. It is observed that each CNW corresponds to a kind of wavy sheet. The CNWs are uniformly grown over the substrate and they form a network structure as seen in Fig. 6.2. The network of CNWs can be easily peeled off the substrate without destroying them, as seen in Fig. 6.2c. In order to characterize the shape of the CNW, its the length, height, and thickness are defined as shown in Fig. 6.3.

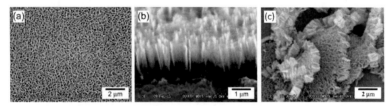

Figure 6.2 SEM images of (a) top view and (b) tilted view of CNWs grown vertically on the quartz substrate by dc-PECVD. (c) A SEM image of CNWs peeled out of the substrate with a knife edge [8].

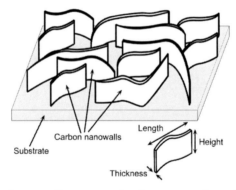

Figure 6.3 Schematic illustration of carbon nanowalls and the definition of their size [8].

6.3 Structural Characterization of CNWS

Various novel properties of carbon nanostructures can be related to their unique structures. Thus, structural characterization is important for their potential applications. Raman spectroscopy is one of the most powerful tools to characterize the structures of carbon nanostructures [35], as similar to TEM. In this section, structural characterization of CNWs by Raman spectroscopy and TEM has been discussed.

6.3.1 Raman Spectroscopy

Figure 6.4 shows SEM images of CNWs obtained by dc-PECVD under different growth conditions such as CH_4/H_2 ratio and substrate temperature for Raman measurements. It is observed that average lengths of CNWs in Fig. 6.4a, b and c are clearly different. The average lengths of CNWs are in the range of 0.5–2.0 μm. The longer the CNWs, the greater is the average height. The average heights of CNWs are in the range of 0.5–3.0 μm. In addition, the density of CNWs increases with their decreasing sizes. On the other hand, it seems that the average thickness of CNWs is independent of the length and height, and is several nanometers. Therefore, the average surface area of CNWs can be controlled by the growth condition.

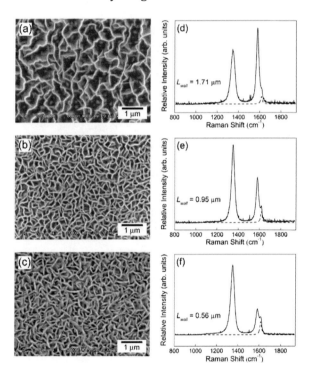

Figure 6.4 SEM images and the corresponding Raman spectra of carbon nanowalls with different sizes. (a), (b), and (c) are SEM images of CNWs with average lengths L_{wall} of 1.71, 0.95, and 0.56 μm, respectively. (d), (e), and (f) are Raman spectra of CNWs corresponding to (a), (b), and (c), respectively. Dashed lines in (d), (e), and (f) represent a Lorentzian fit to a D' band [8].

Raman spectra in CNWs with different average lengths (L_{wall}) corresponding to Figs. 6.4a, b, and c are shown in Figs. 6.4d, e, and 4f, respectively. In all the Raman spectra of CNWs, two main bands are observed at ~1580 and ~1350 cm^{-1}, respectively. The former band corresponds to G band (after graphite), indicating E_{2g} mode of graphite [36–41]. The latter band corresponds to D band (after defect), activated by disorder due to the finite crystallite size [36,42]. In addition, a weak band is observed at ~1620 cm^{-1}. This band corresponds to D' band, that emerges with D band indicating disorder [43]. The D' band appears in graphite-like carbons with relatively low disorder such as microcrystalline graphite and glassy carbon, while it cannot be observed in significantly disordered carbons such as carbon black [39].

Table 6.1 Peak frequencies, band widths, and relative intensities of D, G, and D' bands in Raman spectra of carbon nanowalls with different average lengths L_{wall}

Sample	L_{wall} (µm)	κ_D (cm^{-1})	W_D (cm^{-1})	κ_G (cm^{-1})	W_G (cm^{-1})	$\kappa_{D'}$ (cm^{-1})'	$W_{D'}$ (cm^{-1})	I_D/I_G
1	1.71	1348.7	43.2	1580.0	27.4	1617.0	19.7	0.77
2	0.95	1350.8	39.4	1582.4	31.2	1618.9	18.6	1.58
#3	0.56	1349.2	39.3	1585.5	34.1	1616.9	18.8	2.66

Note: κ_D, κ_G, and $\kappa_{D'}$ are peak frequencies of D, G, and D' bands, respectively. W_D, W_G, and $W_{D'}$ are widths (full widths at half maximum, FWHM) of D, G, and D' bands, respectively. I_D/I_G is the peak intensity ratio of D band to G band. Each band was analyzed by fitting with a Lorentzian line [8].

Each Raman band was analyzed by fitting with a Lorentzian line. The frequencies, bandwidths, and relative intensities of G, D, and D' bands in Figs. 6.4d–f are summarized in Table 6.1 with the corresponding average lengths L_{wall} of the CNWs. As presented in Table 6.1, G-band frequency (ν_G), G-band width (full width at half maximum, FWHM) (W_G) and peak intensity ratio of D band to G band (I_D/I_G) correlate with the L_{wall} of the CNWs, although no strong correlation are observed in the frequency and bandwidth of D and D' bands. Especially, the I_D/I_G intensity ratio strongly depends on the L_{wall} of CNWs. In Fig. 6.5, the I_D/I_G intensity ratio is plotted as a function of L_{wall} for various CNWs obtained under different growth conditions. It is found that the I_D/I_G intensity ratio decreases monotonically with increasing L_{wall} of CNWs according to $I_D/I_G = 1.64/L_{wall}$. This result suggests that the size of CNWs can be estimated from the I_D/I_G intensity ratio.

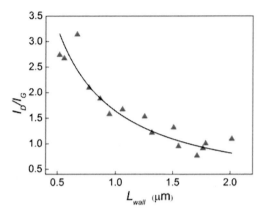

Figure 6.5 The peak intensity ratio of D band to G band (I_D/I_G) in the Raman spectrum as a function of the average length (L_{wall}) of carbon nanowalls [8].

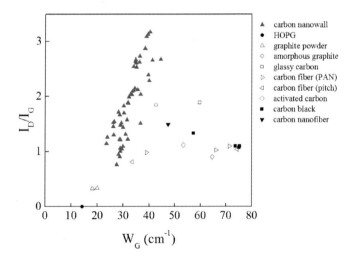

Figure 6.6 Relationship between the peak intensity ratio of D band to G band (I_D/I_G) and the width of G band (W_G) in the Raman spectra of carbon nanowalls, typical graphite-like carbons such as HOPG, graphite powder, amorphous graphite, and glassy carbon, and typical amorphous carbon such as carbon black [8].

To clarify the characteristics of the Raman spectra of CNWs, the Raman spectra of typical graphite-like carbons and amorphous carbons were also measured. For the characterization of the structures of graphite-like carbons, W_G band width and I_D/I_G intensity ratio have been widely used [37,38,41]. It is well known that both W_G and I_D/I_G increase with increasing the disorder in graphite structure [37,41]. In addition, it has been reported that the I_D/I_G intensity ratio is inversely proportional to the size L_a of finite crystallites constituting the graphite-like carbons [36,38]. Therefore, the W_G indicates the degree of graphitization, and the I_D/I_G concerns not only with the degree of graphitization but also the crystallite size.

In Fig. 6.6, the I_D/I_G intensity ratio is plotted as a function of W_G band width for CNWs and typical carbon materials. The HOPG is a typical highly ordered graphite-like carbon and, as expected, exhibits smallest band width ($W_G = 14.3$ cm^{-1}) and lowest intensity ratio ($I_D/I_G = 0$). On the other hand, carbon black is a typical significantly disordered carbon and exhibits larger band width ($W_G \approx 75$ cm^{-1}) and higher intensity ratio ($I_D/I_G \approx 1.2$). Both W_G and I_D/I_G become larger as the degree of disorder increases. As shown in Fig. 6.6, the W_G and I_D/I_G values of graphite-like carbons are expected to stay on straight gray line running from the lower left (HOPG) to the upper right (carbon black) with the degree of disorder. However, the plots of CNWs, amorphous graphite, and glassy carbon appear far above this line as shown in Fig. 6.6. It is well known that amorphous graphite and glassy carbon are composed of small crystallites with a size of several tens Å [44]. Their small crystallite sizes are consistent with the higher I_D/I_G intensity ratios. It should be noted that the characteristic I_D/I_G of CNWs is much higher than those of either glassy carbon or amorphous graphite. It is, therefore, suggested that the CNWs consist of smaller crystallites. In addition, the corresponding W_G is smaller than those of either glassy carbon or amorphous graphite. A narrow W_G means the higher degree of graphitization. To our knowledge, there is no graphite-like carbon with such I_D/I_G–W_G characteristics of CNWs. Thus, it is concluded that CNWs are a new type of graphite-like carbon consisting of small crystallites with a high degree of graphitization.

The crystallite size L_a of graphite-like carbons can be estimated from the I_D/I_G intensity ratio [36]. The empirical formula $L_a = 44/(I_D/I_G)$ has been successfully used for Raman spectra taken with the excitation energies around 514.5 nm [36,38]. Therefore, the above formula was also employed for Raman spectra with 532 nm excitation energy in this study. From these estimations, it is found that the average crystallite sizes L_a in CNWs obtained in this study are in the range of 14 Å (with I_D/I_G = 3.17) to 57 Å (with I_D/I_G = 0.77).

As shown in Fig. 6.5, the I_D/I_G intensity ratio increases with decreasing the size of CNWs. As mentioned above, the I_D/I_G intensity ratio also increases with decreasing the crystallite size of graphite-like carbons [36,38]. These results imply that the CNW size correlates with the crystallite size. Thus, it is suggested thatsmaller the CNW size, smaller is the size of crystallites constituting the CNW.

6.3.2 Transmission Electron Microscopy

Figure 6.7 displays SEM images of the CNWs grown on the quartz substrate by dc-PECVD. As can be seen in the cross-sectional view in Fig. 6.7a, the CNWs grow vertically on the substrate to a height of about 1.4 µm. The top view shown in Fig. 6.7b clearly displays the two-dimensional character of the CNWs. Specifically, the CNWs have the shape of thin sheets with a width of about 10 nm. It should also be noticed that the CNW films are bent and arranged to form a mesh network. These morphological features are consistent with those of CNWs in Fig. 6.2.

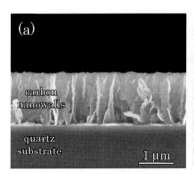
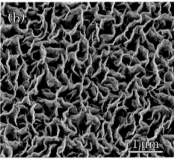

Figure 6.7 SEM images of the CNWs grown on a quartz substrate as seen in (a) a cross-sectional view and (b) a top view [10].

The structure of the CNWs was analyzed by taking electron diffraction patterns using a transmission electron microscope. A detailed analysis of the diffraction patterns by rotating the samples in several directions confirmed that the CNWs had a graphite structure and that the normal direction of each CNW film was basically the [0001] one of the graphite structure. On that basis, the nanostructure of the CNWs using a cross-sectional view was first examined under the condition that the *c* axis was excited in the reciprocal space. Figure 6.8 shows a typical high-resolution image, together with an electron diffraction pattern obtained from one CNW. In the image, we can observe a region with lattice fringes and about 8 nm in width, which is wedged between two amorphous regions with a width of about 2 nm. The formation of these amorphous regions is attributed to the deviation from the CNW-growth condition after switching off of the dc plasma power. As the average distance between the lattice fringes is about 0.34 nm, it is understood that the lattice fringes correspond to the (0002) plane of the graphite structure. That is, the region with the lattice fringes is the graphite region. One important feature of the graphite region is that the (0002) plane is winding due to the introduction of lattice defects such as dislocations. Hence, the [0001] normal direction fluctuates slightly from place to place, as indicated by the thick arrows in Fig. 6.8. This suggests that the CNWs were divided into small graphite regions.

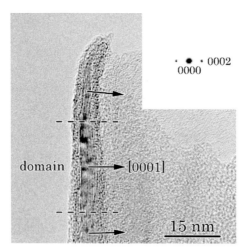

Figure 6.8 A high-resolution image and a corresponding electron diffraction pattern obtained from one CNW sheet. The image was taken under the electron incidence condition that the c^* axis was excited in the reciprocal space. The thick arrows denote the [0001] normal direction of the respective regions [10].

In order to elucidate the reason for the presence of such small regions, the in-plane nanostructure of the CNWs was examined. Figure 6.9a shows an electron diffraction pattern obtained from one CNW. The electron incidence was nearly parallel to the [0001] direction of the graphite structure. In the diffraction pattern, we can see at a glance diffraction spots with sixfold symmetry due to the graphite structure, as well as the first and second halo rings originating from the amorphous regions. These features confirm that the normal direction of the graphite regions is approximately pegged at the [0001] direction. On the other hand, a bright-field image in Fig. 6.9b shows that the CNW nanostructure is characterized by the formation of numerous small regions giving rise to diffraction contrast. The average size of the regions is about 20 nm. The presence of the diffraction contrast is indicative of the variation in the diffraction conditions of the graphite regions even at such a local level. It is thus understood that the CNWs were composed of numerous small graphite regions, dubbed nanographite domains.

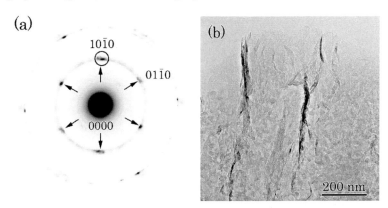

Figure 6.9 (a) An electron diffraction pattern and (b) a bright-field image obtained from one CNW film. The diffraction pattern consists of spots with sixfold symmetry due to the graphite structure. Numerous regions giving rise to diffraction contrast can be observed in the image [10].

Figure 6.10a and b show, respectively, a magnified bright-field image and a corresponding dark-field image taken by using the 1010 fundamental spot (see the circle in Fig. 6.10a) obtained from one CNW film. In the dark-field image, some (not all) nanographite domains with bright contrast can be observed. This indicates that the 1010 fundamental spot originated only from these domains. It should be mentioned here that other nanographite domains gave rise to bright contrast when a dark-field image was taken with another fundamental spot, e.g., the 0110 spot, although the details are not described here. We can thus understand that the normal direction of the nanographite domains fluctuated slightly from place to place, as mentioned above (see Fig. 6.8). The diffraction pattern with sixfold symmetry in Fig. 6.9a is presumed to originate not from the single graphite structure but from the superposition of the slightly deviated 0001 patterns from the numerous nanographite domains. Another important feature of the dark-field image is the appearance of the fringes in one nanographite domain, an example of which is indicated by the circle in Fig. 6.10b. A detailed analysis revealed that the fringes were caused by interference between the diffraction spots around the 1010 fundamental spot (see Fig. 6.9a) and were rotational Moire fringes. It is thus understood that the nanographite domains were composed of slightly rotated graphite sheets along the [0001] direction.

 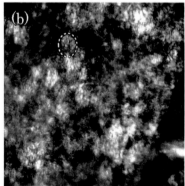

Figure 6.10 (a) A magnified bright-field image of (b). (b) A dark-field image taken by using the 1010 fundamental spot (see the circle in (a) obtained from one CNW sheet. The circles in (a) and (b) indicate one nanographite domain and rotational Moire fringes can be detected in the circle in (b) [10].

On the basis of the experimental results, we illustrate a schematic diagram of the CNW nanostructure in Fig. 6.11. For simplicity, only the nanographite domains in one CNW sheet are depicted in the diagram and two amorphous regions are ignored. In terms of their macroscopic structure, the CNWs have the shape of bent thin films, as shown in the left diagram. It has been reported that CNWs are characterized as being stacks of graphene sheets. However, the presence of the lattice defects in the sheet arrangement caused a local fluctuation in the normal direction in the sheets as well as the rotation of the basal plane of the graphite structure in the domains (see the diagram at the right). Presumably, these defects were introduced during the growth and/or stacking of the graphite sheets. As a result, the local fluctuation gave rise to numerous nanographite domains. This indicates that a nanographite domain with a high degree of graphitization was the constitutional unit of the CNWs examined here. The nanographite domains might correspond to small crystallites mentioned in Raman analysis in Section 1.3.1.

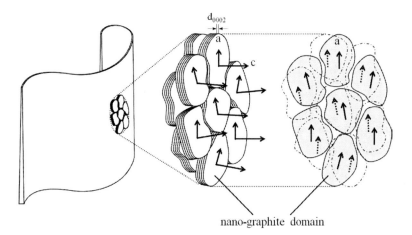

Figure 6.11 A schematic diagram of the nanostructure of one CNW sheet. For simplicity, two amorphous regions are ignored. Small nanographite domains are the constitutional units of the CNW sheet. Characteristics of the domains include the fluctuation of the [0001] normal direction and the rotation of the *ab* basal plane of the graphite structure [10].

Because the domains have the graphite structure, CNWs presumably have a high degree of graphitization. The formation of numerous domains, on the other hand, induces many "domain boundaries" in CNWs, which seem to reduce the average size of the graphite regions. Thus, as shown in Fig. 6.6 [8], the gradient of the index correlation for CNWs can become larger and deviate from that of ordinary graphite-based carbon materials. If the domain size can be controlled by varying the fabrication conditions of CNWs, it should have a significant effect on the physical properties of CNWs.

6.4 Potential Applications of CNWs to Energy Devices

CNWs have been synthesized mainly by PECVD without metal catalysts. The improved dc-PECVD by Tachibana's group has led to the synthesis of a large amount of CNWs [24,28]. The quantity synthesis has enabled to test CNWs for more applications which have been ever difficult with a small amount of samples. In this section, potential applications of CNWs as negative electrodes in lithium ion batteries and as catalyst supports in fuel cells have been discussed.

6.4.1 CNWs as Negative Electrode in Lithium Ion Battery

Lithium ion batteries have been widely developed [45,46] and recently those with high power density have been in demand in many fields. From this point of view, active materials with nanometer-sized structure have been recently focused on for the high-rate use of the battery because they can make the conductive path of lithium ions shorter, so that the internal resistivity of the electrode is effectively decreased to achieve high power of the battery [47–51].

Graphite powder is usually used as an active material in negative electrodes of commercial lithium ion batteries since the highly graphitized carbons generally show good performances on reversible capacity, cycle life, and stably flat potential during charge/discharge [52]. However, nanometer-sized graphite produced by mechanical milling of graphite generally includes a large amount of defects, leading to the irreversible storage of lithium [53]. Therefore the nanometer-sized structure should be made and kept at the moment of carbonization and graphitization processes.

CNW consists of small crystallites or nanometer-sized graphite domains with high degree of graphitization [8,10]. This structure makes it an ideal candidate for negative electrodes in lithium ion batteries with high rate use. In this section, the properties of CNWs as a negative electrode material in lithium ion battery have been discussed in light of the structural features.

6.4.1.1 Preparation of electrode

For preparing electrodes, CNWs synthesized by dc-PECVD method are peeled off from silicon substrate, and used as CNW-as. Some of peeled CNWs are heated up to 1773 K with the heating rate of 10 K min^{-1} under pure nitrogen flow of 200 ml min^{-1}, and kept for 30 min. They are used as CNW-1773. In addition, commercially available artificial graphite KS15 (Lonza Co., average particle size of 8 μm) is used as a reference.

The powders of CNW-as, CNW-1773, and KS15 are mixed with 10 wt% of binder polyvinylidene-fluoride in N-methylpyrrolidone. The mixed slurries with appropriate viscosities are applied on nickel mesh (Nilaco,100 mesh), dried at 423 K and then roll-pressed at 423 K to produce carbon sample electrodes. Conventional three-electrode cells with the carbon sample as a working electrode and two lithium foils as counter and reference ones are constructed in a grove box filled with pure argon.

Lithium insertion properties of three kinds of carbon sample electrodes with CNW-as, CNW-1773, and KS15 are studied by cyclic voltammetry and charge/discharge tests using the three-electrode cells with a commercially available electrolyte from a mixture of ethylenecarbonate and dimethylcarbonate in the ratio of 50/50 containing 1 M lithium tetrafluoroborate.

6.4.1.2 Cyclic voltammogram

Figure 6.12 shows cyclic voltammograms (CVs) of electrode samples with CNW-as and artificial graphite KS15, measured with the scan rate of 1 mVs^{-1}. For both samples, reduction peaks are observed at around 0.6 V in only first cycle. The peaks are caused by the formation of so-called solid electrolyte interface (SEI) due to the electrochemical decomposition of electrolyte on the carbon electrode surface. Moreover, pairs of reduction/oxidation peaks are clearly observed in the range of 0–0.4 V. The peaks correspond to the insertion/extraction of lithium ion, respectively. No other peak of electrochemical reaction except for those related to lithium ion battery as mentioned above is observed. This means that CNWs exhibit the lithium insertion/extraction, as graphitic carbon such as KS15.

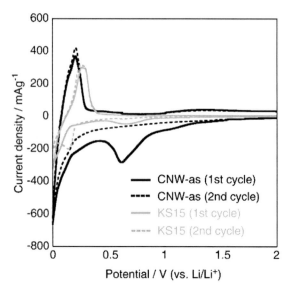

Figure 6.12 Cyclic voltammograms of CNW-as and KS15 electrodes in EC/DEC electrolyte containing 1 M LiBF$_4$. The scan rate is 1 mV s^{-1} [24].

The SEI peak at around 0.6 V in CNW-as is much higher than that of graphite, KS15. This means that the large amount of the electrolyte decomposition to produce SEI layer occurs on CNWs. This can be explained by the structural feature of CNW composed of small graphite domains. The domain size in CNW used in the present experiments is 20 nm, which is much smaller as compared with 170 nm of crystallite size of KS15 [54]. This means that CNWs have a large amount of domain boundaries, i.e. graphite edges, exposed to the electrolyte, although 5 μm of the size of each CNW is almost similar to 8 μm of particle size of KS15. The large exposure of graphite edges in CNW can lead to the large amount of electrolyte decomposition to produce SEI layer.

The lithium insertion/extraction peaks for CNW-as are very sharp, as typical anode property of graphitic carbon such as KS15. In addition, it should be noted that the extraction peak of CNW-as is located at a lower potential than that of KS15. This means that the internal resistivity and/or path length in CNW for lithium extraction are smaller than those in KS15. These can be attributed to the high crystallinity and/or small size of graphite domains in CNW. Thus, CNW can lead to the smooth reaction between insertion and extraction of lithium ions.

6.4.1.3 Charge/discharge curve

Figure 6.13 shows the charge/discharge curves of CNW-as and that heat-treated at 1773 K (CNW-1773), during the first and second cycles measured with constant current density of 50 mA g^{-1}. Note that the charge and discharge processes correspond to the insertion and extraction of lithium ions into the carbon samples.

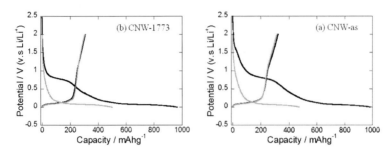

Figure 6.13 Charge/discharge curves of (a) CNW-as and (b) CNW-1773 electrodes with the current density of 50 mA g^{-1} in EC/DEC electrolyte containing 1 M LiBF$_4$. Black and gray lines are those of the first and second cycles, respectively [24].

In CNW-as, only the first charge curve shows two large plateaus at around 1.5 V and 0.8 V. The irreversible capacities estimated from the difference between their first charge and discharge capacities reach to 100 mAh g^{-1} and 750 mAh g^{-1}, respectively. The former plateau is not observed in CNW-1773, whereas the latter one is clearly observed even in CNW-1773. It is known that the thermal annealing at 1773 K can burn off amorphous carbon layer formed on CNWs synthesized by dc-PECVD. Therefore, the former plateau might be attributed to the irreversible storage of lithium in the amorphous carbon layer. On the other hand, the latter plateau corresponds to the large peak at 0.6 V observed in the CV as mentioned above and due to the SEI formation. After the second cycle, same curves are repeated in both the samples. This suggests that the reversible capacity after the second cycle is not affected by thermal treatment at up to 1773 K.

The charge/discharge curves at near 0 V are flat in both samples. The curves correspond to lithium insertion/extraction peaks observed at 0 −0.4 V in the CV as mentioned above, and to reversible capacities. Such flat curves at near 0 V are typical behavior of lithium insertion/extraction making graphite intercalation compounds with lithium.

In addtion, the details of the charge/discharge curves at near 0 V show clear step plateaus which mean staging reactions during deintercalation of lithium from graphene layers, from stage-1 of LiC_6 to graphite via stage-2 of LiC_{12} or LiC_{18} and high stage LiC_x [55]. Thus CNW exhibits clear staging plateaus as typical graphic carbon such as KS15. This staging reaction clearly observed in CNWs is caused by the well developed layer structure of CNWs.

6.4.1.4 High rate property

Moreover, to investigate the high rate property, the charge/discharge curves of CNW-1773 and KS15 were measured with different currrent densities. In these measurements, CNW-1773 was used to exclude the effect of amorphous carbon. Figure 6.14 shows second discharge curves of CNW-1773 and KS15 operated with different current densities such as 100, 200, 500, and 1000 mA g^{-1}. To better clarify the change in the discharge curve, the reversible capacities at different currrent densities were estimated from the curves with respect to the cut off potential at 0.6 V in Fig. 6.14. The reversible capacities for CNW-1773 and KS15 were plotted as a function of the current densities in Fig. 6.15. As clearly shown in Figs. 6.14 and 6.15, CNWs potentially show excellent performances even at large current densities compared with typical graphitic meterials such as KS15. With the small current density of 100 mA g^{-1}, the reversible capacity of CNW-1773 is 260 mAh g^{-1}, which reaches 74% of 350 mAh g^{-1} of KS15 and 70% of theoretical one of graphite, 372 mA g^{-1}.

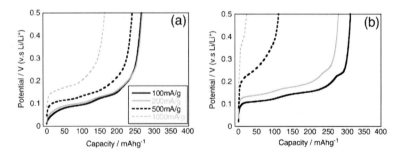

Figure 6.14 Discharge curves of (a) CNW-1773 and (b) KS15 in the second cycle with different current densities. The charge current densities were similar to the discharge ones [24].

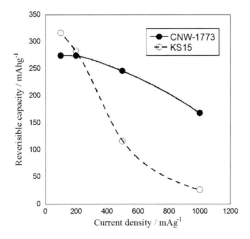

Figure 6.15 Reversible capacity of CNW-1773 and KS15, depending on the current density, obtained from Fig. 6.14.

Reversible capacities of both CNWs and KS15 decrease as the current densities increase. It should be noted that the reversible capacity of CNWs does not drop so much even with large current densities. That is, 70% of the reversible capacity with the low current density of 100 mA g^{-1} is kept even with the high current density of 1000 mA g^{-1}. On the other hand, the reversible capacity of KS15 drastically decreases with large IR drop at large current densities. As a result, the reversible capacity of CNW-1773 becomes about ten times as large as that of KS15 with the high current density of 1000 mA g^{-1}. Such excellent high rate property of CNWs seems to be caused by short range diffusion path of lithium during intercalation in small graphite domains in CNW and large exposure of graphite edges to the electrolyte. In CNW, large amount of lithium can be inserted and extracted at one time, similarly as other nanometer-sized active materials such as carbon nanobeads [56], graphite nanofibers [57,58], pyrolytic carbon thin film [59] for electrode of high-rate lithium ion battery.

More recently, a vertically aligned graphene structure, called CNW, on Ni substrate was used as an electrode directly without being peeled out of the substrate [25]. In this case, to make the CNW electrode, no binder and conductive additives as mentioned in Section 6.4.1.1 are required. As a result, it exhibits higher rate capability. This is due to the effective electrical connenction of CNWs with the substrate. Thus, not only the domain structure but also the vertically aligned morphology of CNWs are useful for further impovement of high rate properties.

6.4.1.5 CNWs as catalyst support in fuel cell

Fuel cells with hydrogen as the fuel represent an environment friendly technology and are attracting considerable interest as a means of producing electricity by direct electrochemical conversion of hydrogen and oxygen into water [60]. Generally, platinum (Pt) particles are used as anode and cathode catalysts in the fuel cells. In addition, carbon materials are used as catalyst supports. Since the activity of a catalyst increases with the increase in the reaction surface area, the diameter of catalyst particles should be reduced to increase the active surface [60 –63]. The catalysts should be supported on a carbon substrate with a high surface area. The support for a fuel cell catalyst must also have sufficient electric conductivity as it can act as a path for the flow of electrons [61,62]. Moreover, the carbon support should have a high percentage of mesopores in the 20–40 nm region to provide a high accessible surface area to catalyst and to monomeric units of the Nafion ionomer and also to boost the diffusion of chemical species [61,63]. In addition to such dispersion effect, the carbon support can potentially influence the activity of the catalysts due to the modification in their electronic characters and geometrical shapes depending on the support [61]. Therefore, new carbon nanostructures with unique morphology and structure have attracted much attention for the improvement of the electrochemical activity of the catalysts [61–64].

The deposition of Pt catalysts on CNWs has been carried out by certain methods such as electro-deposition [26] and supercritical fluid-chemical vapor deposition [27]. For practical applications in fuel cells, not only well-dispersion of Pt catalysts but also high electro-catalytic activity are required. In the following section, the dispersion of Pt catalysts supported on CNWs (Pt/CNW) and their electrochemical properties are shown.

6.4.1.6 Preparation of Pt/CNW

Pt/CNW was prepared by a solution reduction method [65] by the following procedure. First, CNW film was peeled off from the substrate. The peeled CNW film was powdered using agate mortar. One hundred milligram of the CNW powder was suspended in 50 ml distilled water by ultrasonic treatment for 30 min, and the pH of this mixture was adjusted using Na_2CO_3. The solution was heated at 80 °C and then mechanically stirred for 2 h. During the stirring, 25 ml of aqueous solution of hexachloroplatinic acid (H_2PtCl_6) was added slowly into the solution. Note that the mass ratio of CNWs to platinum in this mixture was 4:1. Then formaldehyde was added to the solution, and Pt ion was reduced at 80 °C for 90 min. Following the reduction, the solid was filtered and washed with distilled water, and then dried at 80 °C for 9 h. As a result, Pt/CNW with a Pt loading of 18 wt% was obtained. For comparison, commercially available 45.7 wt% Pt catalysts supported on high surface area carbon (TEC10E50E) provided by Tanaka KIKINZOKU KOGYO K. K., which exhibit good performance, were used as a reference and are called T-Pt/CB hereafter.

6.4.1.7 Physical characterization of Pt/CNW

The dispersion and particle size for Pt/CNW and T-Pt/CB were characterized by TEM and X-ray diffraction (XRD). A typical TEM image of CNW before supporting Pt catalysts is shown in Fig. 6.16a. Similar to Figs. 6.9 and 6.10 [10], diffraction contrasts corresponding to nanographite domains of ~20 nm are clearly observed where a domain is surrounded by a broken line. A TEM image of Pt/CNW is shown in Fig. 6.16b. For comparison, a TEM image of commercially available T-Pt/CB is also shown in Fig. 6.16c. The black round contrasts correspond to the Pt nanoparticles. It is found that Pt nanoparticles are uniformly dispersed on CNWs, as those in T-Pt/CB. Note that the density of Pt/CNW is lower than that of T-Pt/CB because of the lower loading of Pt/CNW where specific surface areas of CNW and CB are about 150 and 800 m^2 g^{-1}, respectively. From histograms of diameters of Pt nanoparticles, it is also found that the particles for Pt/CNW have a relatively narrow size distribution ranging from 1.5 to 7.0 nm, as that for T-Pt/CB. The mean particle size for Pt/CNW is 3.6 nm which is slightly larger than 2.8 nm for T-Pt/CB.

The higher magnification TEM image of Pt/CNW is shown in Fig. 6.16d. It should be noted that Pt nanoparticles are located along the domain boundaries in CNW. Such unique dispersion of Pt nanoparticles is more clearly confirmed in Pt/CNW with a smaller loading of Pt as shown in Fig. 6.16e, where the domain boundaries are highlighted by broken lines. These results indicate that domain boundaries in CNWs can be effectively used as support sites for Pt catalysts.

The loading of Pt catalysts for carbon materials such as carbon nanotubes, carbon nanofibers, and graphene sheets has also been studied by ab initio calculations and molecular dynamics simulations [66–69]. According to the studies, the absorption energies of Pt atom on their defects, such as vacancies and edges, are much larger than those on their perfect surfaces, due to the dangling bonds [66,67]. The strong binding of Pt atom with the defects prevents the metal migration and aggregation. Consequently, the high dispersion of Pt nanoparticle is achievable for such defective carbon materials. The domain boundary in CNW would include graphite edge, vacancies, or dangling bonds [10], although the detailed structure is not yet clear. Thus, the large absorption energy on the domain boundary can lead to the preferred absorption of Pt nanoparticles along the boundary. In addition, the structure of the metal particle depends on the support site, and affects the catalytic properties [68,69]. This also arouses interest in the structure and electro-catalytic activity of Pt nanoparticles supported on the domain boundary in CNW.

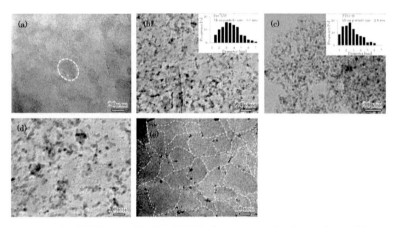

Figure 6.16 TEM images for (a) CNW before supporting Pt catalysts, (b) Pt/CNW with 18 wt%-Pt, and (c) T-Pt/CB with 45.7 wt%-Pt. (b) and (c) include the corresponding histograms of diameters of Pt nanoparticles. (d) and (e) show higher magnification TEM images of Pt/CNWs with 18 wt%-Pt and less than 15 wt%-Pt, respectively [28].

A typical XRD pattern of Pt/CNW is shown in Fig. 6.17. For comparison, an XRD pattern of T-Pt/CB is also included in the figure. All peaks can be assigned to those of Pt crystal with face centered cubic (fcc) structure and graphite with hexagonal structure.

Relatively broad peaks, which are observed at 39.9°, 46.3°, 67.5°, and 81.4°, are attributed to (111), (200), (220), and (311) of fcc Pt crystal, respectively. These peaks are observed in both Pt/CNW and T-Pt/CB. The broad peaks mean that Pt nanoparticles are successfully prepared on CNWs, as those in T-Pt/CB. The mean size of Pt nanoparticles can be calculated from the most intensive (111) peak according to the Scherrer's formula [70]:

$$L = \frac{k\lambda}{B_{(2\theta)}\cos\theta_{max}},\qquad(6.1)$$

where L is the mean size of the Pt particles, λ is the X-ray wavelength which is Cu Kα_1 = 1.5418 Å in this experiment, θ_{max} is the maximum angle of the (111) peak, $B_{(2\theta)}$ is the full width at half maximum for the (111) peak in radians, and k is the coefficient taken here as 1 for the sphere. The mean size of Pt on CNWs is estimated to be 3.5 nm with θ_{max} = 39.98° and $B_{(2\theta)}$ = 0.046 rad, which is slightly large compared with 2.3 nm in T-Pt/C. These particle sizes by XRD are in good agreement with those by TEM as shown in Table 6.2.

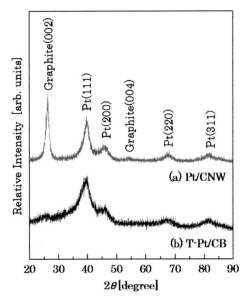

Figure 6.17 XRD patterns for (a) Pt/CNW with 18 wt%-Pt and (b) T-Pt/CB with 45.7 wt%-Pt, taken by using X-ray diffractometer with Cu Kα_1 radiation (λ = 1.5418 Å) operating at 40 kV and 20 mA [28].

Table 6.2 Comparison of particle size, ECSA, GSA, and Pt utilization of Pt/CNW and T-Pt/CB [28]

Catalyst	Pt loading (wt%)	Particle size (TEM) (nm)	Particle size (XRD) (nm)	ECSA $(m^2\,g^{-1}\text{-Pt})$	GSA $(m^2\,g^{-1}\text{-Pt})$	Pt utilization (%)
Pt/CNW	18	3.6	3.5	53.4	80.1	66.7
T-Pt/CB	45.7	2.8	2.3	69.8	121.9	57.3

Moreover, in XRD for Pt/CNW, a sharp and intense peak is clearly observed at 26.3°, as shown in Fig. 6.17. This peak is assigned to (002) of hexagonal graphite. The corresponding graphite (004) peak is also observed at 54.1°. These peaks are originated from CNWs [71]. The sharp peaks mean that CNWs have high degree of graphitization. On the other hand, in T-Pt/CB, graphite peaks are hardly observed, or exhibit very weak and broad profiles. Such spectral feature implies that the carbon support in T-Pt/CB has amorphous-like structure with very small crystalline regions. Therefore, it is expected that CNW has much high electrical conductivity compared with the carbon support in T-Pt/CB. Actually the resistivity of CNW is $\sim 10^{-5}$ Ωm which is much lower than $\sim 10^{-2}$ Ωm of the carbon support in T-Pt/CB [14]. The high conductivity is required for the carbon support so that it can act as a path for the flow of electrons in fuel cell.

6.4.1.8 Electrochemical characterization of Pt/CNW

Electrochemical properties of the catalysts were measured by means of a thin film electrode technique. A glassy carbon electrode with an area of 0.212 cm^2 was used as the substrate, on which 20 μl of an ink of the catalyst was applied. The catalyst layer on the glassy carbon disk was prepared as follows. A mixture containing 5.0 mg of electrocatalysts, 5 ml of mixture of ethanol and 2-propanol, and 30 μl of Nafion solution (5 wt%) were ultrasonically blended for 20 min to obtain a homogeneous ink. Then the ink was spread on the surface of a clean glassy carbon electrode. After the ethanol and 2-propanol volatilization, the electrode was heated at 80 °C for 30 min.

The electrochemical measurements were carried out by using a standard three-electrode cell. The glassy carbon electrode loaded with the catalysts as mentioned above was used as the working electrode. A Pt wire served as the counter electrode and a reversible hydrogen electrode (RHE) was used as the reference electrode. CVs were recorded from 0.05 to 1.20 V vs. RHE at a scan rate of 50 mV s^{-1} in a 0.1 M HClO$_4$. The electrolyte was saturated with nitrogen in order to expel oxygen into this solution. During the experiments, the temperature was kept at 25 °C.

Figure 6.18 shows the CV of Pt/CNW. The hydrogen absorption and desorption peaks clearly appear in the range of 0.05–0.4 V vs. RHE. It can be seen that Pt/CNW produces higher currents in the hydrogen region. This would be due to the higher dispersion of Pt/CNW which is the result of small Pt nanoparticle size on CNWs. The electrochemical active surface areas (ECSA) of the catalysts are estimated from the coulombic charge for the hydrogen adsorption and desorption in the CV according to the equation [72,73]:

$$\text{ECSA} = \frac{Q_H}{[\text{Pt}] \times 0.21}, \qquad (6.2)$$

where [Pt] is the Pt loading in the electrode, 0.21 mC cm^{-2} is the charge required to oxidize a monolayer of H$_2$ on bright Pt, and Q_H is the charge for hydrogen desorption whose value is calculated as the mean value between the amounts of charge transfer during the electro-adsorption and desorption of H$_2$ on Pt sites. The calculated results are also shown in Table 6.2. The Pt/CNW exhibits relatively high ECSA of 53.4 m^2 g^{-1}-Pt with Q_H = 1.91 mC cm^{-2} and [Pt] = 0.017 mg cm^{-2}, although the value is still low compared with 69.8 m^2 g^{-1}-Pt for commercially available T-Pt/CB with good performance.

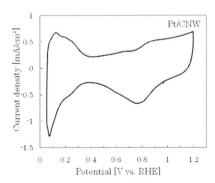

Figure 6.18 A typical cyclic voltammogram of Pt/CNW with 18 wt%-Pt [28].

Moreover, assuming that all Pt particles have mean diameter obtained from XRD as mentioned above, the geometric surface areas (GSA) of these catalysts can be calculated from the equation [74]:

$$GSA = \frac{6}{\rho \times d},\tag{6.3}$$

where ρ is the density of Pt, that is 21.4 g cm^{-3} and d is the mean diameter of Pt nanoparticles. The GSA for Pt/CNW is 80.1 m^2 g^{-1}-Pt with d = 3.5 nm, while the value for T-Pt/CB is 121.9 m^2 g^{-1}-Pt, as shown in Table 6.2. From the above two areas, ECSA and GSA, it is also possible to estimate the Pt utilization efficiency (μ) from the equation:

$$\mu = \frac{ECSA}{GSA}.\tag{6.4}$$

It should be noted that the Pt/CNW exhibits higher μ of 66.7% than 57.3% for T-Pt/CB as shown in Table 6.2. Such high Pt utilization could be attributed to the high electric conductivity of CNW and the improvement of electronic properties of Pt on the domain boundaries, in addition to the high dispersion of Pt nanoparticles.

6.5 Summary

This chapter has shown that each CNW exhibits the domain structure that consists of small crystallites or nanographite domains with high degree of graphitization. The CNWs show quick electrochemical response for lithium insertion/extraction, which is expected for high rate use of lithium ion battery. Such quick response is also an evidence for small graphite domains in CNWs. In addition, the CNWs serve as catalyst support in fuel cells where Pt nanoparticles are well-dispersed along the domain boundaries in CNWs. The Pt particles on the domain boundaries also exhibit relatively high electro-catalytic activity. Thus, the domain structure of CNWs is useful for negative electrodes in high rate lithium ion batteries and catalyst supports in fuel cells. The electrode properties might be further improved since the size of graphite domains in CNWs can be controlled by the growth condition. Such unique structure and morphology of CNWs can lead to further applications.

Acknowledgments

The author acknowledges the contributions to the work presented here by all past and present collaborators, Mr. S. Kurita, Mr. A.

Yoshimura, Mr. H. Yoshimura, Mr. N. Kitada, Mr. K. Nukada, Mr. S. C. Shin, Mr. K. Kobayashi, Mr. H. Nakai, Mr. T. Matsuo, Dr. O. Tanaike, Dr. M. Tanimura, Dr. A. Ishihara, Prof. K. Ota, and Prof. K. Kojima. Especially the author thanks Mr. S. C. Shin for preparing figures in this chapter. This work was partially supported by a Strategic Research Project of Yokohama City University and a KAKENHI (No. 21560691) from the Ministry of Education, Culture, Sports, Science and Technology of Japan.

References

1. Chen J.-H., Jang C., Xiao S., Ishigami M., Fuhrer M.S. (2008) Intrinsic and extrinsic performance limits of graphene devices on SiO_2, *Nat Nanotechnol*, **3**, 206–209.

2. Balandin A.A., Ghosh S., Bao W., et al. (2008) Superior thermal conductivity of single-layer grapheme, *Nano Lett*, **8**, 902–907.

3. Lee C., Wei X., Kysar J.W., Hone J. (2008) Measurement of the elastic properties and intrinsic strength of monolayer graphene, *Science*, **321**, 385–388.

4. Stoller M.D., Park S., Zhu Y., An J., Ruoff R.S. (2008) Graphene-based ultracapacitors, *Nano Lett*, **8**, 3498–3502.

5. Wu Y., Qiao P., Chong T., Shen Z. (2002) Carbon nanowalls grown by microwave plasma enhanced chemical vapor deposition, *Adv Mater*, **14**, 64–67.

6. Hiramatsu M., Shiji K., Amano H., Hori M. (2004) Fabrication of vertically aligned carbon nanowalls using capacitively coupled plasma-enhanced chemical vapor deposition assisted by hydrogen radical injection, *Appl Phys Lett*, **84**, 4708–4710.

7. Wang J.J., Zhu M.Y., Outlaw R.A., et al. (2004) Free-standing subnanometer graphite sheets, *Appl Phys Lett*, **85**, 1265–1267.

8. Kurita S., Yoshimura A., Kawamoto H., et al. (2005) Raman spectra of carbon nanowalls grown by plasma-enhanced chemical vapor deposition, *J Appl Phys,* **97**, 104320-1-5 .

9. Tanaka K., Yoshimura M., Okamoto A., Ueda K. (2005) Growth of carbon nanowalls on a SiO_2 substrate by microwave plasma-enhanced chemical vapor deposition , *Jpn J Appl Phys*, **44**, 2074–2076 .

10. Kobayashi K., Tanimura M., Nakai H., et al. (2007) Nanographite domains in carbon nanowalls, *J Appl Phys*, **101**, 094306-1-4.

11. Ni Z.H., Fan H.M., Feng Y.P., Shen Z.X., Yang B.J., Wu Y.H. (2006) Raman spectroscopic investigation of carbon nanowalls, *J Chem Phys*, **124**, 204703 -1-5.

12. Takeuchi W., Ura M., Hiramatsu M., Tokuda Y., Kano H., Hori M. (2008) Electrical conduction control of carbon nanowalls, *Appl Phys Lett*, **92**, 213103-1-3.

13. Teii K., Shimada S., Nakashima M., Chuang A.T.H. (2009) Synthesis and electrical characterization of n-type carbon nanowalls, *J Appl Phys*, **106**, 084303-1-6.

14. Yamada S., Yoshimura H., Tachibana M. (2010) Anomalous Anderson weak localization of graphite thin film fabricated by plasma-enhanced chemical vapor deposition, *J Phys Soc Jpn*, **79**, 054708-1-4.

15. Zhang C., Wang Y., Huang L., Wu Y. (2010) Electrical transport study of magnetomechanical nanocontact in ultrahigh vacuum using carbon nanowalls, *Appl Phys Lett*, **97**, 062102-1-3.

16. Zhao X., Outlaw R.A., Wang J.J., Zhu M.Y., Smith G.D., Holloway BC (2006) Thermal desorption of hydrogen from carbon nanosheets, *J Chem Phys*, **124**, 194704-1-6.

17. Kinoshita I., Hayashi S., Yoshimura H., Nakai H., Tachibana M. (2008) Ultraviolet photoelectron spectroscopy study of electronic states and deuterium adsorption on carbon nanowalls, *Chem Phys Lett*, **450**, 360–364.

18. Kita Y., Hayashi S., Kinoshita I., et al. (2010) First-principles calculation and transmission electron microscopy observation for hydrogen adsorption on carbon nanowalls, *J Appl Phys*, **108**, 013703 -1-4.

19. Teii K., Nakashima M. (2010) Synthesis and field emission properties of nanocrystalline diamond/carbon nanowall composite films, *Appl Phys Lett*, **96**, 023112-1-3.

20. Stratakis E., Giorgi R., Barberoglou M., et al. (2010) Three-dimensional carbon nanowall field emission arrays , *Appl Phys Lett*, **96**, 043110-1-3.

21. Zhu M.Y., Outlaw R.A., Bagge-Hansen M., Chen H.J., Manos D.M. (2011) Enhanced field emission of vertically oriented carbon nanosheets synthesized by C_2H_2/H_2 plasma enhanced CVD, *Carbon*, **49**, 2526–2531.

22. Hojati-Talemi P., Simon G.P. (2011) Field emission study of graphene nanowalls prepared by microwave-plasma method, *Carbon*, **49**, 2875–2877.

23. Takeuchi W., Kondo H., Obayashi T., Hiramatsu M., Hori M. (2011) Electron field emission enhancement of carbon nanowalls by plasma surface nitridation, *Appl Phys Lett*, **98**, 123107-1-3.

24. Tanaike O., Kitada N., Yoshimura H., Hatori H., Kojima K., Tachibana M. (2009) Lithium insertion behavior of carbon nanowalls by dc plasma CVD and its heat-treatment effect, *Solid State Ionics*, **180**, 381–385.

25. Xiao X., Liu P., Wang J.S., Verbrugge M.W., Balogh M.P. (2011) Vertically aligned graphene electrode for lithium ion battery with high rate capability, *Electrochem Commun*, **13**, 209–212.

26. Giorgi L., Dikonimos T.M., Giorgi R., Lisi N., Salernitano E. (2007) Electrochemical properties of carbon nanowalls synthesized by HF-CVD, *Sens Actuators B*, **126**, 144–152.

27. Machino T., Takeuchi W., Kano H., Hiramatsu M., Hori M. (2009) Synthesis of platinum nanoparticles on two-dimensional carbon nanostructures with an ultrahigh aspect ratio employing supercritical fluid chemical vapor deposition process, *Appl Phys*, **2**, 025001-1-3.

28. Shin S.C., Yoshimura A., Matsuo T., et al. (2011) Carbon nanowalls as platinum support for fuel cells, *J Appl Phys*, **110**, 104308-1-4.

29. Chuang A.T.H., Robertson J., Boskovic B.O., Koziol K.K.K. (2007) Three-dimensional carbon nanowall structures, *Appl Phys Lett*, **90**, 123107 -1-3.

30. Sato G., Morio T., Kato T., Hatakeyama R. (2006) Fast growth of carbon nanowalls from pure methane using helicon plasma-enhanced chemical vapor deposition, *Jpn J Appl Phys*, **45**, 5210–5212.

31. Jain H.G., Karacuban H., Krix D., Becker H.W., Nienhaus H., Buck V. (2011) Carbon nanowalls deposited by inductively coupled plasma enhanced chemical vapor deposition using aluminum acetylacetonate as precursor, *Carbon*, **49**, 4987–4995.

32. Bo Z., Yu K., Lu G., Wang P., Maoa S., Chen J. (2011) Understanding growth of carbon nanowalls at atmospheric pressure using normal glow discharge plasma-enhanced chemical vapor deposition, *Carbon*, **49**,1849–1858.

33. Itoh T., Shimabukuro S., Kawamura S., Nonomura S. (2006) Preparation and electron field emission of carbon nanowall by Cat-CVD, *Thin Solid Films*, **501**, 314–317.

34. Lisi N., Giorgi R., Re M., et al. (2011) Carbon nanowall growth on carbon paper by hot filament chemical vapour deposition and its microstructure, *Carbon,* **49**, 2134–2140.

35. Jorio A., Dresselhaus M.S., Saito R., Dresselhaus G. (2011) *Raman Spectroscopy in Graphene Related Systems*, Wiley-VCH, Weinheim, Germany.

36. Tuinstra F., Koenig J.L. (1970) Raman spectra of graphite, *J Chem Phys*, **53**, pp. 1126–1130.

37. Chieu T.C., Dresselhaus M.S., Endo, M. (1982) Raman studies of benzene-derived graphite fibers, *Phys Rev B*, **26**, 5867–5877.

38. Knight D.S., White W.B. (1989) Characterization of diamond films by Raman spectroscopy, *J Mater Res*, **4**, 385–393.

39. Cuesta A., Dhamelincourt P., Laureyns J., Martinez-Alonso A., Tascon J.M.D. (1994) Raman microprobe studies on carbon materials, *Carbon*, **32**, 1523–1532.

40. Schwan J., Ulrich S., Batori V., Ehrhardt H., Silva S.R.P. (1996) Raman spectroscopy on amorphous carbon films, *J Appl Phys*, **80**, 440–447.

41. Katagiri G. (1996) Raman spectroscopy of graphite and carbon materials and its recent application, *TANSO (Carbon)*, **175**, 304–313.

42. Nemanich R.J., Solin S.A. (1979) First- and second-order Raman scattering from finite-size crystals of graphite, *Phys Rev B*, **20**, 392–401.

43. Vidano R., Fischbach D.B. (1978) New lines in the Raman spectra of carbons and graphite, *J Am Ceram Soc*, **61**, 13–17.

44. Nathan M.I., Smith J.E. Jr., Tu K.N. (1974) Raman spectra of glassy carbon, *J Appl Phys*, **45**, 2370.

45. Winter M., Besenhard J.O., Spahr B.E., Novak P. (1998) Insertion electrode materials for rechargeable lithium batteries, *Adv Mater*, **10**, 725–763.

46. Tarascon J.M., Armand M. (2001) Issues and challenges facing rechargeable lithium batteries, *Nature*, **414**, 359–367.

47. Bueno P.R., Leite E.R. (2003) Nanostructured Li ion insertion electrode. 1. Discussion on fast transport and short path for ion diffusion, *J Phys Chem B*, **107**, 8868–8877.

48. Taberna L., Mitra S., Poizot P., Simon P., Tarascon J.M. (2006) High rate capabilities Fe_3O_4-based Cu nano-architectured electrodes for lithium-ion battery applications, *Nat Mater*, **5**, 567 –573.

49. Hosono E., Fujiwara S., Honma I., Ichihara M., Zhou H. (2006) Fabrication of nano/micro hierarchical Fe_2O_3/Ni micrometer-wire structure and

characteristics for high rate Li rechargeable battery, *J Electrochem Soc*, **153**, A1273–A1278.

50. Wook S.W., Dokko K., Kanamura K. (2007) Preparation and characterization of three dimensionally ordered macroporous $Li_4Ti_5O_{12}$ anode for lithium batteries , *Electrochim Acta*, **53**, 79–82.

51. Watanabe T., Ikeda Y., Ono T., et al. (2002) Characterization of vanadium oxide sol as a starting material for high rate intercalation cathodes, *Solid State Ionics*,**151**, 313–320.

52. Ogumi Z., Inaba M. (1998) Electrochemical lithium intercalation within carbonaceous materials: Intercalation processes, surface film formation, and lithium diffusion, *Bull Chem Soc Jpn*, **71**, 521–534.

53. Disma F., Aymard L. , Dupont L., Tarascon J.M. (1996) Effect of mechanical grinding on the lithium intercalation process in graphites and soft carbons, *J Electrochem Soc*, **143**, 3959–3972.

54. Iwashita N., Inagaki M. (1993) Relations between structural parameters obtained by X-ray powder diffraction of various carbon materials, *Carbon*, **31** , 1107–1113.

55. Ohzuku T., Iwakoshi Y., Sawai K. (1993) Formation of lithium–graphite intercalation compounds in nonaqueous electrolytes and their application as a negative electrode for a lithium ion (shuttlecock) cell, *J Electrochem Soc* , **140**, 2490–2498.

56. Wang H., Abe T., Maruyama S., Iriyama Y., Ogumi Z., Yoshikawa K. (2005) Graphitized carbon nanobeads with an onion texture as a lithium-ion battery negative electrode for high-rate use, *Adv Mater*, **17**, 2857–2860.

57. Doi T., Fukuda A., Iriyama Y., et al. (2005) Low-temperature synthesis of graphitized nanofibers for reversible lithium-ion insertion/extraction, *Electrochem Commun*, **7**, 10–13.

58. Shiraishi S., Ida Y., Oya A. (2003) Application of thin carbon fibers prepared by polymer-blend technique for lithium-ion battery negative electrode, Electrochemistry, **71**, 1157–1159.

59. Ohzawa Y., Mitani M., Li J., Nakajima T. (2004) Structures and electrochemical properties of pyrolytic carbon films infiltrated from gas phase into electro-conductive substrates derived from wood, *Mater Sci Eng* B, **113**, 91–98.

60. Larminie J., Dicks A. (2003) *Fuel Cell Systems Explained*, 2nd ed, John Wiley & Sons, New York.

61. Antolini E. (2009) Carbon supports for low-temperature fuel cell catalysts, *Appl Catal B: Environ*, **88**, 1–24.

62. Dicks A.L. (2006) The role of carbon in fuel cells, *J Power Sources*, **156**, 128–141.

63. Joo S.H., Choi S.J., Oh I., et al. (2001) Ordered nanoporous arrays of carbon supporting high dispersions of platinum nanoparticles, *Nature*, **412**, 169–172.

64. Kim T.W., Park I.S., Ryoo R. (2003) A synthetic route to ordered mesoporous carbon materials with graphitic pore walls, *Angew Chem Int Ed*, **42**, 4375–4379.

65. Keck L., Buchanan J., Hards G. (1991) US Patent, 5, 068, 161.

66. Kong K.J., Choi Y.M., Ryu B.H., Lee J.O., Chang H.J. (2006) Investigation of metal/carbon-related materials for fuel cell applications by electronic structure calculations, *Mater Sci Eng C*, **26**, 1207–1210.

67. Kim S.J., Park Y.J., Ra E.J., et al. (2007) Defect-induced loading of Pt nanoparticles on carbon nanotubes, *Appl Phys Lett*, **90**, 023114-1-3.

68. Sanz-Navarro C.F., Åstrand P.O., Chen D., et al. (2008) Molecular dynamics simulations of the interactions between platinum clusters and carbon platelets, *J Phys Chem A*, **112**, 1392–1402.

69. Kim G.B., Jhi S.H. (2011) Carbon monoxide-tolerant platinum nanoparticle catalysts on defect-engineered graphene, *ACS Nano*, **5**, 805–810.

70. Radmilović V., Gasteiger H.A., Ross P.N. (1995) Structure and chemical composition of a supported Pt–Ru electrocatalyst for methanol oxidation, *J Catal*, **154**, 98–106.

71. Yoshimura H., Yamada S., Yoshimura A., Hirosawa I., Kojima K., Tachibana M. (2009) Grazing incidence X-ray diffraction study on carbon nanowalls, *Chem Phys Lett*, **482**, 125–128.

72. Pozio A., Francesco M.D., Cemmi A., Cardellini F., Giorgi L. (2002) Comparison of high surface Pt/C catalysts by cyclic voltammetry, *J Power Sources*, **105**, 13–19.

73. Maillard F., Martin M., Gloaguen F., Leger J.M. (2002) Oxygen electroreduction on carbon-supported platinum catalysts. Particle-size effect on the tolerance to methanol competition, *Electrochim Acta*, **47**, 3431–3440.

74. Stonehart P. (1992) Development of alloy electrocatalysts for phosphoric acid fuel cells (PAFC), *J Appl Electrochem*, **22**, 995–1001.

Chapter 7

Raman and Infrared Spectroscopic Characterization of Graphene

Da Zhan and Zexiang Shen
Division of Physics and Applied Physics, School of Physical and Mathematical Sciences,
Nanyang Technological University, 21 Nanyang Link, Singapore 63371, Singapore
zexiang@ntu.edu.sg

7.1 Introduction

Graphene attracts tremendous interest in the scientific community owing to its unique electronic, optical, and mechanical properties. In the previous chapters, the transport properties characterized by electrical measurement and the chemical properties for application in energy storage have been introduced. In this chapter, we introduce the optical studies of graphene using Raman, infrared (IR), and visible spectroscopy. Graphene can present good optical contrast under specially constructed substrates despite the fact that it is only one-atom thick. Raman and visible/IR spectroscopic studies, which are based on using the optical microscope to locate the exact sample position, can be easily carried out in air ambient with high efficiency and high throughput. Raman spectroscopy is one of the

---

Two-Dimensional Carbon: Fundamental Properties, Synthesis, Characterization, and Applications
Edited by Yihong Wu, Zexiang Shen, and Ting Yu
Copyright © 2014 Pan Stanford Publishing Pte. Ltd.
ISBN 978-981-4411-94-3 (Hardcover), 978-981-4411-95-0 (eBook)
www.panstanford.com

most commonly used methods for probing various properties of graphene, such as thickness and stacking geometry, doping, defects, edge chirality, and strain. These properties will be introduced in Sections 7.2.1–7.2.5. Despite the fact that IR spectroscopy is not used as common as Raman spectroscopy, it is significant for characterizing the chemical species as well as probing the low-energy regime of electronic band structure in graphene. This will be introduced briefly in Section 7.3.

7.2 Raman Spectroscopic Features of Graphene

Even for pristine graphene, the electronic properties show strong dependence on thickness, stacking geometry, and edge chirality. Furthermore, graphene's electronic properties can be intentionally tuned by introducing defects, doping, and applying strain. Raman spectroscopy provides a fast and easy-to-operate technique that can characterize the above. Normally, Raman spectroscopy analyzes the features of lattice vibrational modes and hence it is very sensitive to the atomic arrangement and phonon structures, and less sensitive to electronic structure. However, for graphene and carbon based materials in general, Raman scattering is very sensitive to the electronic energy levels due to the strong resonance effect. Our discussion below focuses on the electronic properties of graphene instead of the atomic structures.

Graphene and its derivative structures normally contain two prominent Raman active peaks, called the G and the 2D bands, located at ~1580 and 2700 cm^{-1}, respectively (red curve in Fig. 7.1). The G band arises from the E_{2g} vibrational symmetry corresponding to the zero-momentum phonon at Γ point (in-plane transverse optical phonon branch) [1]. The 2D band is the second order of D band which originates from the breathing mode of A_{1g} symmetry at around K point. Remember that the first order D band is silent in Raman measurement for the perfect hexagonal sp^2-based carbon crystal structure, and it can only be observed in the presence of disorders/defects as shown in the blue curve in Fig. 7.1 (indicated by the green circle). The G, 2D, and D bands have rich physics and can practically guide us to analyze the detailed electronic properties of graphene. In addition, some other weak Raman active bands can also be used to assist in analyzing graphene's properties. In the following sections, we will introduce Raman spectroscopic studies of graphene's thickness, stacking geometry, doping, defects, edge chirality, and strain.

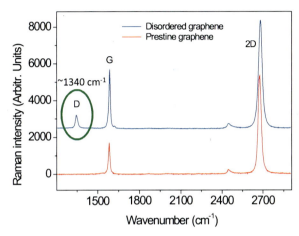

Figure 7.1 Raman spectra for pristine (red) and defected graphene (blue). Reproduced with permission from [49]. Copyright 2010 American Chemical Society.

7.2.1 Identify the Thickness and Stacking Geometry of Graphene

It has been mentioned previously that the electronic structure of graphene is thickness dependent. As single layer graphene is only 3.4 Å thickness, the identification of graphene's thickness is by no means an easy task [2,3]. An easy, fast, and definitive technique is highly desirable.

As Raman spectroscopy of graphene is closely related to its electronic structure, Raman scattering provides a very simple way to identify the number of layers of Bernal stacking graphene (graphene layers arranged in ABAB sequence; it is the most commonly observed stacking geometry in nature) by analyzing its 2D band features. As the 2D Raman band arises from double resonant process, it carries rich information about the electronic band structure. Figure 7.2a shows the typical Raman spectra at 2D band region for graphene with different thickness [4]. It can be clearly seen that the single layer graphene (SLG) presents only one sharp single Lorentzian peak, and this peak broadens with the increase in the number of layers. This is because the SLG has only one conduction and one valence band at low-energy regime [4–8]. Therefore, only one possible pathway can fulfill the double resonant Raman scattering process, which means that the momentum of the selected D phonon is fixed under a monochromatic laser as the excitation source, and hence the corresponding energy (frequency) of D phonon is also fixed; consequently, the 2D band presents only one single Lorentzian peak for SLG [4, 9, 10]. In contrast, the bilayer graphene (BLG) has two parabolic conduction as well as two parabolic valence bands at low-energy regime [6–8]. Considering the Raman selection rule, there are four possible pathways that can fulfill the double resonant Raman scattering process to activate the 2D band even under the excitation of a monochromatic laser (Fig. 7.2b) [4, 9, 10]. These D phonons which scattered in these four pathways possess different momenta and energies, thus the 2D band of the BLG does not present single Lorentzian peak anymore, instead, it presents a broadened asymmetric peak which can be fitted by four subpeaks (Fig. 7.2c) [4, 9, 10]. Similarly, with further increase in the thickness of Bernal stacking graphene, the electronic band structure splits into more subbands at low-energy regime, and the possible pathways that can fulfill the double resonant scattering processes increase accordingly, leading a further increase in the full width of half maximum (FWHM) of the 2D band (Fig. 7.2d). It is worth noting that the FWHM of the 2D band of graphene is variable for few-layer graphene (FLG) with same thickness. In addition to Bernal stacking, rhombohedral stacking and turbostratic FLG also exist. Rhombohedral FLG along the z-direction are arranged in the form of ABC..., and the turbostratic graphene are stacked in twisted angle (not integer times of 60°) between the adjacent layers. The graphene's layer–layer electronic coupling is also very strong for Rhombohedral stacking FLG, and the splitting of electronic band structure at the low-energy regime also results in the broadening of the 2D band FWHM, and the influence is even stronger than that of Bernal stacking. For example, the FWHM of ABA Bernal-stacked trilayer graphene is in the range of 55–59 cm^{-1} [11], while in ABC stacking, the FWHM is in the range of 64–70 cm^{-1} [12, 13]. For turbostratic FLG, the interlayer electronic coupling is relatively weak due to the twisted angle, and its electronic structure is similar to that of the SLG at low-energy regime, which normally possesses only one single linear dispersive band in conduction and valence bands, respectively.

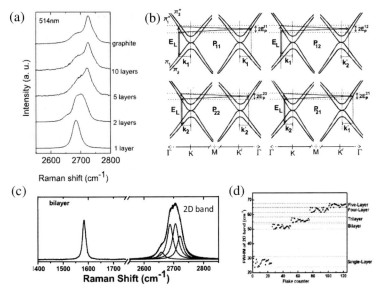

Figure 7.2 (a) Evolution of Raman spectra of graphene with the number of layers (using 514 nm excitation laser). (b) Schematic electronic band dispersion of bilayer graphene and the four different pathways for fulfilling the double resonant process. (c) The Raman spectrum of a bilayer graphene, where the 2D band is a broad peak fitted by four subbands. (d) The statistical data of FWHM of 2D band with respect to the number of layers for AB-stacked bilayer graphene. Reproduced with permission from [4] (a), [10] (b and c), and [11] (d). Copyright 2006 American Physical Society (a), 2007 American Physical Society (b and c), and 2010 WILEY-VCH Verlag GmbH & Co. KGaA, Weinheim (d).

The broadening effect of the 2D peak for turbostratic FLG is normally very weak and its FWHM is comparable with that of SLG [14,15]. Therefore, the stacking geometry can be studied by comparing the FWHM of Raman 2D band. It is worth mentioning that the electronic structure of turbostratic FLG is very much dependent on the twisting angle between the adjacent layers. Recently, Kim et al. measured the Raman spectra of the BLG with misorientated angles in range from 0° to 30° in step of 1°, where the twisting angle is confirmed by high resolution transmission electron microscope (HRTEM) [16]. It is found that at the peak position,

FWHM and normalized intensity of 2D band show, to some extent, dependence on the twist angle, particularly in the range between 1° and 8°. However, the difference is not obvious for angle in the range between 13° and 30° [16]. It is worth noting that for the twisted BLG with angle less than 8°, the FWHM of 2D band is comparable to that of AB-stacked BLG indicating the strong electronic coupling between the graphene layers.

Other Raman active bands of weaker intensity can also be used to identify the stacking geometry of graphene. For example, the band with Raman shift of $\sim$1750 cm^{-1}, called the M band, is only activated for AB-stacked BLG, and it is absent for twisted BLG of which the two layers are incommensurate [17, 18]. At the high frequency region, the 2D+G band, which is always Raman active with Raman shift of $\sim$4250 cm^{-1}, presents an obvious FWHM dependence on thickness for Bernal stacking FLG [19]. At extremely low frequency region, the C band located at $\sim$30–40 cm^{-1} has been recently reported for FLG [20]. The C band arises from the $E_{2g}^{(1)}$ vibration mode of graphene lattice. For SLG, the C band is inactive, but it is Raman active for AB-stacked FLG and graphite. Furthermore, the C band shows a peak position dependence for FLG, the peak position gradually shifts to higher frequency when the layer number increases, and the peak position difference is obviously observed between the FLG with layer numbers 2, 3, 4, and 5, indicating that the layer number of FLG can also be identified by the peak position of the C band [20].

Here, we introduce another very important method that can be used to identify the thickness accurately regardless of its stacking geometry, which is the optical contrast spectroscopy [21]. When graphene samples are deposited onto the Si substrate covered by a 300 nm thick SiO$_2$ layer, the graphene shows optical contrast compared with the pure substrate at visible wavelengths. The contrast spectrum shows an intense peak at the wavelength of $\sim$550 nm as shown in Fig. 7.3a. Interestingly, both the peak position and the value are not sensitive to the stacking geometry, but are only dependent on the thickness. Therefore, the number of graphene layers can be accurately confirmed by comparing the peak value of the optical contrast spectrum as concluded in Fig. 7.3b. It is worth noting that for some special angles of twisted BLG, the peak position and value in the contrast spectrum do show some difference compared to the AB-stacked BLG [22].

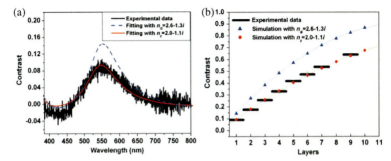

Figure 7.3 (a) The contrast spectrum of single layer graphene on SiO$_2$/Si substrate, the thickness of SiO$_2$ layer is 300 nm. The experimental data, the simulation result using $n = 2.0 - 1.1i$ (n is the refractive index of graphene), and the simulation result using $n = 2.6 - 1.3i$ are plotted by black line, red line, and blue dashed line, respectively. (b) The contrast peak at ~550 nm as a function of graphene layer number simulated by using both n_G (blue triangles) and n_z (red circles), and experiment data (thick black lines), respectively. Reproduced with permission from [21]. Copyright 2007 American Chemical Society.

7.2.2 Probing Doped Charges in Graphene

Naturally, the pristine graphene is a gapless material with zero charge carriers. Normally, the conductivity of graphene can be tuned by doping charge carriers (either electrons or holes) both electrically and chemically. No matter doped by electrons or holes, the conductivity of graphene can be enhanced dramatically. Graphene samples can be naturally doped even under air-ambient environment, and the doping effect is particularly obvious for SLG. The most common sources under natural environment are the SiO$_2$ substrate and O$_2$ [23, 24], which give rise to hole (p-type) doping [23-26]. Traditionally, the doped charge carrier density can be quantitatively characterized by the Hall measurement. However, Hall measurement needs complicated process to fabricate device, which is not easy for many small samples of micrometer size. Moreover, the device fabrication process itself may induce defects in the graphene sample. Raman spectroscopy provides a fast, easy to use, and non-destructive method to probe the doping level of graphene [27]. Doping shifts the Fermi energy and consequently results in two major effects: (a) equilibrium lattice parameter changes with a consequent stiffening/softening effect for the phonons; (b) the onset of effects beyond the adiabatic Born-Oppenheimer approximation that modify the phonon dispersion closes to the Kohn anomalies (KAs) [28, 29]. These two effects make Raman spectroscopy an ideal tool for studying the properties of doped graphene.

High dielectric-constant layer based electrical gating and molecular doping can be used to dope graphene intentionally. The doping level can be tuned in a wide range by using the gating method, while the molecular doping method is a static electric interaction between graphene and the molecule, and the doping level is fixed. Therefore, it is necessary to use the electrical gating method to approach the quantitative investigation of graphene's Raman features as a function of doping level.

Figure 7.4 shows the evolution of Raman spectra of graphene as a function of gating voltage [27]. Note that the +0.6 eV voltage approaches the graphene's neutral point which indicates that the graphene sample is naturally hole doped. The bias voltage below and above 0.6 eV correspond to the hole and electron doping range, respectively. The degree of doping level of graphene can be indirectly deduced by converting the electrical gate voltage. It can be expressed by Eq. 7.1 [27].

$$V_G = \frac{E_F}{e} + \varphi \tag{7.1}$$

where the V_G, E_F, and e represent the gating voltage, Fermi level of graphene, and the elementary electric charge, respectively. ϕ is the electrostatic potential difference between graphene, which is created by the applied gating voltage and determined by the geometrical capacitance C_G. ϕ and C_G can be expressed by Eqs. 7.2 and 7.3 [27].

$$\varphi = \frac{ne}{C_G} \tag{7.2}$$

$$C_G = \frac{\varepsilon\varepsilon_0}{d} \tag{7.3}$$

Here ε and d are the dielectric constant and the thickness of the dielectric layer, respectively, and ε_0 is the permittivity in free space.

Therefore, the evolution of the peak position of the G band as a function of doping level can be obtained accordingly, as shown in Fig. 7.4b. It can be seen that both electron and hole doping induce

Raman Spectroscopic Features of Graphene | 161

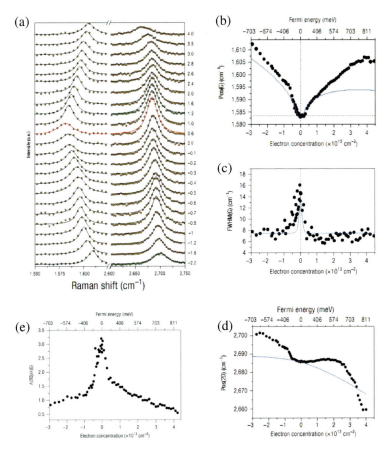

Figure 7.4 (a) Electrical gating voltage dependent Raman spectra of graphene, the top gate voltage varies from −2.2 V to 4.0 V. The dots are the experimental data, the black lines are fitted Lorentzians, and the red line corresponds to the neutral point (Dirac point). (b) Position of the G band as a function of electron concentration (bottom x-axis) and Fermi level (top x-axis). (c) FWHM of G band as a function of electron concentration (bottom x-axis). (d) The position of 2D band as a function of electron concentration. The solid blue lines are the theoretical predictions. The solid lines shown in (a), (b), and (c) are the adiabatic density functional theory calculation data. (e) The intensity ratio of 2D and G bands (I_{2D}/I_G) as a function of electron concentration (bottom x-axis) and Fermi level (top x-axis), respectively. Reproduced with permission from [27]. Copyright 2008 Nature publishing group.

stiffening of the G phonon, also shown in Fig. 7.4a. As the G band is from the $E_{2g}^{(2)}$ vibrational mode at Γ point in the phonon band structure, the doping induced G phonon stiffening effect is from the nonadiabatic removal of the Kohn anomaly near the Γ point [29]. Furthermore, both electron and hole doping reduce the FWHM of the G band (Fig. 7.4c), and it reaches a constant when the G band position is shifted to $\sim$1590 cm^{-1} [27, 30]. Because doping makes the Fermi level shift away from the neutral point (Dirac point), the shifted Fermi level blocks the phonon decay in electronic band to form an electron–hole pair, and thus gives rise to the narrowing for the G band. In other words, we can say that the FWHM of the G band is broadened for pristine graphene because the channel is opened for G phonon to decay into the electron–hole pair.

From Figs. 7.4b and c, both the G band position and the FWHM as function of doping level show symmetric curve with respect to the neutral point. Therefore, using the features of the G band alone, we cannot determine whether the doping is of electron or hole if graphene is doped with unknown molecules. However, the doping type can be identified if the position of the 2D band is considered. As shown in Fig. 7.4d, the position of the 2D band shows different dependence for hole and electron doping; it blueshifts with the hole doping but keeps constant for electron doping and then redshifts when the electron doping is higher than $\sim$2.5$\times$10^{13} cm^{-2} [27].

Furthermore, the intensity variation of the 2D band can also be used to determine the doping level. The intensity of 2D band, $I(2D)$, can be expressed by Eq. 7.4 [31, 32].

$$A(2D) = \frac{8}{3} \left(\frac{e^2}{c} \right)^2 \frac{v_F^2}{c^2} \left(\frac{\gamma_K}{\gamma_{ee} + \gamma_K + \gamma_\Gamma} \right)^2 \tag{7.4}$$

where e, c, and v_F are the electron charge, the speed of light and the Fermi velocity, respectively. They are all the constants. γ_{ee} is the electron–electron collision rate, and γ_K and γ_Γ are emission rate of phonons at K and Γ points, respectively. It has been proved that both γ_K and γ_Γ are not sensitive to doping [32], thus the electron–electron collision rate determines the intensity of the 2D band. When doping increases (regardless of hole or electron doping), the electron–electron collision rate increases, and consequently, the intensity of the 2D band decreases [31–33]. Furthermore, the intensity of the other strong Raman active band, the G band, is not sensitive to doping [30, 32–34]. Therefore, the intensity ratio of

2D and G bands (I_{2D}/I_G) as a function of doping level is normally considered as the variation of normalized 2D band intensity. It should be pointed out that the G band intensity of graphene can be enhanced when the doping induced downshifted Fermi energy approaches to half of the excitation laser energy [35, 36], but this effect can be ignored in most cases as the most commonly used laser energy for characterizing graphene is in the range of 1.96–2.41 eV, which is far beyond twice the maximum shifted Fermi energy of graphene by using the traditional doping method [27, 30, 33]. Figure 7.4e shows the experimentally measured I_{2D}/I_G as a function of shifted Fermi energy (top x-axis) and charge concentration (bottom x-axis). The intensity of 2D band shows a decreasing trend with increase in the doping level (for both electron and hole doping). It can be found that the normalized 2D band intensity decreases with both electron and hole doping, which is in agreement with the theoretical results [31–33].

Recently, it was reported that strong chemical doped graphene can be realized by intercalation of Br_2 and $FeCl_3$ [37–39]. Both Br_2 and $FeCl_3$ are electron-negative molecules which induce strong hole doping effect to the adjacent graphene layers. The Fermi energy shift can be as high as > 0.81 eV, because the 2D band under excitation energy of 1.96 eV was fully quenched. 2D band under excitation of 1.96 eV laser [8, 39]. Furthermore, the configuration of the intercalated FLG can be clearly identified by analyzing the G band features. For example, for $FeCl_3$ intercalated trilayer graphene, the fully intercalated compound can be arranged in two forms: "F/Gr/F/Gr/F/Gr/F" (Fig. 7.5a) and "Gr/F/Gr/F/Gr" (Fig. 7.5b), where F and Gr represent the $FeCl_3$ and graphene layers, respectively. The Raman spectrum for the former compound shows a single Raman G peak which is shifted to ~1625 cm^{-1} [39], while for the latter compound it consists of two splitted G peaks located at ~1613 cm^{-1} and ~1623 cm^{-1}, respectively [38]. For the former configuration, all the three graphene layers are doped to the same level as each graphene layer is flanked by two $FeCl_3$ layers, hence it is reasonable to expect that the G band signals from all three graphene layers are identical and all are blueshifted to the same position (1625 cm^{-1}). While for the latter case, the top and bottom graphene layers are doped less than the middle graphene layer as they are only flanked on one side by $FeCl_3$. As a result, the peak position of G band for the top and bottom layers is shifted less than the middle graphene layer; consequently, a splitted G band is observed.

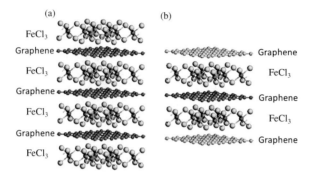

Figure 7.5 The crystal structures of two forms of FeCl$_3$-based full intercalated trilayer graphene compound. (a) Each of the graphene layer is flanked on both sides by FeCl$_3$ layers. (b) Each of the top and bottom graphene layer is flanked on one side by FeCl$_3$ layer, and the middle graphene layer is flanked on both sides by FeCl$_3$ layer. Reproduced with permission from [38]. Copyright 2010 WILEY-VCH Verlag GmbH & Co. KGaA, Weinheim.

7.2.3 Defects in Graphene Probed by Raman Spectroscopy

For a single crystalline electronic material, the existence of defects would adversely influence its electronic performance. For graphene, it is well known that the electronic structure and transport properties would be affected strongly even when the defect concentration is small [40–43]. Despite the fact that the sp^2 hybridized carbon is one of the most stable materials, the crystalline quality of pristine graphene is easily affected, to some extent, as it is only one-atom thick. Normally, Raman spectroscopy is not sensitive to defects for most materials unless the damage is heavy. Nevertheless, the case for graphene is different. For pristine graphene, the first order of D band is normally Raman inactive although the second order 2D band always shows a strong Raman peak [9, 44]. However, the D band can be observed when defects are present. The D band can be used to estimate the defects and domain size of graphene.

The activation of D band needs to satisfy the double resonant Raman scattering process. As shown in Fig. 7.6a, at around K point, an electron in valence band is excited by the excitation photon energy E_L (from A to B). Then this excited electron is scattered by a D phonon (possessing momentum k_D) from energy valley K to another energy valley K' (from B to C). When defects are present, the electron is elastically scattered back to energy valley K by an exchange momentum $-k_D$ (from C to D) and finally it recombines with the hole in the valence band (from D to A). This defect-assisted double resonant Raman process activates the D band. Defects in graphene can be divided into two types, intrinsic and extrinsic defects [43]. The electron and ion irradiation normally creates intrinsic defects in graphene [45, 46] such as point defects and one-dimensional defects. Extrinsic defects are generated by chemical reactions, e.g. hydrogenation [47–49], fluorination [50, 51], and oxidization [52, 53]. These reactions generate covalent bonds between carbon atoms and hydrogen or oxygenated functional groups, leading to the co-existence of sp^2- and sp^3-carbon.

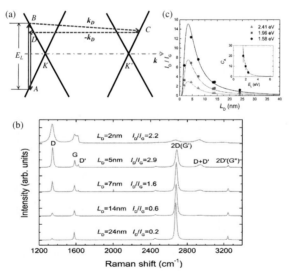

Figure 7.6 (a) The schematic view of double resonant Raman process for activating a D phonon in graphene. (b) Raman spectra of five graphene samples bombarded by Ar$^+$ in different doses. The laser energy is 2.41 eV. (c) I_D/I_G as a function of L_d measured by three different excitation lasers. The solid lines are the fitting data as described elsewhere [56]. Reproduced with permission from [56] (b and c). Copyright 2011 American Chemical Society (b and c).

Raman and Infrared Spectroscopic Characterization of Graphene

For characterizing the intrinsic defects in graphene, it is worth mentioning that the intensity ratio of D and G bands (I_D/I_G) was first used to estimate the crystallite area of 3D graphite [54]. The relationship between the crystallite size (L_a) and I_D/I_G can be expressed by Eq. 7.5 [54].

$$L_a(nm) = (2.4 \times 10^{-10})\lambda_l^4\left(\frac{I_D}{I_G}\right)^{-1} \tag{7.5}$$

where λ_l is the wavelength of the excitation laser. It is worth noting that the I_D/I_G is more sensitive to excitation laser energy than crystallite size. Under the excitation of 1.92 eV laser, it is found that D band intensity is very weak when the crystallite size of graphite is larger than 490 nm. On the other hand, in 2D graphene, the density of defects can also be reflected by another parameter, which is the average distance between inter-defects (L_d) [55, 56]. Cancado et al. used Ar+ ions (kinetic energy of 90 eV) to create defects in graphene, and the density of defects was controlled by the irradiated ion dose. Using excitation laser energy of 2.41 eV, it can be seen in Fig. 7.6b that when L_a decreases from 24 nm to 5 nm, the intensity of D, D', and D+G bands increases gradually (D' and D+G bands are other two defect-assisted Raman bands with Raman shift of ~1620 cm^{-1} and ~2930 cm^{-1}, respectively), while the FWHMs of both D and 2D bands are almost unchanged. When L_a further decreases to 2 nm, both the D and the 2D bands broaden dramatically, and the 2D and D+G bands start to merge into each other. These results indicate that the graphene samples are all "stage 1" disordered graphene (i.e. nanocrystalline graphene) when the crystallite size varies in the range from L_d = 24 nm to 5 nm [56, 57]; and the graphene samples start to change from "stage 1" disordered graphene to "stage 2" disordered graphene when L_d decreases from 5 nm to 2 nm, indicating that graphene with L_d = 2 nm contains low sp^3-carbon. It should be pointed out that both "stage 1" and "stage 2" have also been used in graphite intercalation compounds with different meanings [58]. Furthermore, the laser energy dependent I_D/I_G as a function of L_d have also been investigated by Cancado et al., which have been plotted in Fig. 7.6c [56]. It can be seen that the I_D/I_G as a function of L_d shows similar trend regardless of the excitation laser energy, and it always shows an increasing trend with the decreasing excitation laser energy for graphene samples with the same L_d.

For extrinsic defects in graphene, both hydrogenation and fluorination of graphene can achieve stoichiometric graphene-based derivatives. For hydrogenated and fluorinated graphene, the planar sp^2-networks are transformed to structures that contain distorted sp^3-carbon despite the fact that graphite is one of the most chemically inert materials. Luo et al. reported the Raman-based investigation of the thickness-dependent reversible hydrogenation of graphene layers. Figures 7.7a, b, and c show the Raman spectra of pristine, hydrogenated, and dehydrogenated graphene samples, respectively. After the hydrogenation process, it can be seen that the defect-related Raman bands, D, D', and D+G bands, are all presenting observable intensities (Fig. 7.7b). This indicates that the hydrogenation process gives rise to the formation of sp^3-carbon as well as breaking of the translation symmetry of C=C sp^2 bonds [59, 60]. Interestingly, as shown in Fig. 7.7c, all the defect-related Raman bands are almost eliminated after annealing the graphene samples at 500 °C in vacuum, indicating that the hydrogenation process is largely reversible [59]. For fluorinated graphene, the evolution of Raman spectra with respect to the fluorination time shows difference from that of the hydrogenated graphene. The fluorination of graphene can be divided into three stages. At the first stage (reaction time less than a few hours), the intensity of defect-related Raman bands increases with the fluorination time (Fig. 7.7d) [50]. At the second stage (reaction time between 10 and 20 h), the D and G bands broaden dramatically, and the Raman spectrum looks like that of graphene oxide (graphite oxide) [61]. At the third stage (reaction time more than 30 h), all the Raman active bands disappear gradually, and finally they are fully quenched [50]. The three stages of fluorinated graphene show different Raman features describing the degree of fluorination. Particularly, the fully fluorinated graphene is considered to be obtained as reflected by the disappearance of all Raman bands. In addition to hydrogenated and fluorinated graphene, graphene oxide (GO) is another graphene-derivative which contains large amount of extrinsic defects as the functional groups are chemically attached to it. GO has been extensively studied in the past years and its chemical properties can be used for practical applications [62, 63]. However, different from hydrogenated and fluorinated graphene, the feature of Raman active bands of GO is not sensitive to the oxidation degree, meaning the Raman spectrum always shows similar shapes (similar to that of graphene fluorinated at the second stage). Even for the reduced GO for which the electrical conductivity is 4–6 orders of magnitude higher than that of GO, the D and G bands still show features very similar to that of GO [61, 64]. Therefore, Raman spectroscopy is an ideal tool to probe the information of intrinsic defects in hydrogenated and fluorinated graphene. It can also be used to study graphene oxide although the oxidization degree of GO cannot be inferred.

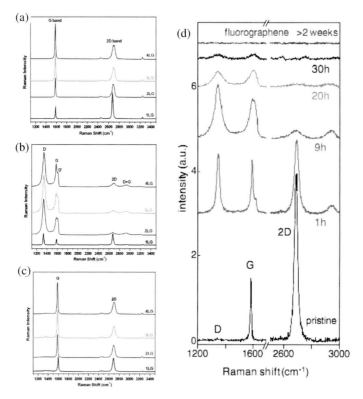

Figure 7.7 (a) Raman spectra of pristine graphene with layer number 1, 2, 3, and 4. (b) Raman spectra of hydrogenated graphene with layer number 1, 2, 3, and 4. (c) Raman spectra of dehydrogenated graphene with layer number 1, 2, 3, and 4. (d) Evolution of Raman spectra for a graphene sample exposed to atomic F with different exposure duration. Reproduced with permission from [59] (a–c), and [50] (d). Copyright 2009 American Chemical Society (a–c) and 2010 WILEY-VCH Verlag GmbH & Co. KGaA, Weinheim (d).

7.2.4 Graphene Edge Chirality Identified by Raman Spectroscopy

Graphene is a 2D material that consists of hexagonal lattice. Graphene can be cut into thin slices to form graphene nanoribbon, the electronic property of which is strongly dependent on its terminating edge structure. The topic on graphene edges has been studied extensively since graphene was isolated and identified experimentally [45, 65–67]. Normally, graphene can be terminated by two types of edges, zigzag or armchair, as shown schematically in Fig. 7.8a. Graphene edges have different chiralities if the angle between the two adjacent edges of graphene is $(2n-1)\times 30°$ (n is an integer), and the two edges have same chirality if the angle between the two adjacent edges is $2n\times 30°$ (n is an integer).

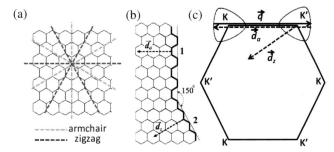

Figure 7.8 (a) Two possible chiralities of graphene. The green dashed lines indicate armchair and the blue dashed lines indicate the zigzag. (b) Schematic illustration of graphene with zigzag and armchair edges with edge angle of 150°. (c) First Brillouin zone of graphene, showing the double resonant mechanism for an armchair edge while the zigzag edge associated wavevector does not fulfill the condition for double resonant process. Part (b) reproduced with permission from [68]. Copyright 2004 American Physical Society.

Atoms on the edges do not have the perfect sp^2-hybridization as half of the atoms are missing, and hence the graphene edges can be considered as one kind of special defects. As aforementioned, the activation of D band of graphene needs the existence of defects. The defects can elastically scatter the electron from one energy valley K' back to the other energy valley K by exchanging momentum $-k_D$. It should be noted that the exchanged momentum has to be along some specific direction in order to fulfill the double resonant process. Figures 7.8b and c show the typical graphene atomic structure and the corresponding first Brillouin zone oriented with respect to the real space lattice, respectively [1, 68]. The bold lines show the edge structures, where edge 1 is armchair and edge 2 is zigzag. The wavevectors of the defects associated with these edges are denoted by d_a (armchair edge) and d_z (zigzag edge). It is clear that only the armchair d_a vector is able to connect two points belonging to the two inequivalent Dirac cones K and K'. The D band should be active for armchair edge as the edge defect associated wavevector can fulfill the double resonant process. On the other hand, the zigzag d_z vector cannot connect the adjacent Dirac cones K and K' and hence the D band is inactive for zigzag edge as the associated defect wavevector cannot fulfill the double resonant Raman scattering process. Furthermore, it is worth noting that the D band intensity for the armchair edge is dependent on the incident laser polarization [68, 69]. The matrix element of creation/annihilation of an electron–hole pair with momenta k and $-k$ (with respect to the Dirac point) by an incident photon with polarization e_{in}, is proportional to $[e_{in} \times k]$. Thus it is maximum intensity when $e_{in} \perp k$. Since a perfect edge allows momentum conservation along its direction, backscattering is possible only at normal incidence. Hence the intensity at the armchair edge is proportional to $\cos^2 \theta_{in} \cos^2 \theta_{out}$, where θ_{in} and θ_{out} represent the angles between the polarizations of incident and scattered photons with the edge, respectively. For Raman measurements where no polarizer is placed in the collection of scattered photons (Raman signal), the Raman signal is independent of θ_{out}.

Experimentally, these phenomena have been observed for both graphite and graphene samples [68, 70]. Figures 7.9a and b show the Raman images constructed by the G and D bands of a graphene sample, respectively [70]. The incident laser is polarized along the horizontal orientation as indicated by a green arrow in each image. Two straight edges in micron-scale are clearly seen in Fig. 7.9a and the angle between these two adjacent edges is 30°. As aforementioned, these two edges have opposite chiralities, one is zigzag and the other is armchair. In Fig. 7.9b, the D band shows very strong intensity along one edge while it is very weak along the other edge. Therefore, it is inferred that the edge with strong D band is the armchair edge, and the other one is the zigzag edge [70]. For the same sample, Fig. 7.9c shows the Raman image of the D band with the incident laser polarization rotated by 90° as indicated by the green arrow. In this image, the edge showing a strong D band in Fig. 7.9b becomes almost non-observable. This is because the angle between the armchair edge and incident laser polarization increased from 15° (Fig. 7.9b) to 75° (Fig. 7.9c). The intensity of D band measured under these two conditions has a difference of $\cos^2 15/\cos^2 75 \approx 14$. Thus, it is reasonable that the D band intensity is much weaker in Fig. 7.9c. It is worth noting that the D band at the "zigzag edge" is not fully quenched (Fig. 7.9b). This is because even the zigzag edge contains a small fraction armchair (and point defects) segments.

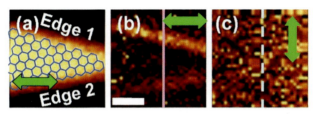

Figure 7.9 (a) Raman image of a graphene sample constructed by the intensity of G band, the schematic blue hexagonal lattice shows the chiralities of the two edges. The chiralities of the two edges are determined by the Raman image constructed by the intensity of D band with horizontal polarization as indicated by the green arrow (b). (c) The Raman image for the same graphene sample constructed by the D band intensity with vertical incident laser polarization as indicated by the green arrow. All images share the same scale bar as indicated in (b) which is 2 μm. Reproduced with permission from [70]. Copyright 2008 American Institute of Physics.

Furthermore, polarized Raman spectroscopy can be used to probe the evolution of graphene edge (armchair, zigzag, and short-range impurities) with thermal annealing [71]. Four pristine graphene samples, each with two adjacent edges forming an angle of 30° and 150°, can be seen in Figs. 7.10Aa, Ba, Ca, and Da. All of them show stark contrast for their two edges' D band intensity as revealed by Raman images as shown in Figs. 7.10Ac, Bc, Cc, and Dc (the polarization of incident laser is represented by the arrow in each image): one edge of each sample shows obvious D peak intensity indicating that it is A-edge (predominant armchair edge), while the other edge of each sample shows very weak D peak indicating that it is Z-edge (predominant zigzag edge). After the samples were annealed in vacuum, it can be seen from the Raman images (Figs. 7.10Ad, Bd, Cd and Dd) that the D band of the Z-edge of each sample becomes Raman active. As the zigzag segments do not present any D band, it indicates that there is a significant change for the Z-edge after thermal annealing. The edge atoms have undergone rearrangement to form armchair segments (along ±30° with respect to the average edge direction (denoted as A-30°)) and the point defects to satisfy double resonance for activating the Raman D band. Normally, the D band intensity induced by point defects does not show laser polarization dependence (as the scattering direction is randomly distributed), but the A-30° induced D band intensity is dependent on the polarization of the incident laser [68]. Therefore, the overall effect of the D band intensity of the annealed Z-edge still shows, to some extent, dependence on the polarization of incident laser. A simple theoretical model can be constructed based on the polarization dependent D band intensity of the annealed Z-edge [71]. The portion of A-30 segments and point defects of each Z-edge annealed at different temperatures can be estimated accordingly, and consequently, the edge dynamic behavior of the single layer graphene can be probed by polarized Raman spectroscopy [71]. It should be noted that although the zigzag edge of single layer graphene is unstable and changes to armchair edge upon annealing, the Z-edges of AB-stacked bilayer graphene can form a closed structure at low-temperature annealing condition [72].

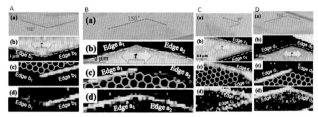

Figure 7.10 (Aa, Ba, Ca, Da) Optical images of four different graphene samples and each of them has two edges with formed angle of either 30° or 150° as indicated by the dashed lines. (Ab, Bb, Cb, Db) Raman images of the G peak intensity of the four pristine graphene samples. The arrows indicate the incident laser polarization. (Ac, Bc, Cc, Dc) Raman images of D peak intensity of the four pristine graphene samples. (Ad, Bd, Cd, Dd) Raman images of D peak intensity of the four pristine graphene samples after annealing treatment at 200 °C, 300 °C, 400 °C, and 500 °C, respectively. Reproduced with permission from [71] (A–D). Copyright: 2011 American Chemical Society (A–D).

7.2.5 Raman Spectroscopy Probes Strain in Graphene

Strain is another important physical parameter that can significantly modify graphene's electronic band structure and thus change its electronic properties. Uniaxial and biaxial are two mostly studied types of strains in graphene. Both of them can be detected by Raman spectroscopy effectively and rapidly [73–76].

Uniaxial strain can cause the asymmetrical distortion of graphene lattice and shift the Dirac cone in the inverse space. Systematic Raman studies have been carried out to probe the uniaxial strain. As shown in Figs. 7.11a and b [73], both peak positions of G and 2D bands show redshift under uniaxial tensile strain regardless of the layer number of graphene [73]. There are two obvious effects for the uniaxial strained graphene samples: (1) The redshift effect for both G and 2D bands of single layer graphene is more sensitive to strain compared to the trilayer graphene; (2) the peak position variation of 2D band is more sensitive to strain compared to the G band [73]. The peak position dependence of both G and 2D bands on uniaxial tensile strain can also be derived from Fig. 7.11, the values are about 14.2 cm^{-1}/% and 27.1 cm^{-1}/%, respectively [73].

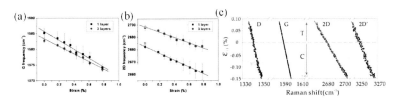

Figure 7.11 (a) The peak positions of G band of single-layer (black squares) and three-layer (red circles) graphene as a function of uniaxial strain. (b) The peak positions of 2D band of single-layer (black squares) and three-layer (red circles) graphene as a function of uniaxial strain. (c) The peak positions of D, G, 2D, and 2D' bands plotted as a function of the biaxial strain, the solid lines are linear fits. Reproduced with permission from [73] (a and b) and [77] (c). Copyright 2008 American Chemical Society (a and b) and 2010 American Chemical Society (c).

Biaxial strain, which can be considered as symmetrical strain in the graphene basal plane, can be realized in graphene by utilizing piezoelectric actuators [77] and creating graphene bubbles in micron-scale [76]. Compared with the uniaxial strain which causes shifting of the relative positions of the Dirac cones and influence the intervalley double-resonance processes (D and 2D peaks), biaxial strain can avoid such Dirac cone perturbations and it does not depend on the Poisson ratio, hence biaxial strain is more appropriate for revealing the strain effects on the multi-phonon involved double resonant processes [76, 77]. Biaxial strain induced Raman peak shift of graphene is shown in Fig. 7.11c. All the Raman bands, D, G, 2D, and 2D' show smooth and linear shift with applied biaxial strain.

Therefore, both the degree of uniaxial and biaxial strains can be quantitatively probed by using Raman spectroscopy using the peak position of the Raman bands, such as the G and 2D bands. As graphene fabricated by CVD and other epitaxial methods contain different degrees of strain, Raman spectroscopy provides a very convenient technique for the characterization of strains.

7.3 Chemical Functional Groups and Energy Gaps of Graphene Probed by Infrared Spectroscopy

Raman spectroscopy can characterize graphene's various properties rapidly and effectively, but still, there are some other properties of graphene that cannot be easily probed by Raman spectroscopy. For example, using Raman spectroscopy, it is difficult to study the concentration of adsorbed chemical functional groups in graphene oxide, or the opened energy gap of graphene. IR spectroscopy is another commonly used spectroscopic tool for characterizing graphene. IR spectroscopy is able to reflect the evolution of GO's chemical structure very clearly. As shown in Figs. 7.12a and b, GO and highly reduced GO show obvious difference in their IR spectra. The vibrational modes of C=C and C=O bonds (shown by the shaded areas in Fig. 7.12) decrease dramatically after GO is highly reduced, inferring that most chemically attached oxygenated functional groups in GO have been effectively removed in the reduction process [78].

Graphene possesses many exotic electronic properties, such as anomalous quantum Hall effect [79, 80], extremely high mobility [81, 82], fine structure, constant defined optical absorption [50], and Klein tunneling [83, 84]. However, pristine graphene cannot be used in field-effect-transistor applications because of the gapless

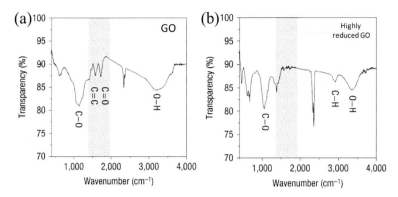

Figure 7.12 IR absorption spectra (400–4000 cm^{-1}) of GO (a) and highly reduced GO (b). Reproduced with permission from [78]. Copyright 2008 Nature Publishing Group.

electronic band structure [8]. In the past years, one of the most active areas of graphene research is to open a bandgap in graphene. By applying a perpendicular electric field on the AB-stacked bilayer graphene [85–87], the inversion symmetry of graphene is broken and consequently opens a bandgap [88]. In AB-stacked bilayer graphene, it has two nearly parallel parabolic conduction bands above two nearly parallel parabolic valence bands at low-energy regime at around the Dirac point [6, 8]. As the energy splitting between the two parallel subbands at conduction or valence band is ~0.4 eV, it is possible to use IR spectroscopy for this study [85]. Thus IR spectroscopy is also a powerful scientific tool for probing the low-energy electronic dispersion in graphene. For inversion symmetry breaking in AB-stacked bilayer graphene induced by a perpendicular electric field, the modified electronic band structure at the low-energy regime is shown in Fig. 7.13a. The opened bandgap between π_1 and π_1^* increases with the increase in the external electric field. Transition I denoted in Fig. 7.13a corresponds to the opened bandgap. It can be seen in Fig. 7.13b, the IR absorption spectra at the energy range of 0.15–0.6 eV resolves the absorption peak clearly. The absorption peak position shifts to higher energy with increase in the electric field. This experimental result is fully consistent with the theoretical results as shown in Fig. 7.13c, indicating that the values of opened bandgaps (Δ) of AB-stacked bilayer graphene under various electric field intensities have been directly probed by the IR spectra.

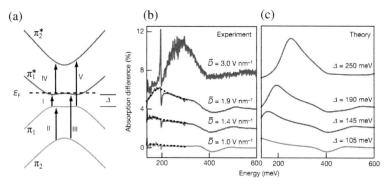

Figure 7.13 (a) Schematic electronic band structure of bandgap opened bilayer graphene and the allowed optical transitions between different subbands. (b) Experimental results of infrared absorption spectra at charge neutral point for different applied average displacement fields. (c) Theoretical results of the gate induced absorption spectra based on a tight-binding model. Reproduced with permission from [86]. Copyright 2009 Nature Publishing Group.

7.4 Summary

In this chapter, it is demonstrated that both Raman and IR spectroscopies are power scientific tools that can let researchers know many important physical and chemical parameters of graphene, including the thickness, stacking geometry, doped charges, defects, edge chirality, strain, chemically attached functional groups, and energy gaps. As a 2D material with only angstrom-scale in z-direction, the electronic band structure of graphene is very easy to be modified by changing any of the above mentioned physical and chemical parameters. Therefore, the understanding of such effective spectroscopic tools for characterizing these parameters of graphene is of great importance, not only for knowing the modified crystal structure of graphene, but also for the modified electronic structure of graphene. Considering the advantages and efficiencies of Raman and IR spectroscopy, these spectroscopic methods will not only be limited to graphene's fundamental research characterization, but can also be used as important examination tools in graphene's industrialization application in future.

References

1. Pimenta M.A., Dresselhaus G., Dresselhaus M.S., Cancado L.G., Jorio A., Saito R. (2007) Studying disorder in graphite-based systems by Raman spectroscopy, *Phys Chem Chem Phys*, **9**, 1276–1291.

2. Berger C., Song Z.M., Li T.B., et al. (2004) Ultrathin epitaxial graphite: 2D electron gas properties and a route toward graphene-based nanoelectronics, *J Phys Chem B*, **108**, 19912–19916.

3. Novoselov K.S., Geim A.K., Morozov S.V., et al. (2004) Electric field effect in atomically thin carbon films, *Science*, **306**, 666–669.

4. Ferrari A.C., Meyer J.C., Scardaci V., et al. (2006) Raman spectrum of graphene and graphene layers, *Phys Rev Lett*, **97**, 187401.

5. Wallace P.R. (1947) The band theory of graphite, *Phys Rev*, **71**, 622–634.

6. Trickey S.B., Mullerplathe F., Diercksen G.H.F., Boettger J.C. (1992) Interplanar binding and lattice-relaxation in a graphite dilayer, *Phys Rev B*, **45**, 4460–4468.

7. Castro Neto A.H., Guinea F., Peres N.M.R., Novoselov K.S., Geim A.K. (2009) The electronic properties of graphene, *Rev Mod Phys*, **81**, 109–162.

8. Zhan D., Yan J.X., Lai L.F., Ni Z.H., Liu L., Shen Z.X. (2012) Engineering the electronic structure of graphene, *Adv Mater (Weinheim, Ger)*, **24**, 4055.

9. Malard L.M., Pimenta M.A., Dresselhaus G., Dresselhaus M.S. (2009) Raman spectroscopy in graphene, *Phys Rep*, **473**, 51–87.

10. Malard L.M., Nilsson J., Elias D.C., et al. (2007) Probing the electronic structure of bilayer graphene by Raman scattering, *Phys Rev B*, **76**, 201401.

11. Hao Y.F., Wang Y.Y., Wang L., et al. (2010) Probing layer number and stacking order of few-layer graphene by Raman spectroscopy, *Small*, **6**, pp. 195–200.

12. Lui C.H., Li Z.Q., Chen Z.Y., Klimov P.V., Brus L.E., Heinz T.F. (2011) Imaging stacking order in few-layer graphene, *Nano Lett*, **11**, 164–169.

13. Cong C.X., Yu T., Sato K., et al. (2011) Raman characterization of ABA- and ABC-stacked trilayer graphene, *ACS Nano*, **5**, 8760–8768.

14. Ni Z.H., Wang Y.Y., Yu T., You Y.M., Shen Z.X. (2008) Reduction of Fermi velocity in folded graphene observed by resonance Raman spectroscopy, *Phys Rev B*, **77**, 235403.

15. Ni Z.H., Liu L., Wang Y.Y., et al. (2009) G-band Raman double resonance in twisted bilayer graphene: Evidence of band splitting and folding, *Phys Rev B*, **80**, 125404.

16. Kim K., Coh S., Tan L.Z., et al. (2012) Raman spectroscopy study of rotated double-layer graphene: Misorientation-angle dependence of electronic structure, *Phys Rev Lett*, **108**, 246103.

17. Rao R., Podila R., Tsuchikawa R., et al. (2011) Effects of layer stacking on the combination Raman modes in graphene, *ACS Nano*, **5**, 1594–1599.

18. Cong C.X., Yu T., Saito R., Dresselhaus G.F., Dresselhaus M.S. (2011) Second-order overtone and combination Raman modes of graphene layers in the range of 1690–2150 cm(-1), *ACS Nano*, **5**, 1600–1605.

19. Li D.F., Zhan D., Yan J.X., et al. (2012) Thickness and stacking geometry effects on high frequency overtone and combination Raman modes of graphene, *J Raman Spectrosc*, DOI: 10.1002/jrs.4156.

20. Tan P.H., Han W.P., Zhao W.J., et al. (2012) The shear mode of multilayer graphene, *Nat Mater*, **11**, 294–300.

21. Ni Z.H., Wang H.M., Kasim J., et al. (2007) Graphene thickness determination using reflection and contrast spectroscopy, *Nano Lett*, **7**, 2758–2763.

22. Wang Y.Y., Ni Z.H., Liu L., et al. (2010) Stacking-dependent optical conductivity of bilayer graphene, *ACS Nano*, **4**, 4074–4080.

23. Ni Z.H., Yu T., Luo Z.Q., et al. (2009) Probing charged impurities in suspended graphene using Raman spectroscopy, *ACS Nano*, **3**, 569–574.

24. Liu L., Ryu S.M., Tomasik M.R., et al. (2008) Graphene oxidation: Thickness-dependent etching and strong chemical doping, *Nano Lett*, **8**, 1965–1970.

25. Ni Z.H., Wang H.M., Luo Z.Q., et al. (2010) The effect of vacuum annealing on graphene, *J Raman Spectrosc*, **41**, 479– 483.

26. Zhang W., Lin C.-T., Liu K.-K., et al. (2011) Opening an electrical band gap of bilayer graphene with molecular doping, *ACS Nano*, **5**, 7517–7524.

27. Das A., Pisana S., Chakraborty B., et al. (2008) Monitoring dopants by Raman scattering in an electrochemically top-gated graphene transistor, *Nat Nanotechnol*, **3**, 210–215.

28. Lazzeri M., Mauri F. (2006) Nonadiabatic Kohn anomaly in a doped graphene monolayer, *Phys Rev Lett*, **97**, 266407.

29. Pisana S., Lazzeri M., Casiraghi C., et al. (2007) Breakdown of the adiabatic Born-Oppenheimer approximation in graphene, *Nat Mater*, **6**, 198–201.

30. Casiraghi C., Pisana S., Novoselov K.S., Geim A.K., Ferrari A.C. (2007) Raman fingerprint of charged impurities in graphene, *Appl Phys Lett*, **91**, 233108.

31. Basko D.M. (2008) Theory of resonant multiphonon Raman scattering in graphene, *Phys Rev B*, **78**, 125418.

32. Basko D.M., Piscanec S., Ferrari A.C. (2009) Electron–electron interactions and doping dependence of the two-phonon Raman intensity in graphene, *Phys Rev B*, **80**, 165413.

33. Casiraghi C. (2009 Doping dependence of the Raman peaks intensity of graphene close to the Dirac point, *Phys Rev B*, **80**, 233407.

34. Basko D.M. (2009) Calculation of the Raman G peak intensity in monolayer graphene: Role of Ward identities, *New J Phys*, **11**, 095011.

35. Kalbac M., Reina-Cecco A., Farhat H., Kong J., Kavan L., Dresselhaus M.S. (2010) The influence of strong electron and hole doping on the raman intensity of chemical vapor-deposition graphene, *ACS Nano*, **4**, 6055–6063.

36. Chen C.F., Park C.H., Boudouris B.W., et al. (2011) Controlling inelastic light scattering quantum pathways in graphene, *Nature*, **471**, 617–620.

37. Jung N., Kim N., Jockusch S., Turro N.J., Kim P., Brus L. (2009) Charge transfer chemical doping of few layer graphenes: Charge distribution and band gap formation, *Nano Lett*, **9**, 4133–4137.

38. Zhan D., Sun L., Ni Z.H., et al. (2010) $FeCl_3$-based few-layer graphene intercalation compounds: Single linear dispersion electronic band structure and strong charge transfer doping, *Adv Funct Mater*, **20**, 3504–3509.

39. Zhao W.J., Tan P.H., Liu J., Ferrari A.C. (2011) Intercalation of few-layer graphite flakes with FeCl(3): Raman determination of Fermi level, layer by layer decoupling, and stability, *J Am Chem Soc*, **133**, 5941–5946.

40. Bolotin K.I., Ghahari F., Shulman M.D., Stormer H.L., Kim P. (2009) Observation of the fractional quantum Hall effect in graphene, *Nature*, **462**, 196–199.

41. Ghahari F., Zhao Y., Cadden-Zimansky P., Bolotin K., Kim P. (2011) Measurement of the $\nu = 1/3$ fractional quantum Hall energy gap in suspended graphene, *Phys Rev Lett*, **106**, 046801.

42. Dean C.R., Young A.F., Cadden-Zimansky P., et al. (2011) Multicomponent fractional quantum Hall effect in graphene, *Nat Phys*, **7**, 693–696.

43. Banhart F., Kotakoski J., Krasheninnikov A.V. (2011) Structural defects in graphene, *ACS Nano*, **5**, 26–41.

44. Thomsen C., Reich S. (2000) Doable resonant Raman scattering in graphite, *Phys Rev Lett*, **85**, 5214–5217.

45. Girit C.O., Meyer J.C., Erni R., et al. (2009) Graphene at the edge: Stability and dynamics, *Science*, **323**, 1705–1708.

46. Liu G.X., Teweldebrhan D., Balandin A.A. (2011) Tuning of graphene properties via controlled exposure to electron beams, *IEEE Trans Nanotechnol*, **10**, 865–870.

47. Ryu S., Han M.Y., Maultzsch J., et al. (2008) Reversible basal plane hydrogenation of graphene, *Nano Lett*, **8**, 4597–4602.

48. Luo Z.Q., Shang J.Z., Lim S.H., et al. (2010) Modulating the electronic structures of graphene by controllable hydrogenation, *Appl Phys Lett*, **97**, 233111.

49. Ni Z.H., Ponomarenko L.A., Nair R.R., et al. (2010) On resonant scatterers as a factor limiting carrier mobility in graphene, *Nano Lett*, **10**, 3868–3872.

50. Nair R.R., Ren W.C., Jalil R., et al. (2010) Fluorographene: A two-dimensional counterpart of teflon, *Small*, **6**, 2877–2884.

51. Jeon K.J., Lee Z., Pollak E., et al. (2011) Fluorographene: A wide bandgap semiconductor with ultraviolet luminescence, *ACS Nano*, **5**, 1042–1046.

52. Stankovich S., Dikin D.A., Dommett G.H.B., et al. (2006 Graphene-based composite materials, *Nature*, **442**, 282–286.

53. Nourbakhsh A., Cantoro M., Vosch T., et al. (2010) Bandgap opening in oxygen plasma-treated graphene, *Nanotechnology*, **21**, 435203.

54. Cancado L.G., Takai K., Enoki T., et al. (2006) General equation for the determination of the crystallite size L-a of nanographite by Raman spectroscopy, *Appl Phys Lett*, **88**, 163106.

55. Lucchese M.M., Stavale F., Ferreira E.H.M., et al. (2010) Quantifying ion-induced defects and Raman relaxation length in graphene, *Carbon*, **48**, 1592–1597.

56. Cancado L.G., Jorio A., Ferreira E.H.M., et al. (2011) Quantifying defects in graphene via Raman spectroscopy at different excitation energies, *Nano Lett*, **11**, 3190–3196.

57. Ferrari A.C., Robertson J. (2000) Interpretation of Raman spectra of disordered and amorphous carbon, *Phys Rev B*, **61**, 14095–14107.

58. Dresselhaus M.S., Dresselhaus G. (2002) Intercalation compounds of graphite, *Adv Phys*, **51**, 1–186.

59. Luo Z.Q., Yu T., Kim K.J., et al. (2009) Thickness-dependent reversible hydrogenation of graphene layers, *ACS Nano*, **3**, 1781–1788.

60. Elias D.C., Nair R.R., Mohiuddin T.M.G., et al. (2009) Control of graphene's properties by reversible hydrogenation: Evidence for graphene, *Science*, **323**, 610–613.

61. Stankovich S., Dikin D.A., Piner R.D., et al. (2007) Synthesis of graphene-based nanosheets via chemical reduction of exfoliated graphite oxide, *Carbon*, **45**, 1558–1565.

62. Wang X., Zhi L.J., Mullen K. (2008) Transparent, conductive graphene electrodes for dye-sensitized solar cells, *Nano Lett*, **8**, 323–327.

63. Stoller M.D., Park S.J., Zhu Y.W., An J.H., Ruoff R.S. (2008) Graphene-based ultracapacitors, *Nano Lett*, **8**, 3498–3502.

64. Zhan D., Ni Z.H., Chen W., et al. (2011) Electronic structure of graphite oxide and thermally reduced graphite oxide, *Carbon*, **49**, 1362–1366.

65. Son Y.W., Cohen M.L., Louie S.G. (2006) Energy gaps in graphene nanoribbons, *Phys Rev Lett*, **97**, 216803.

66. Ritter K.A., Lyding J.W. (2009) The influence of edge structure on the electronic properties of graphene quantum dots and nanoribbons, *Nat Mater*, **8**, 235–242.

67. Jia X.T., Hofmann M., Meunier V., et al. (2009) Controlled formation of sharp zigzag and armchair edges in graphitic nanoribbons, *Science*, **323**, 1701–1705.

68. Cancado L.G., Pimenta M.A., Neves B.R.A., Dantas M.S.S., Jorio A. (2004) Influence of the atomic structure on the Raman spectra of graphite edges, *Phys Rev Lett*, **93**, 247401.

69. Casiraghi C., Hartschuh A., Qian H., et al. (2009) Raman spectroscopy of graphene edges, *Nano Lett*, **9**, 1433–1441.

70. You Y.M., Ni Z.H., Yu T., Shen Z.X. (2008) Edge chirality determination of graphene by Raman spectroscopy, *Appl Phys Lett*, **93**, 163112.

71. Xu Y.N., Zhan D., Liu L., et al. (2011) Thermal dynamics of graphene edges investigated by polarized Raman spectroscopy, *ACS Nano*, **5**, 147–152.

72. Zhan D., Liu L., Xu Y.N., et al. (2011) Low temperature edge dynamics of AB-stacked bilayer graphene: Naturally favored closed zigzag edges, *Sci Rep*, **1**, 12.

73. Ni Z.H., Yu T., Lu Y.H., Wang Y.Y., Feng Y.P., Shen Z.X. (2008). Uniaxial strain on graphene: Raman spectroscopy study and band-gap opening, *ACS Nano*, **2**, 2301–2305.

74. Frank O., Mohr M., Maultzsch J., et al. (2011) Raman 2D-band splitting in graphene: Theory and experiment, *ACS Nano*, **5**, 2231–2239.

75. Huang M.Y., Yan H.G., Heinz T.F., Hone J. (2010) Probing strain-induced electronic structure change in graphene by raman spectroscopy, *Nano Lett*, **10**, 4074–4079.

76. Zabel J., Nair R.R., Ott A., et al. (2012) Raman spectroscopy of graphene and bilayer under biaxial strain: Bubbles and balloons, *Nano Lett*, **12**, 617–621.

77. Ding F., Ji H.X., Chen Y.H., et al. (2010) Stretchable graphene: A close look at fundamental parameters through biaxial straining, *Nano Lett*, **10**, 3453–3458.

78. Li X.L., Zhang G.Y., Bai X.D., et al. (2008) Highly conducting graphene sheets and Langmuir-Blodgett films, *Nat Nanotechnol*, **3**, 538–542.

79. Novoselov K.S., Geim A.K., Morozov S.V., et al. (2005) Two-dimensional gas of massless Dirac fermions in graphene, *Nature*, **438**, 197–200.

80. Zhang Y.B., Tan Y.W., Stormer H.L., Kim P. (2005) Experimental observation of the quantum Hall effect and Berry's phase in graphene, *Nature*, **438**, 201–204.

81. Bolotin K.I., Sikes K.J., Hone J., Stormer H.L., Kim P. (2008) Temperature-dependent transport in suspended graphene, *Phys Rev Lett*, **101**, 096802.

82. Bolotin K.I., Sikes K.J., Jiang Z., et al. (2008) Ultrahigh electron mobility in suspended graphene, *Solid State Commun*, **146**, 351–355.

83. Katsnelson M.I., Novoselov K.S., Geim A.K. (2006) Chiral tunnelling and the Klein paradox in graphene, *Nat Phys*, **2**, 620–625.

84. Young A.F., Kim P. (2009). Quantum interference and Klein tunnelling in graphene heterojunctions, *Nat Phys*, **5**, 222–226.

85. Kuzmenko A.B., van Heumen E., van der Marel D., et al. (2009) Infrared spectroscopy of electronic bands in bilayer graphene, *Phys Rev B*, **79**, 115441.

86. Zhang Y.B., Tang T.T., Girit C., et al. (2009) Direct observation of a widely tunable bandgap in bilayer graphene, *Nature*, **459**, 820–823.

87. Oostinga J.B., Heersche H.B., Liu X.L., Morpurgo A.F., Vandersypen L.M.K. (2008) Gate-induced insulating state in bilayer graphene devices, *Nat Mater*, **7**, 151–157.

88. Xia F.N., Farmer D.B., Lin Y.M., Avouris P. (2010) Graphene field-effect transistors with high on/off current ratio and large transport band gap at room temperature, *Nano Lett*, **10**, 715–718.

Chapter 8

Graphene-Based Materials for Electrochemical Energy Storage

Jintao Zhang[a] and Xiu Song Zhao[a,b]

[a]*Department of Chemical & Biomolecular Engineering,*
National University of Singapore, 4 Engineering Drive 4, Singapore 117576
[b]*School of Chemical Engineering, Faculty of Engineering,*
Architecture and Information Technology, The University of Queensland,
St Lucia, Brisbane, QLD 4072, Australia
george.zhao@uq.edu.au

8.1 Introduction

Graphene, a two-dimensional monolayer of sp^2 hybridized carbon bonded in a hexagonal lattice, is a rising star in the field of material science and physics. A great progress in the study of graphene and related materials in the last few years, including material synthesis and understanding of fundamental properties, has significantly boosted the exploration of promising applications for electrochemical energy storage. Standing as bridges between high-power-output conventional capacitors and high-energy-density batteries, supercapacitors are an ideal electrochemical energy-storage system

---

Two-Dimensional Carbon: Fundamental Properties, Synthesis, Characterization, and Applications
Edited by Yihong Wu, Zexiang Shen, and Ting Yu
Copyright © 2014 Pan Stanford Publishing Pte. Ltd.
ISBN 978-981-4411-94-3 (Hardcover), 978-981-4411-95-0 (eBook)
www.panstanford.com

suitable for rapid storage and release of energy. This chapter aims to summarize the recent research progress towards the design and exploitation of graphene-based materials for supercapacitor applications. Beginning from the research advances on the synthesis of graphene-based materials and the brief description of the principles of energy storage in supercapacitor devices, we also discuss how to evaluate the capacitive performance of supercapacitors to highlight ongoing research strategies. A particular focus is set on the synthesis of graphene-based materials as the active materials for supercapacitors, given the large inherent potential of such materials to maximize the electrochemical performance of supercapacitors.

8.1.1 A Family of Graphene-Based Materials

Carbon materials, including zero-dimensional (0D) fullerenes, one-dimensional (1D) carbon nanotubes (CNTs), two-dimensional (2D) graphene, and three-dimensional (3D) graphite, are of particular interest because of their excellent electrical and mechanical properties as well as unique structures [1–3]. Graphene, a 2D flat monolayer of sp^2 hybridized carbon tightly bonded in a hexagonal lattice, is the basic building block for all the graphitic carbons. As shown in Fig. 8.1, the 0D fullerene, 1D CNT, and 3D graphite or diamond can be formed by wrapping, rolling, and stacking of a graphene sheet, respectively [1–4]. For the electrochemical energy storage applications, the essential characteristics of an electrode material are its high surface area and electrical conductivity. Graphene has a theoretical surface area of ~ 2630 m^2 g^{-1}, which is about two times larger than that of single wall carbon nanotubes (SWCNTs) (~ 1315 m^2 g^{-1}). The high conductivity (~ 64 mS cm^{-1}) of graphene is superior to many other carbon materials [5]. The production and processing of graphene appear to be simpler and more economic in comparison with CNTs and the related carbon materials. For example, the residual metallic impurities are inherent to chemical vapor deposition (CVD) preparation process of CNTs, which have hindered the configuration of reliable energy devices. However, the problem would be precluded in the preparation process of graphene sheets. The transparent conductive graphene sheets have been prepared by a simple transfer process of large-area graphene grown on Cu foils by a CVD process [6]. More importantly, they are capable of storing an electrical double-layer capacitance value of up to 550 F g^{-1}, provided the whole surface area of graphene is fully utilized [7], highlighting the superior characteristic for energy storage application.

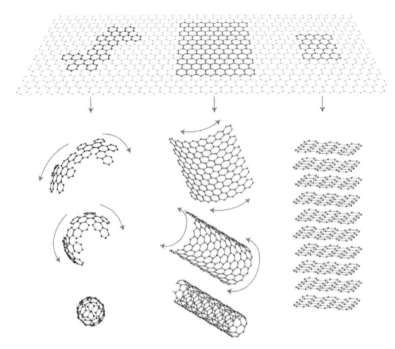

Figure 8.1 Graphene: The mother of all graphitic carbon materials. Graphene (bottom from left to right) Graphite, Nanotube, and Fullerenes. Reprinted by permission from Macmillan Publishers Ltd: Geim, A. K., and Novoselov, K. S. (2007). The rise of graphene, *Nat Mater*, **6**, pp.183–191], copyright @ 2010.

The pristine graphene generally consists of single- and few-(three to nine) layer graphene sheets. Over the past few years, a number of methods have been reported to synthesize graphene sheets, which include mechanical or chemical exfoliation of graphite, epitaxial growth on SiC surface, chemical vapor deposition on various metal surfaces, solvothermal synthesis, total organic synthesis, unzipping carbon nanotubes, etc. [8–11]. However, the uniform growth of single-layer graphene in a large-scale is still a challenge. A family of chemically modified graphene (CMG) consisting of structural and chemical derivates of graphene has been prepared by using chemical methods from various precursors.[12, 13]. Reduced graphene oxide (RGO) sheet is one of the important CMGs for the promising energy storage applications. The most common route to prepare RGO sheets begins with the oxidation of graphite to graphite oxide which consists of a layered structure of graphene oxide [12]. Due to the strong hydrophilicity of graphite oxide, the intercalation of water molecules between the layers occurs easily. The complete exfoliation of graphite oxide produces an aqueous colloidal suspension of graphene oxides (GO), which is readily carried out by sonication. Notably, the resulting GO sheets are electrically insulating owing to the disruption of graphene networks with oxygenate groups. The significant amount of oxygenate groups on the surface of GO sheets suggests that the electrostatic repulsion between the negatively charged GO sheets leads to the stable dispersion of GO sheets in water. The oxidation degree of graphite oxide is varied with the reaction conditions and the graphite precursors [14]. As a result, GO sheets with various levels of oxidation would be produced. The reduction of GO sheets using a reducing regent or thermal treatment result in the formation of RGO sheets which are electrically conductive. Nonetheless, element analysis revealed the existence of a large amount of residual oxygen (atomic ratio of C/O, ~10) on RGO sheets [15], suggesting that RGO is not the same as pristine graphene.

To exploit the potential of graphene-based materials for wide applications, many efforts have been focused on CMGs to improve the diversity of graphene. CMGs have been prepared by structural and chemical modification and functionalization of graphene or RGO sheets. Blending CMG with a second component was also used to form graphene-based composite materials. Graphene-based materials are of special importance because of the intriguing properties from the synergistic effects of their counterparts except for the intrinsic properties from each component. However, information as to how graphene and CMGs are prepared is crucial because the properties of graphene strongly depend on the methods of fabrication. For example, mechanical exfoliation produces few-layered graphene of highest quality, while chemical method is demonstrated to give high throughput and at relatively low costs, which enables technical applications in a variety of fields such as energy conversion and storage materials, catalysis and sensors. Therefore, the effective synthesis strategies of graphene-based materials on a large scale with controlled sizes and layers are of great importance for the applications in different fields.

8.1.2 Electrochemical Energy Storage Systems

Diminishing reserves of fossil fuels and its severe impact on humans and environment, such as climate change and global warming due to emissions from burning of fossil fuels, have been increasingly driving our society towards clean and sustainable energy development. With the fast-growing market for portable electronic devices and the development of hybrid electric vehicles, there has been an ever increasing and urgent demand for environment- friendly high-power energy resources. Transforming natural energy such as wind, tides, and solar energy can generate large amounts of sustainable energy. Hence, the development of energy storage devices is extremely important to store the harvested energy from the natural phenomena with limited control [16]. Batteries and supercapacitors (also named electrochemical capacitors), as the sustainable energy storage devices, have attracted considerable attention [17–19]. On the basis of the specific energy and power capabilities of several energy storage and conversion systems (capacitors, supercapacitors, batteries, and fuel cells) in Fig. 8.2, it is interesting to note that no single energy source can match all power and energy region. Supercapacitors and batteries fill up the gap between conventional capacitors and fuel cells and therefore are ideal electrochemical energy-storage systems.

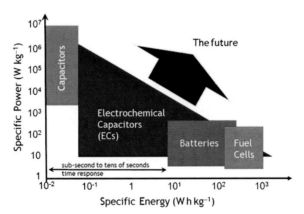

Figure 8.2 Ragone plot of specific energy and power capabilities for various energy storage and conversion devices. Rolison, D. R. and Nazar, L. F. (2011). Electrochemical energy storage to power the 21st century, *MRS Bulletin*, **36**, pp. 486–493. Reprinted with permission.

The common features of these systems are that the process of energy release takes place at the interface between electrode and electrolyte and that the electron and ion transport are separated [20]. Owing to the inherent differences between batteries and supercapacitors with respect to energy storage mechanism and electrode materials, the characteristic performance of supercapacitors sets them apart from batteries. In order to better understand the inherent differences between batteries and supercapacitors as well as the conventional capacitors (electrolytic capacitors), Table 8.1 summarizes the basic characteristics of these energy storage systems [21].

Table 8.1 The basic characteristics of supercapacitor, battery, and electrolytic capacitor

Parameters	Electrolytic capacitor	Supercapacitor	Battery
Storage mechanism	Physical	Physical	Chemical
Charge time	$10^{-6} \sim 10^{-3}$ s	$1 \sim 30$ s	$1 \sim 5$ h
Discharge time	$10^{-6} \sim 10^{-3}$ s	$1 \sim 30$ s	$0.3 \sim 3$ h
Energy density (Wh kg^{-1})	< 0.1	$1 \sim 10$	$20 \sim 100$
Power density (kW kg^{-1})	~ 10	$5 \sim 10$	$0.5 \sim 1$
Charge/ discharge efficiency (%)	~ 100	$75 \sim 95$	$50 \sim 90$
Cycle life	Infinite	> 500,000	$500 \sim 2000$
Max. voltage determinants	Dielectric thickness and strength	Electrode and electrolyte stability window	Thermodynamics of phase reactions
Charge stored determinants	Electrode area and dielectric	Electrode microstructure and electrolyte	Active mass and thermodynamics

In a battery, energy is stored in chemical form as active materials in electrodes whereas energy is released in an electrical form by connecting a load across the terminals of a battery where electrochemical reactions of electrode materials with ions from an electrolyte occur, leading to the conversion of chemical energy to electrical energy [22]. Lithium-ion batteries (LIBs) are the most popular rechargeable batteries. LIB is mainly composed of an anode (negative), a cathode (positive), an electrolyte, and a separator. Research effort has been devoted to developing a wide variety of electrode materials and electrolytes for LIB. Typically, the cathode materials are Li-containing metal oxides with tunnel-structure (e.g., lithium manganese oxide) or layered materials (e.g., lithium cobalt oxide), while the anode materials are insertion-type materials (such as, carbon, $Li_4Ti_5O_{12}$, and TiO_2), alloying-type materials (such as Si and Sn), conversion-type materials (such as iron oxides, nickel oxides, and cobalt oxides) [23, 24]. A solution of lithium salts dissolved in a mixture of two or more organic solvents is the electrolyte for LIB. When a battery is cycled, Li ions are exchanged between anode and cathode. The discharge rate and power performance of batteries are determined by the reaction kinetics of active materials as well as the mass transport. Therefore, battery generally yields a high energy density (150 Wh kg^{-1} is possible for LIBs) and a rather low power rate as well as limited cycle life. Thus, the ever-increasing demand of the power requirements is a great challenge to the capability of battery design [25].

Supercapacitors, also known as electrochemical capacitors or ultracapacitors, are energy storage devices that offer higher specific power density than most batteries and a higher specific energy density in a small package than the conventional capacitors. The configuration of a typical supercapacitor is somewhat similar to a battery and consists of a pair of polarizable electrodes with current collectors, a separator, and an electrolyte (Fig. 8.3) [26]. The fundamental difference between the supercapacitors and the batteries is that energy is physically stored in a supercapacitor by means of ion adsorption at the electrode/electrolyte interface (namely, electrical double-layer capacitors, EDLCs). As a result, the supercapacitors offer the ability to store/release energy in timescales of a few seconds with an extended cycle life [27]. Generally, carbon materials, such as activated carbon and porous carbon, are used as electrode materials for EDLCs [28, 29]. Metal oxides and conducting polymers are also used as electrode materials of supercapacitors, where energy storage is based on the fast and reversible redox reaction with a small amount of charge stored by EDLC. This class of supercapacitor is represented as pseudocapacitor. The electrolytes of supercapacitors can be aqueous or organic. The aqueous electrolytes offer low internal resistance, but the operating voltage limits to about 1.0 V due to the thermodynamic electrochemical window of water (1.23 V). The decomposition voltage of electrolyte determines the operating voltage of the supercapacitor. The organic medium with broad potential window will significantly enhance the energy accumulated in supercapacitors.

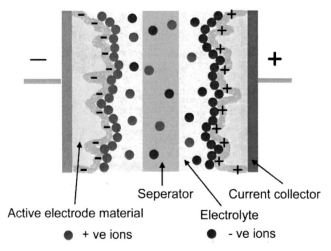

Figure 8.3 Schematic diagram of a two-electrode cell of supercapacitor. Zhang, J., Zhao, X. S. (2012). On the configuration of supercapacitors for maximizing electrochemical performance, *ChemSusChem.*, **5**, pp. 818–841. Reproduced with permission.

8.2 Synthesis of Graphene-Based Materials

8.2.1 Synthesis of Graphene

While theoretical studies on graphene started some 60 years ago [30, 31], it is only when free-standing graphene was first experimentally turned into a reality has it aroused a worldwide resurgence. With the confirmation of its unusual physical properties, the single-layer graphene which is mechanical exfoliation of highly oriented pyrolitic graphite (HOPG), first and foremost yields access to a large amount of interesting physics and electronics [32]. Mechanically exfoliated single-layer graphene was firstly developed by Geim and co-workers using a technique called micromechanical cleavage [33]. Typically, a cellophane tape is used to peel off graphene layers from a graphite flake, followed by pressing the tape against a substrate. Upon removing the tape, a single sheet graphene is obtained. The earlier works provided opportunities to experimentally investigate the electronic structure of nanosized graphene and formed the foundation to develop graphene-based nanoelectronics [32]. However, the low throughput and yield of the mechanical exfoliation method largely limits its applications for mass production.

Liquid exfoliation (exfoliation of graphite in solutions) is predicated an alternative approach to producing high quality of graphene sheets or graphene nanoplates because the pristine properties of graphite are retained after exfoliation [34–36]. Coleman and co-workers [35] demonstrated a scalable method to produce high-quality graphene flakes from graphite powders. By using certain solvents, such as N-methylpyrrolidone (NMP), graphene was dispersed at concentrations of up to 0.01 mg ml^{-1}. The strong interaction between solvent and graphite sidewall was likely to decrease the energetic penalty for exfoliation and subsequent solvation, leading to the exfoliation of graphene sheets.

The direct synthesis of single-layer graphene in large-scale for macroscopic applications is still a challenge. Single-layer graphene sheet is mainly synthesized by epitaxial growth or CVD on various substrates [8, 37–40]. For the epitaxial growth of graphene on silicon carbide (SiC), the high temperature in the range of 1200–1600°C leads to the decomposition of single crystal SiC in vacuum. Excess carbon is left on the surface due to the lower sublimation rate of carbon than that of silicon. The carbon atoms are rearranged to form graphene sheets [41, 42]. Epitaxial growth yields high-quality graphene samples interacting strongly with their substrates, but the isolation of graphene sheets from the substrates would be beneficial to study its properties without the interference of the substrates. The photolithographic techniques and an oxygen plasma etching procedure were employed to prepare free-standing epitaxial graphene sheets [40]. The resulting free-standing graphene sheets paved the way for a variety of mechanical, electronic, and optical experiments to probe the true nature of epitaxial graphene. In order to grow graphene sheets from epitaxial method, the conditions must be carefully controlled to promote crystal growth without seeding additional layers or forming grain boundaries. These strict conditions would limit the use of these techniques. Recently, the thermal splitting of commercial polycrystalline SiC granules has been developed to synthesize graphene sheets [43]. As shown in Fig. 8.4, the carbon species self-assemble into freestanding graphene nanosheets at a high temperature following the breaking of Si–C bonds and the sublimation of Si. This synthetic approach would be easily scaled up since the throughput is only limited by the size of the oven. Nonetheless, one drawback of this process is that the resulting graphene sheets prefer to form few-layer graphene sheets due to their strong π–π stacking interaction.

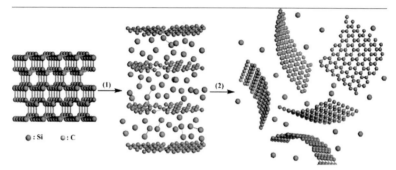

Figure 8.4 Schematic representation of the proposed mechanism of synthesis of freestanding graphene nanosheets from commercial polycrystalline SiC granules. Deng, D., Pan, X., Zhang, H., Fu, Q., Tan, D., Bao, X. (2010). Freestanding graphene by thermal splitting of silicon carbide granules, *Adv. Mater.*, **22**, pp. 2168–2171]. Reproduced with permission.

Alternatively, CVD is a promising method for the synthesis of uniform graphene layers. CVD generally produces graphene layers on metal surfaces by using hydrocarbon gases as precursors at the temperature of about 1000°C under inert atmosphere. Recently, Shen and co-workers reported the synthesis of bi-layer graphene (BLG) homogeneously at a large scale by thermal CVD [44]. The growth mechanism of the second graphene layer was found to be dependent on the purity-controlled catalytic activity of Cu surface. The study provides an effective approach to growing strongly coupled or even AB-stacked BLG on Cu foils at large scale, which is of particular importance for special electronic and photonic device applications based on their degenerate electronic bands.

A method for the synthesis of high-quality graphene layers with large-area was developed by Ruoff and co-workers [45], in which a mixture of methane and hydrogen was used as a precursor for the growth of grapnene sheets on copper foils. The method opens a new route for large-area synthesis of high-quality graphene films. More importantly, the growth of graphene is no longer limited to the use of rigid substrates.The method has been extended to grow graphene sheets on copper thin films (about 300 mm) on a Si substrate, resulting in the facile transfer of the graphene sheets to alternative substrates. These results suggest that the synthesis of high-quality graphene would be achieved by modifying and optimizing the CVD process [46–48]. Indeed, a modified CVD process has been developed for large-scale pattern growth of graphene on a nickel layer with thickness less than 300 nm [49]. Note that the fast cooling rate ($\sim 10°C\ s^{-1}$) is critical in suppressing the formation of multiple layers and for separating graphene layers efficiently from substrate in the later process [50]. The redox reaction of between Fe(III) and Ni is used to etch nickel substrate layers, which is effective in a mild pH range without forming gaseous products or precipitates. Large flexible copper foils have been used in the form of a roll-type substrate fitting inside a tubular furnace to maximize the scale and homogeneity of produced graphene films (Fig. 8.5) [51]. The flexibility of graphene and copper foils allows the efficient etching and transfer processes, leading to a cost- and time-effective roll-to-roll production of large-area and high-quality graphene sheets.

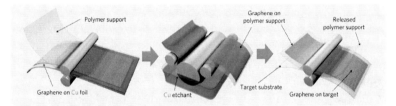

Figure 8.5 Schematic representation of the roll-based production of graphene films grown on a copper foil. Reprinted by permission from Macmillan Publishers Ltd: Bae, S., Kim, H., Lee, Y., Xu, X., Park, J. S., Zheng, Y., Balakrishnan, J., Lei, T., Ri Kim, H., Song, Y. I., Kim, Y. J., Kim, K. S., özyilmaz, B., Ahn, J. H., Hong, B. H., Iijima, S. (2010). Roll-to-roll production of 30-inch graphene films for transparent electrodes, *Nat Nanotech*, **5**, pp. 574–578], copyright @ 2010.

Template CVD route has proven to an effective strategy for synthesizing graphene with controlled shapes [52, 53]. More recently, Cheng and co-workers [54] reported the synthesis of 3D foam-like graphene macrostructures, graphene foams (GFs), by template-directed CVD. The resulting GF was composed of an interconnected flexible network with graphene as the fast transport channel of charge carriers for high electrical conductivity. Integration of individual 2D graphene sheets into macroscopic structures is essential for their applications.

Chemical synthesis from different precursors, such as polycyclic aromatic hydrocarbons (PAH), as a bottom-up method is expected to synthesize graphene with precise control over composition and structure [32]. The major challenge in the synthesis of graphene with large size is the limited dispersibility because the face-to-face interactions (π–π interaction) between the graphene layers increase with increasing size. Müllen and co-workers [55] exhibited a major break-through in the organic synthesis of 2D graphene ribbons with length up to 12 nm. However, it is hard to synthesize large graphene sheet in large-scale by using the precursor of polycyclic aromatic hydrocarbons. Choucair et al. [36] reported the direct chemical synthesis of carbon nanosheets in gram-scale quantities in a bottom-up approach by using the common laboratory reagents (ethanol and sodium) which were reacted to give an intermediate solid. The subsequent pyrolysis yielded a fused array of graphene sheets that was easily dispersed by mild sonication. Although the carbon nanosheets did not appear to consist of single graphene sheets, the production of large quantities of carbon nanosheets would allow for the development of large-scale applications of this unique material. Recently, a novel method for one-pot direct synthesis of N-doped graphene via the reaction between tetrachloromethane (CCl_4) and lithium nitride under mild conditions was developed (Fig. 8.6) [56]. The study revealed that the dichlorocarbene, free $-C=C-$, and $-C=N-$ groups were the likely intermediates for transforming sp^3 hybridized carbon in CCl_4 to sp^2 hybridized carbon. These intermediates would readily couple with each other, dechlorinate to form small domains of sp^2 hybridized carbon containing N, and then grow into N-doped graphene sheets. The resulting N-doped materials are expected to broaden the already widely explored applications of graphene sheets.

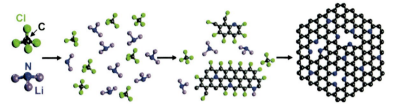

Figure 8.6 Schematic representation of the proposed mechanism for solvothermal synthesis of N-doped graphene via the reaction of CCl$_4$ and Li$_3$N, where gray balls represent C atoms, blue for N, green for Cl, and purple for Li. Reprinted with permission from Deng, D., Pan, X., Yu, L., Cui, Y., Jiang, Y., Qi, J., Li, W.-X., Fu, Q., Ma, X., Xue, Q., Sun, G., Bao, X. (2011). Toward N-doped graphene via solvothermal synthesis, *Chem. Mater.*, **23**, pp. 1188–1193. Copyright @ American Chemical Society.

Unzipping carbon nanotube has been developed to prepare graphene nanoribbon, which can be performed by oxidative treatment of CNTs, cutting CNTs with metal nanoparticles (Ni or Co), plasma etching, and so on [2, 57–61]. Figure 8.7 demonstrates different approaches to unzipping carbon nanotubes to yield graphene nanoribbons with controlled widths. The graphene nanoribbons obtained by these methods are of importance for the fundamental researches. However, only minute quantities of graphene nanoribbons can be produced by these methods, which is inefficient for the energy storage applications.

8.2.2 Synthesis of Chemically Modified Graphene

Chemical conversion of GO to RGO is efficient to produce CMG in large-scale [62]. As schematically illustrated in Fig. 8.8, graphite is firstly oxidized to produce graphite oxide using previously reported methods [13, 14, 63, 64]. The larger interlayer distance makes it readily dispersed in solution with a relatively high stability, leading to the formation of GO suspension [65]. The structural model of GO was firstly proposed on the basis of solid-state ^{13}C NMR spectra [66, 67]. The detailed chemical structure of GO, such as the type and distribution of oxygen-containing functional groups, has been investigated by Ruoff and co-workers, suggesting that the basal plane of GO sheet is decorated with hydroxyl and epoxy (1,2-ether) functional groups with small amount of lactol, ester, acid, and

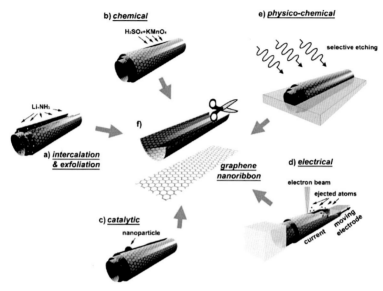

Figure 8.7 A sketch showing the different ways nanotubes could be unzipped to yield graphene nanoribbons (GNRs). (a) Intercalation–exfoliation of MWCNTs, involving treatments in liquid NH_3 and Li, and subsequent exfoliation using HCl and heat treatments; (b) chemical route, involving acid reactions that start to break carbon–carbon bonds (e.g., H_2SO_4 and $KMnO_4$ as oxidizing agents); (c) catalytic approach, in which metal nanoparticles "cut" the nanotube longitudinally like a pair of scissors; (d) the electrical method, by passing an electric current through a nanotube; and (e) physicochemical method by embedding the tubes in a polymer matrix followed by Ar plasma treatment. The resulting structures are either GNRs or graphene sheets (f). Reprinted from Terrones, M., Botello-Médez, A. R., Campos-Delgado, J., López-Urías, F., Vega-Cantú, Y. I., Rodríguez-Macías, F. J., Elías, A. L., Muñoz-Sandoval, E., Cano-Márquez, A. G., Charlier, J.-C., Terrones, H. (2010). Graphene and graphite nanoribbons: Morphology, properties, synthesis, defects and applications, *Nano Today*, **5**, pp. 351–372]. Copyright @ 2010, with permission from Elsevier.

ketone carbonyl groups at the edges of GO [68]. GO is electrically insulating and contains irreversible defects and disorders because of the disruption of conjugated electronic structure by the oxygenate functional groups. Chemical reduction of GO would restore its conductivity at values orders of magnitude below that of pristine graphene [69], leading to the formation of RGO sheets.

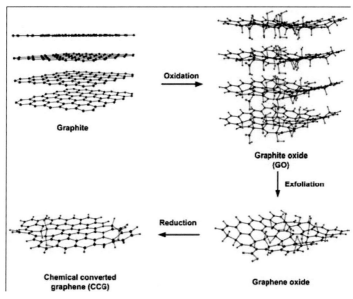

Figure 8.8 Preparation of chemically modified graphene (CMG) by reduction of graphene oxide. Bai, H., Li, C., Shi, G. (2011). Functional composite materials based on chemically converted graphene, *Adv. Mater.*, **23**, pp. 1089–1115]. Reproduced with permission.

Various methods have been developed to reduce GO with hydrazine solution and other reducing agents [15, 35, 69–71]. RGO sheets prepared by chemical reduction methods would be easily deposited on any substrate to form thin films, exhibiting promising applications for ultrathin electrochemical devices. Li et al. [71] reported the reduction of GO by hydrazine under a high pH value. The colloid stability of resulting RGO sheets is relatively high. However, the addition of NaCl and long-term storage resulted in an irreversible agglomeration. In view of the toxicity and unstability of hydrazine, it is desirable to explore green chemistry routes for the reduction of GO [72]. A facile approach was developed to reduce GO by using ascorbic acid as a reducing agent under mild conditions [73, 74]. Dong and co-workers [75] reported that reducing sugars, such as glucose, fructose, and sucrose, can also be used as both reducing agent and stabilizer to synthesize CMG sheets. These reduction methods avoid the usage of harmful chemical reductants and additional capping agents, providing environment-friendly routes to produce CMG. A two-step method consisting of deoxygenation and dehydration processes was used to prepare RGO with good stability. In this method, nearly complete reduction of functional groups on GO surface was easily achieved by deoxygenation with NaBH4 and dehydration with concentrated sulfuric acid, followed by thermal annealing [70]. Only small amounts of impurities were present in the final product (less than 0.5 wt% of sulfur and nitrogen, compared with about 3 wt% with other chemical reductions). This method is particularly effective to restore the π-conjugated structure of graphene, leading to highly soluble and conductive RGO materials. Loh and co-workers [76] demonstrated that the supercritical water under hydrothermal conditions would result in the conversion of GO to RGO by dehydration, providing a simple and clean hydrothermal method to prepare RGO. Fan et al. [77] reported that a stable RGO suspension was easily prepared by heating a GO suspension under storing alkaline conditions at moderate temperature of about 50–90°C. In this process, the alkaline-catalyzed deoxygenation reaction would contribute to the reduction of GO because the higher pH of GO suspension led to the faster reaction.

Some organic solvents allow high temperature processing, resulting in the reduction of GO suspension without the additional reducing agents [78–80]. An effective solvothermal reduction method was employed to prepare RGO suspension with increased conjugation domains and less defect in an organic solvent of *N,N*-dimethylformamide (DMF) [78]. Ruoff and co-workers [80] demonstrated that a wide variety of organic solvent systems would be used to prepare homogeneous colloidal suspensions of RGO. Notably, a microwave-assisted method has proven to readily produce homogeneous colloidal suspensions of RGO in an organic media in seconds [81–84]. Ang et al. [85] proposed a straightforward one-step intercalation and exfoliation method to produce large-sized conductive graphene sheets without using surfactants. The method is based on the rich intercalation chemistry of GO aggregates. The large-sized GO aggregates consisted of multilayer graphene flakes, which were highly oxidized on the outer layer whereas the inner layer consisted of mildly oxidized graphene sheets. Intercalation of such GO aggregates by tetrabutylammonium (TBA) cations via electrostatic attraction and cation–π interaction followed by exfoliation in DMF yielded large-sized conductive graphene sheets with a high yield (>90%). Recently, Compton et al. [86] demonstrated that suspensions of RGO were readily prepared by refluxing GO for a short time in appropriate organic solvents. The C/O ratio of RGO after heat treatment would be modulated according to the boiling point of the solvent, allowing for the functional group composition of RGO nanosheets to be tunable in a well-controlled fashion. For example, hydroxyl and carboxyl groups are readily removed via a thermal treatment above 155°C, while epoxides are more resilient, being removable only when the temperature is higher than 200°C, as in the case of N-methylpyrrolidone (NMP) reflux.

Thermal exfoliation and reduction of graphite oxide is a very effective approach to producing RGO, which utilizes the heat treatment to quickly remove the oxygenate functional groups from graphene oxide surfaces. A detailed analysis of the thermal expansion mechanism of graphite oxide to produce RGO sheets was demonstrated by Aksay's group [87, 88]. For the success of this process, it is essential that the decomposition rate of the oxygenate groups, such as epoxy and hydroxyl groups, on graphite oxide exceeds the diffusion rate of the evolved gases, thus yielding pressure that exceeds the interlayer forces (van der Waals force) holding the graphene sheets together. A comparison of the Arrhenius dependence of the reaction rate against the calculated diffusion coefficient based on Knudsen diffusion suggested the critical temperature for exfoliation of graphite oxide to occur is about 550°C [88]. However, the treatment temperatures for full exfoliation of graphite oxide to RGO sheets experimentally were normally above 1000°C. Lv et al. [89] developed a novel exfoliation approach (as shown in Fig. 8.9). The exfoliation process is realized at a very low temperature (200°C) which is far below the proposed critical exfoliation temperature, by introducing a high vacuum to the exfoliation process.

The arc discharge method has been extensively used for producing fullerenes and multiwalled, single-walled, and double-walled carbon nanotubes (CNTs) [90]. In combination with solution-phase dispersion and centrifugation techniques, a hydrogen arc discharge exfoliation method was used to synthesize RGO sheets from graphite oxide. Temperature could be instantaneously increased to more than 2000°C during the arc discharge process, leading to the efficient exfoliation and deoxygenation of GO [91]. This method is effective to eliminate the defects and heal the exfoliated RGO sheets. As a convenient and rapid heating source, microwave irradiation was used to prepare exfoliated RGO powder from graphite oxide in a short time (few seconds to min) [72, 81, 92, 93]. Thermal exfoliation of solid graphite oxide to RGO sheets generates several specific properties: (i) RGO sheets with a high ratio of C/O and relatively good electrical

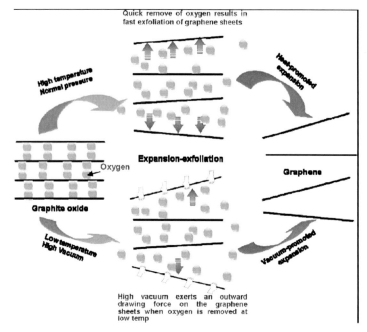

Figure 8.9 Schematic representation of chemical exfoliation of graphene. (Top) High-temperature (above 1000 °C) exfoliation under an atmospheric pressure; (bottom) low-temperature (as low as 200 °C) exfoliation under high vacuum, where high vacuum introduces a negative pressure surrounding the graphene layers. Reprinted with permission from Lv, W., Tang, D.-M., He, Y.-B., You, C.-H., Shi, Z.-Q., Chen, X.-C., Chen, C.-M., Hou, P.-X., Liu, C., Yang, Q.-H. (2009). Low-temperature exfoliated graphenes: Vacuum-promoted exfoliation and electrochemical energy storage, *ACS Nano*, **3**, pp. 3730–3736]. Copyright @ American Chemical Society.

conductivity would be obtained. However, the RGO sheets obtained are usually a mixture of one to few layers. The dispersibility in water is low due to the low content of oxygenate groups. Post treatments, such as solution-phase dispersion by sonication and centrifugation techniques, are essential to obtain a single-layer RGO suspension. (ii) The obtained RGO sheets suffer from relatively high structural defects, such as vacancies and topological defects caused by the release of gases (such as carbon dioxide) during deoxygenation. These defects disrupt the band structure and partly degrade the electronic properties. The quality of RGO sheets would benefit from removing structural defects or even healing defects.

Other synthesis methods of RGO performing under various conditions (electrochemical, photocatalytic, laser, plasma, and so on) have also been developed, resulting in their different properties. Recently, systematic studies were carried out to evaluate the reduction methods of GO by several factors in terms of dispersibility, reduction degree, defect repair degree, and electrical conductivity [94]. An ideal reduction method is not only to remove the oxygen containing functional groups but also to repair the defect to obtain high-quality RGO. The study revealed that the two-step method appeared to be the best one for RGO synthesis. The resulting RGO has better reduction degree, defect repair degree, and electrical conductivity, but relatively weak dispersibility and tedious preparation process still need to improve. The systemic comparison would be helpful to further understand the mechanism of reduction and help select or develop more effective reduction methods. Table 8.2 summarizes the basic characteristics of graphene-related materials prepared by various methods.

Table 8.2 Comparisons of different methods of synthesis of graphene-related materials

Methods	Number of layers	Electronic quality of layers	Nature of produced graphene	Throughput	Precursor
Mechanical exfoliation	S/M	High	G	Low	Graphite
Liquid-phase exfoliation	S/M	High	G	High	Graphite
Epitaxial growth by thermal deposition	S/M	High	G	Low	Silicon carbide
Chemical vapor deposition on transition metals	S/M	High	G	Low	Hydrocarbon

				High	Graphite oxide
Chemical reduction of graphene oxide liquid suspension	S/M	Low	RGO		
Thermal reduction of graphene oxide	S/M	Relative high	RGO	High	Graphite oxide
Solvothermal synthesis	S/M	—	G/N-doped G	High	Ethanol/CCl$_4$
Unzipping carbon nanotubes	S/M	High	RGO	Low	Carbon nanotubes

Note: S, single layer; M, multilayer; G, graphene; RGO, reduced graphene oxide.

8.2.3 Synthesis of Graphene-Based Composite Materials

Graphene with high surface area and high conductivity suggests its use as a significant support for metal oxides and conducting polymers. The incorporation of the second component is the key procedure for preparing graphene-based composite materials. However, pristine graphene usually has a poor solubility in both polar and apolar solvents. Therefore, special attention is focused on the process of inducing the second component to avoid aggregation of graphene sheets. GO with good solubility is an important precursor for the in-situ synthesis of graphene-based materials because a large amount of oxygenate functional groups on the surface of GO are efficient to offer the compatibility with the second component. A salt containing the metal ions is mixed with GO and then converted to the corresponding oxide or hydroxide via hydrolysis or redox reaction, forming a GO/metal compound composite. The subsequent reduction of GO leads to the formation of RGO-based composite materials. For example, $FeCl_2/FeCl_3$ is added to a GO solution. Ions of Fe^{2+} and Fe^{3+} are easily adsorbed on the surface of GO via an electrostatic interaction. Then, ammonia is added quickly to convert the ions to Fe_3O_4 nanoparticles. Finally, GO/Fe_3O_4 composite is reduced by hydrazine, forming a RGO/Fe_3O_4 composite [95]. A similar process was used to prepare GO/MnO_2 via the oxidation of adsorbed Mn^{2+} on GO surface [96]. However, the subsequent reduction was inefficient for the preparation of RGO/MnO$_2$ because MnO$_2$ would get dissolved away in the reduction process. A facile one-step approach to producing RGO-metal oxide composite materials by using GO as a reactant has been developed by several groups [97–101]. UV-assisted photocatalytic reduction and hydrothermal treatment of GO were proven to be useful to prepare RGO–metal oxide composite materials [97, 98]. Cao et al. reported the preparation of RGO–CdS nanocomposite by a one-step method. In this method, GO is reduced solvothermally to RGO in dimethyl sulfoxide (DMSO) at 180°C. In the presence of cadmium acetate, DMSO, as a source of sulfur, results in the formation of CdS quantum dots on the surface of RGO [99]. As demonstrated in Fig. 8.10, GO sheets can be reduced by Sn^{2+} or Ti^{3+} ions to in situ form RGO–SnO$_2$ and RGO–TiO$_2$ composite materials consisting of RGO and metal oxide nanoparticles. During this redox reaction, GO is reduced to RGO while Sn^{2+} and Ti^{3+} are oxidized and hydrolyzed to SnO$_2$ and TiO$_2$, depositing on the surface of RGO [100, 101]. The present method offers several advantages over the previously reported ones, including (i) an extra reducing agent, such as toxic hydrazine, is not required for the reduction of GO to RGO, (ii) in situ growth of metal oxides leads to the formation of uniform nanoparticles on individual RGO sheets, and (iii) the process can be carried out under mild conditions.

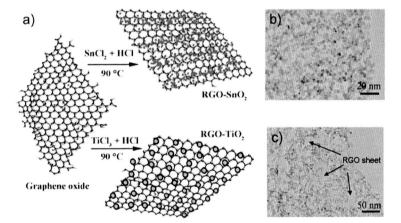

Figure 8.10 A schematic illustration showing the preparation of samples RGO–SnO$_2$ and RGO–TiO$_2$ (a). TEM images of RGO–SnO$_2$ (b) and RGO–TiO$_2$ (c). Zhang, J., Xiong, Z., Zhao, X. S. (2011). Graphene–metal–oxide composites for the degradation of dyes under visible light irradiation, *J. Mater. Chem.*, **21**, pp. 3634–3640]. Reproduced by permission of The Royal Society of Chemistry.

The influence of oxidation degree of graphene sheets on the nanocrystal gro(wth process was investigated recently [102]. CMG made by exfoliation reintercalation-expansion method (GS, oxygen content: ~5%) and GO (with oxygen content: ~20%) were used as model materials. Ni(OH)$_2$ nanoparticles were coated onto both GS and GO sheets by hydrolyzing Ni(CH$_3$COO)$_2$ at 80°C in a DMF/water mixed solvent. After a hydrothermal reaction at 180°C, these small particles on GS sheets diffused across graphite lattice and recrystallized into single-crystalline nanoplates. However, this phenomenon is not observed in the GO/Ni(OH)$_2$ composite (Fig. 8.11) [102]. These results imply that the strong interaction between GO with larger content of oxygen and metal oxide nanoparticles possibly prevent latter from recrystallizing to single-crystalline structure.

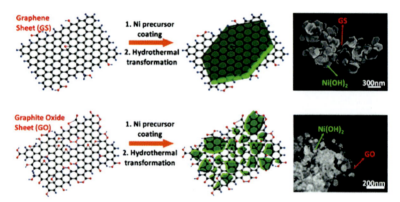

Figure 8.11 Schematic illustration of Ni(OH)$_2$ nanocrystal growth on (top) RGO sheets (GS) and (bottom) GO. Dark gray balls, C atoms; blue balls, H atoms; red balls, O atoms; green plates, Ni(OH)$_2$, remained as densely packed nanoparticles pinned by the functional groups and defects on the GO surface. Reprinted with permission from Wang, H., Robinson, J. T., Diankov, G., Dai, H. (2010). Nanocrystal growth on graphene with various degrees of oxidation, *J. Am. Che. Soc.*, **132**, pp. 3270–3271. Copyright @ American Chemical Society.

Chemical modification is an effective technique to increase the processibility of graphene as well as the compatibility and interfacial bonding force in the composite. Dai and co-workers investigated the atomic layer deposition (ALD) of metal oxide on pristine and functionalized graphene [103]. For pristine graphene, ALD coating was only actively grown on the edges and defect sites, where dangling bonds or surface groups reacted with ALD precursors which afforded a simple method to decorate and probe single defect sites in graphene planes. The functionalization of graphene surface with perylene tetracarboxylic acid (PTCA) led to the densely packed surface groups on graphene, resulting in a uniform ultrathin ALD coating on the modified graphene over a large area. The proof-study demonstrates that the surface modification of graphene sheets isimportant for preparing uniform graphene-based composite materials. Anionic sulfate surfactants have been used to stabilize RGO sheets aqueous suspensions and facilitate the self-assembly of in situ grown TiO_2 nanocrystals with RGO sheets [104]. The use of surfactants not only addresses the hydrophobic/hydrophilic incompatibility problem, but also provides molecular templates for controlled nucleation and growth of nanostructured inorganics (Fig. 8.12). An approach akin to the pioneering studies was based on the ternary self-assembly of surfactant micelles on RGO sheets and their hybrid nanostructures binding with metal cations to form an ordered nanocomposite. After removal of surfactants, metal oxides were crystallized between RGO sheets, producing a new class of nanocomposites in which alternating layers of RGO sheets and metal oxide nanocrystals were assembled into layered nanostructures. Alternatively, the structures of metal oxides would be controlled by self-assembly with various surfactants [105]. The robust method led to the formation of a new class of layered nanocomposites containing stable, ordered alternating layers of nanocrystalline metal oxides, such as SnO_2, NiO, and MnO_2 and RGO sheets.

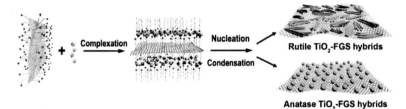

Figure 8.12 Anionic sulfate surfactant mediated stabilization of graphene and growth of self-assembled TiO_2–FGS hybrid nanostructures. Reprinted with permission from Wang, D., Choi, D., Li, J., Yang, Z., Nie, Z., Kou, R., Hu, D., Wang, C., Saraf, L. V., Zhang, J., Aksay, I. A., Liu, J. (2009). Self-assembled TiO_2–graphene hybrid nanostructures for enhanced Li-ion insertion, *ACS Nano*, **3**, pp. 907–914]. Copyright @ American Chemical Society.

A variety of insulating polymers including polystyrene (PS) [106], poly(vinyl alcohol) (PVA) [107, 108], poly(sodium 4-styrensulfonate) (PSSNa) [109], poly(acrylic acid) (PAA) [110], etc. have been incorporated into RGO or GO to form composite materials. However, conducting polymers (CP) are quite different from the insulating polymers. The conjugated backbones of CP provide them with unique electrical and optical properties. It is noteworthy that CPs are conductive in their doped states while insulating in their neutral states. Furthermore, CPs are usually brittle and weak in mechanical strength, which is a big drawback for supercapacitor applications. The incorporation of CMG into CP is attractive for combining the properties of both the components to improve the properties of resulting composites. For the electrochemical applications, most published works were focused on composite materials of CMG and CPs (CMG/CPs) including polyaniline (PANi) [111–113], polypyrrole (PPy) [114–116], and poly(3,4-ethylenedioxythiophene) (PEDOT) [117]. In situ polymerization is the most widely applied method for preparing CMG/CP composites. The basic process is to polymerize CP monomers in RGO or GO dispersion with an oxygenating agent [118]. In order to produce doped CPs with good conductivity for electrochemical applications, the polymerization would be processed by adjusting the pH values of the reaction systems carefully.

The poor solubility of CPs, such as PEDOT, is the dominant factor for limiting the application of in situ polymerization method. Recently, Xu et al. [117] synthesized sulfonated RGO sheets (SRGO). They claimed that the sulfonate groups would increase the solubility of RGO sheets and act as dopants of PEDOT. Therefore, the polymerization of 3,4-ethylenedioxythiophene (EDOT) in SRGO suspensions led to the formation of SRGO/PEDOT composites. Electrochemical polymerization has been employed to synthesize RGO/CP composites. For example, PANi and PPY were deposited onto RGO films by the electrochemical polymerization of aniline and pyrrole monomers, respectively [111, 114]. The superior feature of electrochemical polymerization is that the polymerization process is easily controllable by adjusting the applied potential, current as well as polymerization time without using an additional oxidant. However, difficulty in a large-scale synthesis is the main drawback because polymerization only occurs on the electrode surface.

8.3 Graphene-Based Materials as Supercapacitor Electrodes

8.3.1 Energy Storage in Supercapacitor

Supercapacitor can store energy by either ion adsorption at the electrode–electrolyte interface (electrical double layer capacitors, EDLC) or fast and reversible faradic reactions (pseudocapacitors) [27]. Both mechanisms can sometimes function simultaneously depending on the nature of electrode materials. EDLC is based on the theory of electrical double layer (EDL), which is only an organization of charges at the interface of conductor and electrolyte formed by the electrostatic attraction. Helmholtz firstly proposed the model of EDL in 1853 [119]. The model was further developed by Gouy and Chapman, proposing that the two charged layers with opposite charges built up in the interface between the electrode and the electrolyte. Later, Stern combined the Helmholtz model with the Gouy–Chapman model to explicitly recognize two layers of ion distribution: the compact layer (also named Stern layer) and the diffuse layer (Fig. 8.13). The double-layer capacitance is made of contributions from the compact layer and the diffuse layer [18]. For the EDL type of supercapacitor, the specific capacitance, C (F g^{-1}), of each electrode is generally assumed to follow that of a parallel-plate capacitor:

$$C = \frac{\varepsilon_r \varepsilon_0}{d} A. \tag{8.1}$$

in which ε_r (a dimensionless constant) is the relative permittivity, ε_0 (F m^{-1}) is the permittivity of a vacuum, A (m^2 g^{-1}) is the specific surface area accessible to the electrolyte ion, and d (m) is the effective thickness of EDL (also known as Debye length). The nature of EDLC is the charge accumulation on the surface of the electrode materials. Therefore, the high surface area of active materials with a good electrical conductivity is the fundamental issue to obtain high capacitances. Activated carbons (ACs) are extensively investigated as electrode materials for EDL capacitors. It is traditionally believed that the specific capacitance of AC is determined by the accessible surface area because many pores in carbon materials are smaller than 0.5 nm and not accessible to hydrate ions (0.6–0.76 nm) [120]. Recent studies have demonstrated an anomalous increase in capacitance for pore sizes below 1 nm because the partial desolvation of hydrated ions occurs in micropores [121, 122]. However, the phenomenon is only observed on some special kind of carbon materials. Therefore, further theoretical and experimental studies on the in-depth understanding of the EDL charge storage mechanism in the nano-confined spaces would be helpful to design novel carbon materials with fine pore-size control and dimensional control for supercapacitor applications.

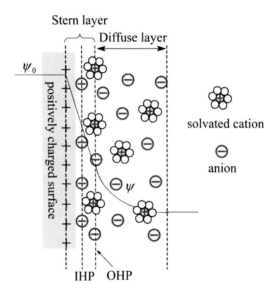

Figure 8.13 The EDL structure based on Stern model formed at a positively charge porous electrode surface. The IHP refers to the distance of closest approach of specially adsorbed ions (generally anions) and OHP refers to that of the non-specially adsorbed ions. The OHP is also the plane where the diffuse layer begins. Zhang, L. L., Zhao, X. S. (2009). Carbon-based materials as supercapacitor electrodes, *Chem. Soc. Rev.*, **38**, pp. 2520–2531. Reproduced by permission of The Royal Society of Chemistry.

Different from EDLCs, the pseudocapacitance is faradic in origin, involving fast and reversible redox reactions of electro-active species at or near the electrode surface. The promising materials providing a pseudocapacitive property are metal oxides such as hydrous ruthenium oxide [123, 124], manganese oxide [125, 126], and conducting polymers including PANi, PPy, and PEDOT [127]. Hydrous ruthenium oxide (RuO$_2$ ·xH$_2$O) is a typical example of a material giving pseudocapacitive property [123]. As demonstrated in Fig. 8.14, the charge storage process is a reversible redox reaction of ruthenium oxide involving insertion and extraction of protons at or near the electrode surface according to the following reaction [128]:

$$Ru^{IV}O_2 + \delta H^+ + \delta e^- \Leftrightarrow Ru^{IV}_{1-\delta}Ru^{III}_{\delta}O_2H_{\delta}. \tag{8.2}$$

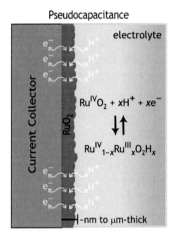

Figure 8.14 Schematic representation of charge storage via the process of pseudocapacitance. Long, J. W., Bélanger, D., Brousse, T., Sugimoto, W., Sassin, M. B., Crosnier, O. (2011). Asymmetric electrochemical capacitors-Stretching the limits of aqueous electrolytes, *MRS Bulletin*, **36**, pp. 513–522]. Reprinted with permission.

In a proton-rich electrolyte (e.g., H$_2$SO$_4$), the faradic charges can be reversibly stored and delivered through the redox transitions of the oxyruthenium groups,

i.e., Ru(IV)/Ru(III). On the basis of the mean electron transfer numbers, the theoretical specific capacitance of RuO$_2$·xH$_2$O is estimated to range from ca. 1300 to 2200 F g^{-1}. A very high specific capacitance of 1300 F g^{-1} was reported for a pure bulk nanostructured RuO$_2$·xH$_2$O electrode [129]. The high cost of noble metal oxides and the difficulties in large scale production are the major problems

Raman and Infrared Spectroscopic Characterization of Graphene

for their practical applications. Therefore, great efforts have been focused on reducing the mass of noble metal oxide and/or improving their utilization by synthesizing composite materials and optimal structures. Manganese oxides have a high theoretical capacitance of about 1370 F g^{-1} [125]. Additional advantages in low cost, low environmental impact, and safety have made them promising electrode materials. Nonetheless, manganese oxide has some inherent shortcomings, including the poor electronic conductivity and a slow redox reaction that limit the power capability. However, the nanoscale manganese oxide coatings on conductive substrates, such as graphene, would benefit to the high redox utilization and facile kinetics. Thus, a number of carbon materials, including carbon nanotubes, nanofoams, and graphene have been used for the configuration of manganese oxide modified architectures, exhibiting enhanced performances in comparison with the pure manganese oxides.

8.3.2 Basic Principles and Techniques for Evaluation of Supercapacitor

To evaluate the capacitive performances, cyclic voltammetry (CV), galvanostatic charge/discharge (GCD), and electrochemical impedance spectroscopy (EIS) are the most common techniques used. A CV curve generally presents a rectangular shape when the capacitance simply originates from EDL and there are no Faradaic reactions between the active materials and the eletrolyte. The specific capacitance (C_{sp}), capacitance per unit mass, is estimated from the current at the middle point of potential range (I) and scan rate (v) according to the equation $C_{sp} = I/mv$, where m is the mass of active material [29]. The pesucocapacitive behaviors usually lead to the presence of redox peaks with a derivation from rectangle shape. Thus, the average capacitance is calculated using the voltammetric charge integrated from CV curves according to Eq. (8.3) [22, 130]:

$$C_{sp}\,(F/g) = \frac{Q}{2mV} = \frac{1}{2mVv} \int_{V_-}^{V_+} I(V)dV. \tag{8.3}$$

in which Q is the total charge obtained by the integration of positive and negative sweeps in a CV curve, m is the mass of the active material in the two electrodes, v the scan rate, and $(V = V_+ - V_-)$ represents the potential window. For a two-electrode supercapacitor, two electrodes are set across a separator, and the potential difference between the electrodes is monitored and controlled. Each electrode/electrolyte interface represents a capacitor in parallel with a resistor. For this reason, the complete cell is considered as two capacitors in series. Therefore, the specific capacitance (C_T) for a two-electrode cell is one-fourth of the capacitance of single electrode measured in a three-electrode system, theoretically.

For the GCD technique, the potential is generally linear response to the charge/discharge time (dV/dt = constant) during a constant current operation so that the state-of-charge (SOC) can be exactly pinpointed. In contrast, most batteries exhibit a relatively constant operating voltage because of the thermodynamics of the reactants. As a result, the energy E stored in the battery is proportional to the voltage V, whereas the energy stored in the supercapacitor is proportional to the square of voltage squared [131–133]. On the basis of the GCD curve, the specific capacitance (C_{sp}) can be calculated according to Eq. (8.4)

$$C_{sp} = \frac{I}{mdV/dt}, \tag{8.4}$$

where I (A) is the discharge current, m (g) is the mass of active materials, t (s) is the discharge time, and V is the potential during the discharge process after IR drop. Hence, dV/dt is the slope of discharge curve. Recently, Ruoff and co-workers [134] recommended to calculate the specific capacitance by using two datum points from the discharge curve with $dV/dt = (V_{max}-\frac{1}{2}V_{max})/(t_2-t_1)$, especially for the case of nonlinear response between potential and time resulted from the pseudocapacitive reaction. Here, t_2 and t_1 are the discharge times at the points of maximum potential (V_{max}) and half of the voltage.

Energy density and power density are two important parameters to evaluate the performance of a supercapacitor device. Energy density is the capacity to do work whereas power density is how fast the energy is delivered. The standard way to obtain energy density and power density is based on the specific capacitance (C_T) of a two-electrode system, although the capacitance obtained from a three-electrode system is used to calculate the specific energy density and power density. The three-electrode system will largely overestimate the performance of electrode materials [134]. For example, the equivalent series resistance (ESR) of a two-electrode system is totally different from that of a three-electrode system. The maximum energy stored (E_{max}, Wh kg^{-1}) and power delivered (P_{max}, W kg^{-1}) for such a two-electrode supercapacitor are given in Eqs. (8.5) and (8.6), respectively [27, 134]:

$$E_{max} = \frac{1}{2}C_T V^2 \qquad (8.5)$$

$$P_{max} = \frac{V^2}{4R_s}, \qquad (8.6)$$

where V is the cell voltage (in V). The cell voltage is determined by the thermodynamic stability of the electrolyte solution. The specific capacitance of a supercapacitor depends extensively upon the electrode materials. ESR comes from various types of resistance associated with the intrinsic electronic properties of the electrode matrix and electrolyte solution, mass transfer resistance of ions in the matrix, contact resistance between current collector and electrode. Hence, for a supercapacitor to perform well, it must simultaneously satisfy the requirement of having large capacitance value, high operating cell voltage, and minimum ESR. In the fundamental research, the transformed equations are usually used to calculate the maximum energy density and power density, respectively [135].

$$E_{max} = \frac{0.5C_T V^2}{3.6}. \qquad (8.7)$$

$$P_{max} = \frac{E_{max} \times 3600}{t}. \qquad (8.8)$$

Electrochemical impedance spectroscopy (EIS) is a powerful tool to evaluate the supercapacitor frequency behavior and ESR. Normally, EIS is conducted at the open-circuit voltage (OCV) by applying a small amplitude of alternative potential (5–10 mV) in a range of frequency (generally 0.01–10000 Hz). The resistance (Z) is defined as $Z=Z'+jZ''$, where Z' and Z'' are the real part and the imaginary part of the impedance, respectively. The capacitance is calculated from the imaginary part (Z'') of the collected EIS data according to the following equation:

$$C = \frac{-1}{2\pi f Z'' m}, \qquad (8.9)$$

where f (in Hz) is the frequency, and m is the mass of electrode materials.

The impedance can be modeled as a function of angular frequency (ω) which is equal to $2\pi f$ [136, 137].

$$Z(\omega) = Z'(\omega) + jZ''(\omega) = \frac{1}{j\omega C(\omega)}. \tag{8.10}$$

Therefore, the capacitance can be written as a function of angular frequency:

$$C(\omega) = \frac{-Z''(\omega)}{\omega|Z(\omega)|^2} - j\frac{Z'(\omega)}{\omega|Z(\omega)|^2} = C'(\omega) - jC''(\omega). \tag{8.11}$$

leading to

$$C'(\omega) = \frac{-Z''(\omega)}{\omega|Z(\omega)|^2} \quad and \quad C''(\omega) = \frac{Z'(\omega)}{\omega|Z(\omega)|^2}. \tag{8.12}$$

Figure 8.15a presents a typical example showing the real part of capacitance ($C(\omega)'$) change vs. frequency. When the frequency decreases, $C(\omega)'$ sharply increases, and then tends to be less frequency dependent. The low frequency value of $C(\omega)'$ corresponds to the capacitance of a supercapacitor that is measured during constant-current discharge. Figure 8.15b presents the evolution of imaginary part of the capacitance ($C(\omega)''$) vs. frequency. The imaginary part of capacitance goes through a maximum at a frequency f_0, defining a time constant as $t_0 = 1/f_0$. The time constant is described as a characteristic relaxation time of the whole system (the minimum time needed to discharge all the energy from the device with an efficiency of greater than 50%). Thus, the smaller value indicates the higher rate capability [137, 138].

The other important form for EIS is to plot the real part of impedance (Z') against the imaginary part (Z''), named Nyquist diagram. Figure 8.15c shows a typical EIS (Nyquist plot) recorded in two-electrode cell using AC as both electrode materials. The impedance spectrum exhibits a semicircle over the high frequency range, followed by a linear part in the low frequency region. It should be noted that the large semicircle observed is indicative of a high charge-transfer resistance, contributing to the poor electrical conductivity of materials, whereas a more vertical line corresponds to an electrode close to an ideal capacitor [137]. The quantitative data for these parameters can be obtained by fitting the impedance spectra using the electrical equivalent circuit in Fig.16c [139]. In this

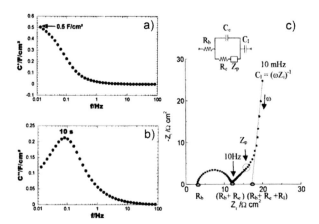

Figure 8.15 Evolution of the real part (a) and imaginary capacitance (b) vs. frequency for 4 cm² cell assembled with two electrodes containing 15 mg cm⁻² of activated carbon in acetonitrile (AN) with tetraethylammonium tetrafluoroborate (Et₄NBF₄). (Taberna, P. L., Simon, P., Fauvarque, J. F. (2003). Electrochemical characteristics and impedance spectroscopy studies of carbon-carbon supercapacitors, *J. Electrochem. Soc.*, **150**, pp. A292-A300. Reproduced by the permission of the Electrochemical Society.) A Nyquist plot of EIS (10 mHz to 10 kHz) recorded in two-electrode mode and the equivalent circuit for impedance analysis (c) [139]. (Fabio, A. D., Giorgi, A., Mastragostino, M., Soavi, F. (2001). Carbon-poly(3-methylthiophene) hybrid supercapacitors, *J. Electrochem. Soc.*, **148**, pp. A845–A850]. Reproduced by the permission of the Electrochemical Society.)

circuit, C_L is the limit capacitance and Z_p is the Warburg impedance. The double-layer capacitance (C_e) is usually substituted with the constant phase elements (CPEs) in order to better fit the high-frequency capacitive loop.

$$CPE = \frac{1}{Q(j\omega)^n}. \tag{8.13}$$

Here, Q is defined as the frequency independent constant relating to the surface electroactive properties and ω is the angular frequency. The value of exponent n varies in the range of -1 to 1. When $n = -1$, CPE represents an inductor; when $n = 0$, it represents a pure resistor, whereas a pure capacitor is represented with a phase angle of $-90°$ at $n = 1$; CPE corresponds to the Warburg impedance at $n = 0.5$ [140].

Generally, R_b is the combinational resistance of ionic resistance of electrolyte, intrinsic resistance of substrate, and contact resistance at the active material/current collector interface [141]. R_e is the charge-transfer resistance caused by double-layer capacitance on particle surfaces. The sum of R_b and R_e is the main contributor to ESR, limiting the specific power of a supercapacitor. In the presence of a pesudocapacitive material, Faradic reactions also contribute to the resistance. *It's worth noting that the analysis and understanding of EIS should be conducted carefully on a case-by-case basis, especially for the pseudocapacitive materials with complex kinetics of electrode process because a perfect semicircle is not obtained usually* [142, 143].

Alternatively, ESR values are usually determined from a linear fit to the IR drop values (IR_{drop}) obtained from GCD curves at different current densities [94, 144, 145].

$$IR_{drop} = a + bI. \tag{8.14}$$

in which a represents the difference between the applied potential and the charged potential of a capacitor, b is twice of the value of ESR (R_s), and I is the discharge current. The maximum power density follows:

$$P_{max} = \frac{V^2}{4R_s} = \frac{(4-a)^2}{2b}. \tag{8.15}$$

On the basis of the above discussion on the various techniques, we have proposed the following recommendations for the evaluation of capacitive performances.

1. The mass loading of test electrodes should be comparable. While extremely thin films containing minute amount of the material are preferable for investigating the intrinsic capacitance, such a low mass loading may potentially overestimate the capacitive performances with respect to both energy and power densities. If thin films have to be used, it is more recommended to evaluate the performance by considering the geometrical area instead of the gravimetric value.

2. When a specific capacitance is reported, the scan rate for CV or current density for GCD should be specified. This is because the specific capacitance obtained from CV or GCD data varies with scan rates or current densities. A high specific capacitance may be obtained at a low scan rate or current density, but the rate capability is poor under these experimental conditions.

3. A two-electrode cell is essential to estimate the energy density, the power density, and the cycle life of a supercapacitor cell. Three-electrode system would overestimate the energy and power densities, resulting from the overstatement of specific capacitance and underestimated ESR.

8.3.3 Graphene and Reduced Graphene Oxide as Supercapacitor Electrodes

Progress towards supercapacitor technologies can benefit from the significant advancements of nanostructured electrode materials. Graphene is predicted to be a promising electrode material for supercapacitors because of its high electrical conductivity, high surface area, great flexibility, and excellent mechanical properties. When used as electrodes for supercapacitors, physically separated graphene sheets that were vertically grown on a metal substrate showed an exceptional frequency response [146]. Zhao et al. [147] employed the CVD method to synthesize carbon nanosheets composed of 1–7 graphene layers on carbon fibers and carbon papers, respectively (Fig. 8.16). The outer surfaces of extended graphene sheets were exposed to the electrolyte and made available for forming electrical double layers (Fig. 8.16b). It was found that such carbon nanosheets yielded a capacitance value of 0.076 F cm^{-2} based on the geometric testing area in a H_2SO_4 solution. A total capacitance was estimated to be 1.49×10^4 F according to a virtual supercapacitor device rolled in a sandwich pad with a given dimension (Fig. 8.16c). Despite the intense interests and continuously growing number of publications, practical applications of the pristine graphene sheets have not yet been explored. This is mainly due to the difficulty in the production of high-quality graphene in a large-scale. The performance of pristine graphene sheets is yet to be significantly improved to compete with the traditional porous carbon materials in energy storage devices. To harness the excellent properties of graphene for macroscopic applications, both large-scale synthesis and integration of graphene sheets with single- and few-layer to advanced multifunctional structures are required.

While CVD and mechanical exfoliation produce graphene with the highest quality, the methods are neither high throughput nor high yield. RGO can be prepared in large-scale and at relatively low costs through chemical conversion from GO. RGO sheets can partly restore

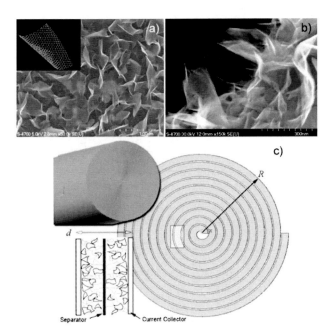

Figure 8.16 (a) SEM image of carbon nanosheets (top view). The inset figure shows a schematic diagram of a single graphene sheet. (b) SEM image of carbon nanosheets (cross-section view) shows carbon nanosheets about 0.6 μm tall and less than 1 nm thick. (c) A virtual supercapacitor cell containing carbon nanosheets as the electrode material. A rolled sandwiched-pad forms the supercapacitor. The sandwich-pad contains two conductive electrodes as current collectors. It has one insulating layer as an ion permeable separator. Carbon nanosheets are filled in as electrode material. Left corner inset shows the cross-section schematic of the pad (screening zone in the middle of the figure). Reprinted from Zhao, X., Tian, H., Zhu, M., Tian, K., Wang, J. J., Kang, F. and Outlaw, R. A. (2009). Carbon nanosheets as the electrode material in supercapacitors, *J. Power Sources*, **194**, pp. 1208–1212. Copyright @ 2009, with permission from Elsevier.

the structure as well as the conductivity of graphene. Additionally, RGO sheets can adjust themselves to be accessible to different types of electrolyte ions, which enable technical applications in a variety of fields, such as energy conversion and storage devices. RGO sheets have been one of the most widely used electrode materials for supercapacitors due to their facile preparation process, large productivity, low cost, and potential for functionalization.

The initial studies [148] have shown that the specific capacitances of RGO can reach 135 F g^{-1}, 99 F g^{-1}, and 75 F g^{-1} in aqueous, organic, and ionic liquid electrolytes, respectively. A coin-size symmetric supercapacitor has been fabricated with RGO sheets prepared by using a gas-based hydrazine reduction, giving remarkable results in terms of specific capacitance of 205 F g^{-1}, energy density of 28.5 Wh kg^{-1}, and power density of 10 kW kg^{-1} [149], significantly higher than those of CNT-based supercapacitors [150]. A similar supercapacitor using curved RGO sheets as electrodes was fabricated in an ionic liquid electrolyte. The curved morphology of RGO sheets enabled the formation of mesopores accessible to ionic liquids with a capable operating voltage >4 V, benefiting to the high performance with respect to a specific energy density of 85.6 kW kg^{-1} at room temperature and 136 kW kg^{-1} at 80°C. The high energy densities are comparable to that of Ni metal hydride battery [7]. The key point to success is the ability to make full use of the intrinsic surface by preparing curved RGO sheets that would not restack face-to-face, demonstrating that the facile strategy to achieve capacitive performance in graphene-based materials is to find an effective way to prevent graphene sheets from sticking to one another.

RGO powers obtained by thermal reduction exhibited a moderated capacitance of 117 F g^{-1}, resulting from the large proportion of inaccessible surface area [151]. Several research groups have modified the process of thermal exfoliation to synthesize RGO sheets for supercapacitors [152, 153]. The RGO sheets exhibited improved capacitance as high as 230 F g^{-1} by controlled thermal exfoliation at a low temperature (300°C) [153]. Vacuum-promoted exfoliation process resulted in the formation of RGO sheets at a low temperature of 200°C. The specific capacitance as high as 279 F g^{-1} has been achieved at a scan rate of 10 mV s^{-1}, which is much larger than that of high-temperature exfoliated samples [89]. These varied electrochemical performances would be ascribed to the different surface chemistry of RGO sheets in terms of oxygen content and specific surface area. With the assistance of microwave heating, the exfoliation and reduction of graphite oxide was achieved within 1 min. When used as electrode materials in a supercapacitor cell, specific capacitance of 191 F g^{-1} was demonstrated in a KOH electrolyte. Recently, improvement was achieved by chemical activation of RGO sheets [154]. The simple activation of microwave exfoliated graphite oxide resulted in the formation of a porous carbon with continuous 3D network structure. Surprisingly, the BET surface area of the porous carbon was up to 3100 m^2 g^{-1}, even larger than the theoretical value of graphene sheets. A packaged supercapacitor device using the activated RGO sheets as both electrodes exhibited an energy density of above 20 Wh kg^{-1} in organic electrolyte, which is four times higher than the AC-based supercapacitors, and nearly equal to that of the lead acid batteries [155]. After 10,000 GCD cycles at a current density of 2.5 A g^{-1}, 97% of its initial specific capacitance was retained. The excellent performance opens the possibility to produce supercapacitor electrodes based on this form of RGO, aimed to target a wide range of applications. Moreover, the facile processes used to prepare the RGO electrode material are readily scalable to industrial levels.

A self-assembled RGO hydrogel (RGOH) was prepared by a convenient one-step hydrothermal method. The RGOH had a 3D network consisting of ultrathin RGO walls and cross-linking sites formed by regional π–π stacking of graphene sheets. The unique structure endows the RGOH with high mechanical strength and conductivity. The specific capacitances of RGOH at scan rates of 10 and 20 mV s^{-1} were calculated to be 175 and 152 F g^{-1}, respectively, much higher than that of the supercapacitor based on RGO agglomerate particles tested under the same condition (100 F g^{-1}, scan rate = 20 mV s^{-1}).

With the advent of atomically thin and flat layers of graphene, new designs for thin film energy storage devices with good performance have become possible. An "in-plane" fabrication approach for ultrathin supercapacitors based on electrode materials comprised of pristine graphene or RGO sheets has been developed [156]. In the case of stacked geometry, the electrochemical surface area is incompletely utilized, because some of the regions are inaccessible to the electrolyte ions (Fig. 8.17a). In contrast, the in-plane design offers effective routes for the electrolyte ions to enhance interaction with all the graphene layers (Fig. 8.17b), leading to a full utilization of their high surface area. All solid-state supercapacitors with the 2D in-plane structure were fabricated by using a polymer-gel (PVA-H_3PO_4) electrolyte over the RGO thin films. The favorable in-plane design exhibited a high specific capacitance up to 247.3 F g^{-1} (390 μF cm^{-2}). The mass of RGO electrode (about 0.283 μg) is very minute, resulting in the difficulty in the accurate evaluation. Therefore, it is more reasonable to evaluate the performance by geometrical area instead of the gravimetric value.

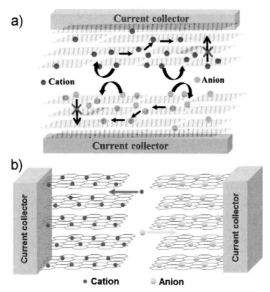

Figure 8.17 Schematic depiction of the stacked geometry used for the fabrication of supercapacitor devices (a) and the operating principle in case of the in-plane supercapacitor device (b). Reprinted with permission from Yoo, J. J., Balakrishnan, K., Huang, J., Meunier, V., Sumpter, B. G., Srivastava, A., Conway, M., Mohana Reddy, A. L., Yu, J., Vajtai, R., Ajayan, P. M. (2011). Ultrathin planar graphene supercapacitors, *Nano Lett.*, **11**, pp. 1423–1427. Copyright @ American Chemical Society.

The effect of nitrogen doping on the capacitive performances of RGO sheets was recently evaluated by fabricating symmetric supercapacitors. Using nitrogen doped RGO sheets produced by a simple plasma process as electrode materials, the supercapacitors yielded a specific capacitance of about 282 F g^{-1}, about 4 times larger than those of the control case with pristine RGO (69 F g^{-1}). Moreover, the supercapacitors demonstrated excellent cycle life (>200000), high power capability (~80 kW kg^{-1}) and energy density (~48 Wh kg^{-1}). According to the integrated analysis using X-ray photoelectron spectroscopy (XPS) in the microscopic and bulk scale resolutions and the ionic binding energy calculation, especially, the scanning photoemission microscopy with a capability of probing local nitrogen–carbon bonding configurations within a single sheet of RGO, the origin of improved capacitance is ascribed to a certain N-configuration at basal planes [157]. This work is highly significant in that it successfully revealed the N-doped sites (at basal planes) for the improved performances, which is useful to guide the design of novel graphene-based materials.

A recent study revealed that GO exhibited higher capacitance (up to 189 F g^{-1}) than RGO sheets. They claimed that an additional pseudo-capacitance effect of the attached oxygen-containing functional groups on GO basal planes was ascribed to the enhanced capacitive performance [158]. The conclusion however was made according to the samples prepared in the work only because the pseudocapacitance and the electrical conductivity of RGO sensitively depended on the oxygen content and the π-conjugated structure of graphene [159]. Our previous study has revealed that the atomic ratio of O/C of RGO sheets has significant impact on their capacitive performances [160]. A specific capacitance as high as 218 F g^{-1} for the RGO electrode with about 10.0% O/C ratio was achieved. A thermal treatment resulted in a dramatic reduction of O/C ratio to 3.5%, its surface area slightly decreased from 801 to 787 m^2 g^{-1}. However, the specific capacitance of RGO electrode significantly decreased to 95 F g^{-1} after the thermal treatment. These findings help to explain the previously reported unusual capacitance of GO sheets. Thus, the formation of partially oxidized graphene sheets with optimal oxygen content is likely the cause for the improved pseudocapacitive performances. For example, a graphite oxide consisted of multilayered graphite flakes, which were oxidized on the outer layers while the inner layers consisted of pristine or mildly oxidized graphene sheets. The exfoliation of this kind graphite oxide produced conductive graphene sheets. The percentage of C–C bonds increased from 55% after 1 h to 81% after 2 days. The increase in the C–C component and the corresponding decrease in the C–O (epoxide, ether, and hydroxyl groups) and C=O (carbonyl and carboxyl) components with reaction time indicated the generation of large domains of π-conjugated structures. The heating of graphite oxide in dimethylformamide (DMF) might result in the reduction of GO to some extent during the exfoliation process. Nonetheless, the exfoliation of the less oxidized inner sheets in which the oxygen functionalities were situated mainly at the graphene edge planes led to the appreciable conductivity of the product [85]. Recent report demonstrated that graphite oxide intercalated with an insulting polymer, poly(sodium 4-styrensulfonate) (PSS–GO) showed a high performance of EDLC compared to that of the pristine graphite oxide (Fig. 8.18) [109]. Specific capacitance of the PSS–GO reached 190 F

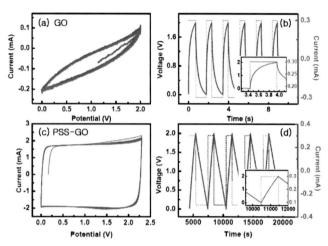

Figure 8.18 Cyclic voltammetry and galvanostatic charge/discharge curves of graphite oxide (a, b) and PSS–GO (c, d). The scan rate was 10 mV s^{-1} and cycles of 10 times in CV, and the current density of charge–discharge curves was 0.1 A g^{-1} with a discharge voltage of 2.0 V. Reprinted with permission from Jeong, H.-K., Jin, M., Ra, E. J., Sheem, K. Y., Han, G. H., Arepalli, S., Lee, Y. H. (2010). Enhanced electric double layer capacitance of graphite oxide intercalated by poly(sodium 4-styrensulfonate) with high cycle stability, *ACS Nano*, **4**, pp. 1162–1166]. Copyright @ American Chemical Society.

g^{-1}, and the energy density was much improved to 38 Wh kg^{-1} with a power density of 61 W kg^{-1}. Cycle test showed that the specific capacitance decreased by only 12% after 14860 cycles, exhibiting excellent cyclic stability.

The graphite oxide used in this study consisted of multilayered graphite flakes, which were oxidized on the outer layers while the inner layers consisted of pristine or mildly oxidized graphene sheets. Therefore, the intercalation of polymer into GO sheets led to the wide interlayer distance and simple pore structures accommodating fast ion kinetics, resulting in the high EDLC performance of PSS–GO composites.

Table 8.2 summarizes the basic characteristics and capacitive performances of RGO prepared by various methods. Since the pseudocapacitance and the electrical conductivity of RGO sheets depend sensitively on the content of the oxygenate groups and the domains of π-conjugated structures of RGO sheets, the careful adjustment on surface chemistry of RGO would benefit to the capacitive performance of RGO-based materials for supercapacitor applications [13, 161, 162].

Table 8.2 A comparison of the capacitive properties of different CMG materials

Methods	C_{sp} (F g^{-1})	Electrode configuration	Electrolyte	SSA (m^2 g^{-1})	Reference
Microvave assisted exfoliation of Go	191	2E	KOH	463	[81]
Vacuum-promoted exfoliation of Go	279	2E	KOH	382	[89]
Chemical reduction of GO	135	2E	KOH	705	[148]
Chemical reduction of GO	205	2E	KOH	320	[149]
Thermal exfoliation of Go	117	2E	H_2SO_4	925	[151]
Hydrothermal reduction of Go	175	2E	KOH	-	[163]
Thermal exfoliated Go (300°C)	232	3E	KOH	404	[153]
Thermal exfoliation of Go (900°C)	91	3E	KOH	737	[153]

Note: Go: graphite oxide; GO: graphene oxide; 2E: two-electrode cell; 3E: three-electrode cell.

8.3.4 Graphene Based Composite Materials as Supercapacitor Electrodes

The agglomeration of RGO sheets not only decreases the surface area, but also increases the difficulty of ion diffusion to the inner surface of RGO sheets, deteriorating the capacitive performance of RGO-based materials. Thus, the experimentally observed capacitances are mainly limited by the agglomeration of graphene sheets and do not reflect the intrinsic capacitance of an individual graphene sheet. To minimize the restacking of RGO sheets, some attempts have been made to combine RGO with the second materials, such as CNTs, porous carbon, and carbon spheres. Zhao and co-workers have developed an approach to prepare three-dimensional (3D) carbon-based architectures consisting of mesoporous carbon spheres intercalated between RGO sheets [160]. In the preparation process, the electrostatic interaction between colloidally dispersed GO sheets and positively charged mesoporous silica spheres (MSS) led to the formation of a MSS–GO composites. The MSS were then used as templates for replicating mesoporous carbon spheres (MCS) via a CVD process, during which the GO sheets were reduced to RGO. Removal of the silica spheres left behind a 3D hierarchical porous carbon architecture with slightly crumpled RGO sheets intercalated with MCS. A symmetric supercapacitor was fabricated by using the 3D hierarchical porous carbon architecture as electrode materials, exhibiting a higher power density in comparison with that using the pristine RGO sheets and comparable energy density. Recently, a flexible RGO/multiwalled carbon nanotubes (RGO/MWCNTs) film was prepared by a flow-directed assembly from the complex dispersion of graphite oxide and pristine MWCNTs followed by the use of gas-based hydrazine to reduce the GO into RGO sheets. The MWCNTs in the obtained composite film not only efficiently increased the basal spacing but also bridged the defects for electron transfer between RGO sheets, increasing the electrolyte/electrode contact area and facilitating the transportation of electrolyte ion and electron into the inner region of electrode. The RGO/MWCNT film yielded a specific capacitance of 265 F g^{-1} at 0.1 A g^{-1} and displayed an excellent specific capacitance retention of 97% after 2000 continuous charge/discharge cycles [153]. The results indicated that the freestanding RGO/MWCNT film had a potential application in flexible energy storage devices. 3D CNT/RGO sandwich structures with CNT pillars grown in between the RGO layers have been prepared by CVD [164]. The unique structure endowed the high-rate transportation of electrolyte ions and electrons throughout the electrode matrix, resulting in a specific capacitance of 385 F g^{-1} at 10 mV s^{-1}. Notably, the presence of redox peaks of cobalt hydroxide suggested that the capacitance would be contributed to pseudocapacitance from cobalt hydroxide, which was from the conversion of cobalt catalysts used for CNT growth.

A flexible thin film was fabricated recently via vacuum filtration of the mixed dispersions of RGO sheets and PANi fibers, exhibiting a sandwiched layered structure with alternate PANi fibers and RGO sheets [165]. Supercapacitor device based on the conductive and flexible composite films exhibited a specific capacitance of 210 F g^{-1} at a discharge rate of 0.3 A g^{-1}. The energy density was above 18 Wh kg^{-1}, but the stability needed to be improved further (20% capacitance lost during 800 charge/discharge cycles). Li and co-workers [166] have demonstrated that water, the very "soft" matter, could serve as an effective "spacer" to prevent the restacking of RGO sheets. In contrast to the common expectation that RGO sheets would restack to graphite when stacked face-to-face, the hydrated RGO sheets remained significantly separated when combined together in a nearly parallel manner. The resultant multilayered RGO film provided a highly open pore structure, allowing the electrolyte solution to easily access to the surface of individual sheets. These superior features would make it possible to combine ultrahigh power density and high energy density in RGO-based supercapacitors and to allow the devices operational at high rates. As expected, the supercapacitor based on the hydrated RGO films exhibited a maximum power density of 414.0 kW kg^{-1} at a discharge current of 108 A g^{-1}, 1 to 3 orders of magnitude higher than the freeze-dried or thermally annealed RGO films. When an ionic liquid was used as the electrolyte, the operation voltage for the RGO film based supercapacitor was increased to 4 V, offering a specific capacitance up to 273.1 F g^{-1}, an energy density and the maximum power density of up to 150.9 Wh kg^{-1} and 776.8 kW kg^{-1}, respectively. Note that the high energy density is comparable to that of lithium ion batteries [167]. This simple and effective method provides an amazing strategy for addressing the key challenge that has limited the performances of RGO sheets as supercapacitor electrodes.

Many studies have focused on the synthesis and characterization of RGO/GO-based composite materials that are prepared by following the basic procedure (Fig. 8.19) [100, 168, 169]. The composite materials of RGO or GO with various metal oxides, such as nickel hydroxide [170, 171], ruthenium oxide [172], and manganese dioxide (MnO_2) [173, 174] have been prepared in order to investigate their capacitive properties. The full potential of composite materials are expected to be realized by the synergic effect of both the components [165, 170, 171, 173–175], including (i) the deposition of second components on RGO sheets effectively prevents the latter from agglomeration, which makes the surfaces of RGO sheets easily accessible for electrolyte ions, eventually improving the EDL capacitances of RGO sheets. (ii) The intimate interaction between pseudocapacitive materials and RGO sheets with good electrical conductivity provides an effective route for charge transfer, leading to rapid redox reactions of metal oxides or conducting polymers supported on RGO sheets. Thus, enhanced capacitive performances would be achieved.

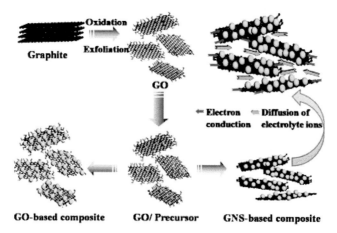

Figure 8.19 Schematic illustration for the synthesis and the structural advantages of GO or/and RGO-based composites. Yuan, C. Z., Gao, B., Shen, L. F., Yang, S. D., Hao, L., Lu, X. J., Zhang, F., Zhang, L. J., Zhang, X. G. (2011). Hierarchically structured carbon-based composites: Design, synthesis and their application in electrochemical capacitors, *Nanoscale*, **3**, pp. 529–545. Reproduced by permission of The Royal Society of Chemistry.

Direct growth of nickel hydroxide on RGO sheets with different oxidation degrees was achieved. The composite RGO sheets with Ni(OH)$_2$ nanoplates exhibited the largest capacitance of 935 F g^{-1} [170]. More importantly, the studies revealed that the electrochemical performances of these composite materials were dependent on the quality of graphene substrates, and the morphology and crystallinity of the nanomaterials grown on top. A GO and manganese oxide

(GO–MnO$_2$) composite has been successfully prepared by making full use of the surface oxygenate groups of GO as the center for anchoring manganese ions and the subsequent nucleation and growth [96]. The GO–MnO$_2$ composite materials exhibited a moderate capacitance of 197.2 F g^{-1}, smaller than even that of the nanostructured MnO$_2$ (211.2 F g^{-1}) [96]. The results exhibited that the preparation of GO–MnO$_2$ composites with a small amount of GO would improve the high dispersion of nanostructured MnO$_2$, leading to a convenient interaction between the electrolyte ions and the electrode materials. However, the incorporation of GO into MnO$_2$ would not lead to the synergic effect because the insulating GO will largely reduce the conductivity of composite materials, eventually deteriorating the capacitive performances [176]. In contrast, high capacitances and good rate performances were obtained when used the conductive RGO sheets as the starting materials. With the assistance of microwave heating, the redox reaction between carbon and permanganate ions has been empolyed to deposit manganese dioxide on RGO sheets [174]. The composite displayed a specific capacitance of about 310 F g^{-1} at a low scan rate of 2 mV s^{-1}, which was almost three times higher than that of pure graphene (104 F g^{-1}) and MnO$_2$ (103 F g^{-1}). We have development a novel assembly method to prepare composite material of functionalized RGO (f-RGO) and MnO$_2$ sheets (f-RGO-MnO$_2$) by using the electrostatic interation between the positively charged f-RGO and the negatively charged MnO$_2$ nanosheets [173]. The obtained f-RGO-MnO$_2$ exhibited enhanced capacitive performances in comparison with the single counterparts, contributing to the synergic effects. Cheng and co-workers reported that a sandwich membrane with alternate RGO sheets and nickel oxide nanoparticles (RGO/NiO) was prepared by a multi-step assembly approach. The RGO/NiO membrane was capable of assembling into a CR2032 coin cell as electrode materials without any binders or conductive additives for two-electrode measurements [171]. The supercapacitor cell had a much better capacitive performance (150–220 F g^{-1}, 0.1 A g^{-1}) than those of pristine RGO membranes. The immobilization of oxide nanoparticles between adjacent RGO sheets led to the construction of ordered channels for ion transport, improving the rate capability. Moreover, the layered sandwich structure also acted as an ideal strain buffer to accommodate volume changes of the nanoparticles, and thus had a better structural stability in the electrochemical charge/discharge process. The effect of Ru content in the RGO–RuO$_2$ composites on their capacitive performances has been investigated by the same group [165]. Benefits from the combined advantages of RGO sheets and RuO$_2$ were that the supercapacitors exhibited high specific capacitance (~570 F g^{-1} for 38.3 wt% Ru loading), excellent electrochemical stability (~97.9% retention after 1000 cycles), high energy density (20.1 Wh kg^{-1}) as well as power density (10 kW kg^{-1}).

Conducting polymers offer high pseudocapacitances, but their poor stability is a big drawback for the supercapacitor applications. Graphene sheets provide a great opportunity to improve the stability of conducting polymers [177]. Murugan and co-workers demonstrated a microwave-assisted solvothermal process to produce RGO nanosheets without the need for highly toxic chemicals. The authors investigated the energy storage properties of thus-prepared RGO nanosheets and associated PANi composites. The composite with 50 wt% RGO displayed both EDL capacitance and pseudocapacitance with an overall specific capacitance of 408 F g–1 [178]. Wang et al. prepared a RGO polyaniline composite paper (GPCP) by in situ anodic electropolymerization of aniline monomer as a PANi film on graphene paper [111]. The obtained composite paper combined flexibility, conductivity, and electrochemical activity and exhibited a gravimetric capacitance of 233 F g–1 and a volumetric capacitance of 135 F cm–3. However, the agglomerated paperlike structure resulted in the inhomogeneous coating of PANi on the 3D structure of RGO paper. Composite films of sulfonated RGO (SRGO) and PPy were deposited on Pt foil by an electrochemical polymerization process [114]. With the assistance of surfactant, dodecylbenzene sulfonic acid (DBSA), the amorphous PPy component formed thick coatings on SRGO surfaces. Meanwhile, the gaps between SRGO sheets were also filled by PPy, resulting in a low surface area. An electrode material consisting of fibrillar PANi doped with GO sheets was synthesized via in situ polymerization of monomer in the presence of GO. The obtained nanocomposite with a small amount of GO (mass ratio of aniline/GO, 100:1) exhibited a higher specific capacitance of 531 F g–1 at the current density of 0.2 A g–1 in comparison with 216 F g–1 of individual PANi [176]. A dilute polymerization at a low temperature (–10°C) has been used to construct hierarchical nanocomposites by combining one-dimensional (1D) conducting PANi nanowires with 2D GO nanosheets (PANi–GO composite) [179]. PANi nanowire arrays were aligned vertically on GO substrate, contributing to an optimized ionic transport pathway. Consequently, PANi–GO nanocomposite showed higher electrochemical capacitance and better stability than each individual component. This study introduces a facile method to construct a hierarchical nanocomposite using 1D and 2D nanocomponents and may guide the way for designing new composite materials at the nanoscale level. A series of RGO or GO and PANi nanofiber composites has been prepared by using the in situ polymerization method [113]. PANi fibers were adsorbed on the RGO surface and/or filled between the RGO sheets. The results revealed that good capacitive performance would be achieved by doping either RGO with a small amount of PANi or bulky PANi with a small amount of RGO. The RGO–PANi composite displayed a specific capacitance of as high as 480 F g^{-1} at a current density of 0.1 A g^{-1}. For the sample of GO–PANi, the electrical conductivity of the resulting nanocomposites would be increased by the chemical reduction of GO and subsequent re-doping process. However, the complex post processing usually resulted in the conducting polymers in a partially agglomerated form on RGO sheets. Yan et al. [112] also reported that RGO nanosheet/polyaniline (RGO/PANi) composite was synthesized using the direct polymerization method (see Fig. 8.20). Although the aggregation of RGO/PANi was observed in the as-prepared composite materials, dispersion of PANi was greatly improved because only aggregated PANi fibers were synthesized without the presence of RGO sheets. A specific capacitance as high as 1,046 F g^{-1} was achieved on the RGO/PANi composite at a low scan rate of 1 mV s^{-1} compared to 115 F g^{-1} for pure PANi, exhibiting the superior feature. These nanocomposite strategies open up the possibilities to combine graphene with other redox pseudocapacitive materials like polyaniline, MnO_2, and RuO_2 to enhance the energy density of supercapacitors.

As described in Section 8.3.2, optimizing the electrolytes to broaden the electrochemical window would lead to improved performances in terms of high energy density and power density. Aqueous electrolytes generally work in a maximum voltage window of 1.0 V (the thermodynamic window of water is 1.23 V) with a relatively low ESR. A smart asymmetric configuration can extend the operating voltages approaching $\sim$ 2 V, which offers great possibilities to explore the upper-right quadrant of Ragone plot (Fig. 8.2). The key to fabricate an ASC is to couple two kinds of electrode

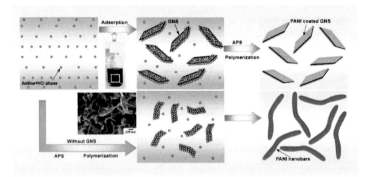

Figure 8.20 A scheme illustrating the synthesis of RGO/PANi composite. Reprinted from Yan, J., Wei, T., Shao, B., Fan, Z., Qian, W., Zhang, M., Wei, F. (2010). Preparation of a graphene nanosheet/polyaniline composite with high specific capacitance, *Carbon*, **48**, pp. 487–493. Copyright @ 2010, with permission from Elsevier.

materials in the same cell [128, 180]. The great innovation of ASC is to increase the maximum operation voltage of aqueous electrolyte in a cell by making full use of the different potential windows of the two electrodes, leading to the improved energy density and power density [128]. Activated carbon is generally used as one of the electrodes while a transition metal oxide such as manganese oxide is employed as the other electrode.

Great progress in the configuration of ASCs has already been achieved with graphene-based materials. As a typical example in Fig. 8.21, the operating potential windows for MnO_2 nanowire/RGO composite (MGC) and RGO are in the positive and negative ranges in aqueous electrolyte of Na_2SO_4, respectively. Consequently, the potential window of the asymmetric cell is enlarged to 2 V due to the compensatory effect of opposite over-potentials [165]. Such an ASC demonstrated that the maximum energy density was up to 30.4 Wh kg−1, which was much higher than those of symmetric supercapacitors based on RGO||RGO (2.8 Wh kg^{-1}) and MGC||MGC (5.2 Wh kg^{-1}). A similar ASC coupling RGO/MnO_2 composite as positive electrode with activated carbon nanofibers (ACN) as negative electrode in a neutral aqueous electrolyte could be cycled reversibly in the voltage range of 0–1.8 V [181]. The ASC exhibited the maximum energy density of up to 51.1 Wh kg^{-1}, much higher than that of asymmetric MnO_2||CNT cell (29.1 Wh kg^{-1}), highlighting the important role of RGO sheets.

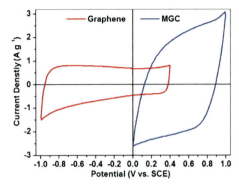

Figure 8.21 CV curves of graphene and MGC electrodes performed in a three-electrode cell in a 1 M Na$_2$SO$_4$ solution at a scan rate of 10 mV s^{-1}. Reprinted with permission from Wu, Z. S., Ren, W., Wang, D. W., Li, F., Liu, B., Cheng, H. M. (2010). High-energy MnO$_2$ nanowire/graphene and graphene asymmetric electrochemical capacitors, *ACS Nano*, **4**, pp. 5835–5842]. Copyright © 2010, American Chemical Society.

Additionally, the ASC of RGO/MnO$_2$||ACN exhibited excellent cycling stability with 97% specific capacitance retained after 1000 cycles. Zhang et al. developed an ASC using RGO sheets modified with ruthenium oxide (RGO–RuO$_2$) or polyaniline (RGO–PANi) as the positive and negative electrodes, respectively [182]. Benefit from the broad potential window was that an energy density of 26.3 Wh kg^{-1} was achieved, which was about two times higher than those of the SSCs based on RGO–RuO$_2$ (12.4 Wh kg^{-1}) and RGO–PANi (13.9 Wh kg^{-1}) electrodes. In addition, a power density of 49.8 kW kg^{-1} was obtained at an acceptable energy density of 6.8 Wh kg^{-1}. Dai and co-workers [183] grew Ni(OH)$_2$ nanoplates and RuO$_2$ nanoparticles on RGO sheets in order to maximize the specific capacitances of these materials. By pairing up a Ni(OH)$_2$/RGO electrode with a RuO$_2$/RGO electrode, an ASC operating in aqueous solutions at a voltage of ~1.5 V was successfully fabricated. An energy density of ~48 Wh kg^{-1} at a power density of ~0.23 kW kg^{-1}, as well as a power density of ~21 kW kg^{-1} at an energy density of ~14 Wh kg^{-1} have been achieved. These studies suggest that pairing up metal oxide/RGO or CPs/RGO composite materials for ASCs represents a new approach to enhancing the performances of supercapacitor devices. Cui and co-workers [184] have demonstrated that solution-exfoliated graphene nanosheets (about 5 nm in thickness) were conformably coated from solution on the three-dimensional, porous textiles to facilitate the access of electrolytes to those materials. With the controllable electrodeposition of pseudocapacitive MnO_2 nanomaterials, the graphene/MnO_2-based textile yielded a high capacitance. Interestingly, ASCs with graphene/MnO_2-textile as the positive electrode and SWNTs-textile as the negative electrode were successfully fabricated in an aqueous Na_2SO_4 electrolyte. These devices exhibited promising characteristics with respect to an operation voltage of 1.5 V, a maximum power density of 110 kW kg^{-1}, an energy density of 12.5 Wh kg^{-1}, and an excellent cycling performance of ~95% capacitance retention over 5000 cycles. The energy density and power density of the ACS was relative lower as compared to the previously reported all-paper supercapacitors [185]. However, this shortage would be technically overcome using ultrathin textile and/or thick graphene/MnO_2 electrodes. With these feasible solutions, the outstanding performances of the hybrid electrodes are expected to be fully exploited for practical applications.

8.4 Summary and Perspectives

The research within the field of graphene-based materials is currently undergoing an exciting development, driven by advancements in the exploitation and design of novel graphene-based materials. Moreover, the demand for portable, flexible, and space-effective energy sources will further drive more researches towards the development of graphene-based materials. On the configuration of supercapacitor devices, graphene-based materials with various microtextures and wide availabilities represent very attractive materials. Especially, the potential for integrated structures containing the three essential components (electrodes, spacer, and electrolyte) of the electrochemical device provides the significant opportunity to design a variety of compact devices such as flexible and ultrathin supercapacitors. Furthermore, the improved energy storage in supercapacitors would be achieved by designing composite materials to combine graphene sheets together with pseudo-capacitive materials (metal oxides and conducting polymers). The dual functions of graphene sheets in terms of high EDLC and superior conducting agent would be beneficial to the design of high power and energy supercapacitors. Widening the cell operating voltage by using proper organic electrolyte or ionic liquid is a very effective way to enhance both energy density and power density of supercapacitors. From the view of environmental friendly and cost-effective points, the superior features of aqueous electrolyte have boosted the configuration of asymmetric supercapacitors by using graphene-based composite materials. Benefiting from the enlarged operating voltage of asymmetric supercapacitor fabricated with graphene-based composite materials, the enhanced energy densities have been achieved. However, it must bear in mind that with the improvement on energy density, the principal advantages of supercapacitors which are high rate capability and long cycle life must not be sacrificed.

As the promising electrode materials for supercapacitor applications, novel materials and structures based on graphene sheets are expected to complement or replace the electrode materials in certain aspects of supercapacitor applications achieving higher energy density, higher power as well as higher level of reliability. With the rapidly growing demand, it will continue to generate much research activities towards the development of graphene-based materials. It is anticipated that the revolutionary breakthrough in the configuration of graphene-based materials for supercapacitors will make inroads into a wider range of technological applications.

8.5   Acknowledgements

We thank the Ministry of Education (Singapore) for financial support under a Tier 2 grant (MOE2008-T2-1-004). The Australian Research Council (ARC) is also acknowledged for the financial support under the ARC Future Fellowship Program (FT100100879).

References

1. Geim A.K., Novoselov K.S. (2007) The rise of graphene, *Nat Mater*, **6**, 183–191.
2. Terrones M., Botello-Médez A.R., Campos-Delgado J., et al. (2010) Graphene and graphite nanoribbons: Morphology, properties, synthesis, defects and applications, *Nano Today*, **5**, 351–372.
3. Soldano C., Mahmood A., Dujardin E. (2010) Production, properties and potential of graphene, *Carbon*, **48**, 2127–2150.

4. Wu Y.H., Yu T., Shen Z.X. (2010) Two-dimensional carbon nanostructures: Fundamental properties, synthesis, characterization, and potential applications, *J Appl Phys*, **108**,. 071301.

5. Brownson D.A.C., Kampouris D.K., Banks C.E. (2011) An overview of graphene in energy production and storage applications, *J Power Sources*, **196**, 4873–4885.

6. Li X., Zhu Y., Cai W., et al. (2009) Transfer of large-area graphene films for high-performance transparent conductive electrodes, *Nano Lett*, **9**, 4359–4363.

7. Liu C., Yu Z., Neff D., Zhamu A., Jang B.Z. (2010) Graphene-based supercapacitor with an ultrahigh energy density, *Nano Lett*, **10**, 4863–4868.

8. Allen M.J., Tung V.C., Kaner R.B. (2010) Honeycomb carbon: A review of graphene, *Chem Rev*, **110**, 132–145.

9. Zhang L.L., Zhou R., Zhao X.S. (2010) Graphene-based materials as supercapacitor electrodes, *J Mater Chem*, **20**, 5983–5992.

10. Rao C.N.R., Sood A.K., Subrahmanyam K.S., Govindaraj A. (2009) Graphene: The new two-dimensional nanomaterial, *Angew Chem Int Ed*, **48**, 7752–7777.

11. Guo S., Dong S. (2011) Graphene nanosheet: Synthesis, molecular engineering, thin film, hybrids, and energy and analytical applications, *Chem Soc Rev*, **40**, 2644–2672.

12. Park S., Ruoff R.S. (2009) Chemical methods for the production of graphenes, *Nat Nano*, **4**, 217–224.

13. Bai H., Li C., Shi G. (2011) Functional composite materials based on chemically converted graphene, *Adv Mater*, **23**, 1089–1115.

14. Hummers W.S., Offeman R.E. (1958) Preparation of graphitic oxide, *J Am Chem Soc*, **80**, 1339–1339.

15. Stankovich S., Dikin D.A., Piner R.D., et al. (2007) Synthesis of graphene-based nanosheets via chemical reduction of exfoliated graphite oxide, *Carbon*, **45**, 1558–1565.

16. Cook T.R., Dogutan D.K., Reece S.Y., Surendranath Y., Teets T.S., Nocera D.G. (2010) Solar energy supply and storage for the legacy and nonlegacy worlds, *Chem Rev*, **110**, 6474–6502.

17. Long J.W., Dunn B., Rolison D.R., White H.S. (2004) Three-dimensional battery architectures, *Chem Rev*, **104**, 4463–4492.

18. Novak P., Muller K., Santanam K.S.V., Hass O. (1997) *Chem Rev*, **97**, 207.

19. Rolison D.R., Nazar L.F. (2011) Electrochemical energy storage to power the 21st century, *MRS Bulletin*, **36**, 486–493.

20. Winter M., Brodd R.J. (2004) What are batteries, fuel cells, and supercapacitors?, *Chem Rev*, **104**, 4245–4270.

21. Pandolfo A.G., Hollenkamp A.F. (2006) Carbon properties and their role in supercapacitors, *J Power Sources*, **157**, 11–27.

22. Burke A. (2000) Ultracapacitors: Why, how, and where is the technology, *J Power Sources*, **91**, 37–50.

23. Cheng F., Liang J., Tao Z., Chen J. (2011) Functional materials for rechargeable batteries, *Adv Mater*, **23**, 1695–1715.

24. Liu C., Li F., Ma L.-P., Cheng H.-M. (2010) Advanced materials for energy storage, *Adv Mater*, **22**, E28–E62.

25. Miller J.R., Simon P. (2008) Electrochemical capacitors for energy management, *Science*, **321**, 651–652.

26. Zhang J., Zhao X.S. (2012) On the configuration of supercapacitors for maximizing electrochemical performance, *ChemSusChem*, **5**, 818–841.

27. Zhang L.L., Zhao X.S. (2009) Carbon-based materials as supercapacitor electrodes, *Chem Soc Rev*, **38**, 2520–2531.

28. Jampani P., Manivannan A., Kumta P.N. (2010) Advancing the supercapacitor materials and technology frontier for improving power quality, *Electrochem Soc Interface*, **19**, 57–62.

29. Simon P., Gogotsi Y. (2008) Materials for electrochemical capacitors, *Nat Mater*, **7**, 845–854.

30. Wallace P.R. (1947) The band theory of graphite, *Phys Rev*, **71**, 622–634.

31. McClure J.W. (1956) Diamagnetism of graphite, *Phys. Rev.*, **104**, 666–671.

32. Wu J., Pisula W., Müllen K. (2007) Graphenes as potential material for electronics, *Chem Rev*, **107**, 718–747.

33. Novoselov K.S., Geim A.K., Morozov S.V., et al. (2004) Electric field effect in atomically thin carbon films, *Science*, **306**, 666–669.

34. Lotya M., Hernandez Y., King P.J., et al. (2009) Liquid phase production of graphene by exfoliation of graphite in surfactant/water solutions, *J Am Chem Soc*, **131**, 3611–3620.

35. Hernandez Y., Nicolosi V., Lotya M., et al. (2008) High-yield production of graphene by liquid-phase exfoliation of graphite, *Nat Nanotechnol*, **3**, 563–568.

36. Choucair M., Thordarson P., Stride J.A. (2009) Gram-scale production of graphene based on solvothermal synthesis and sonication, *Nat Nanotechnol*, **4**, 30–33.

37. Reina A., Jia X., Ho J., et al. (2008) Large area, few-layer graphene films on arbitrary substrates by chemical vapor deposition, *Nano Lett*, **9**, 30–35.

38. de Heer W.A., Berger C., Wu X., et al. (2007) Epitaxial graphene, *Solid State Commun*, **143**, 92–100.

39. Sutter P.W., Flege J.-I., Sutter E.A. (2008) Epitaxial graphene on ruthenium, *Nat Mater*, **7**, 406–411.

40. Shivaraman S., Barton R.A., Yu X., et al. (2009) Free-standing epitaxial graphene, *Nano Lett*, **9**, 3100–3105.

41. Forbeaux I., Themlin J.M., Charrier A., Thibaudau F., Debever J.M. (2000) Solid-state graphitization mechanisms of silicon carbide 6H-SiC polar faces, *Appl Surf Sci*, **162–163**, 406–412.

42. Charrier A., Coati A., Argunova T., et al. (2002) Solid-state decomposition of silicon carbide for growing ultra-thin heteroepitaxial graphite films, *J Appl Phys*, **92**, 2479.

43. Deng D., Pan X., Zhang H., Fu Q., Tan D., Bao X. (2010) Freestanding graphene by thermal splitting of silicon carbide granules, *Adv Mater*, **22**, 2168–2171.

44. Luo Z., Yu T., Shang J., et al. (2011) Large-scale synthesis of bi-layer graphene in strongly coupled stacking order, *Adv Func Mater*, **21**, 911–917.

45. Li X., Cai W., An J., et al. (2009) Large-area synthesis of high-quality and uniform graphene films on copper foils, *Science*, **324**, 1312–1314.

46. Srivastava A., Galande C., Ci L., et al. Novel liquid precursor-based facile synthesis of large-area continuous, single, and few-layer graphene films, *Chem Mater*, **22**, 3457–3461.

47. López V., Sundaram R.S., Gómez-Navarro C., et al. (2009) Chemical vapor deposition repair of graphene oxide: A route to highly conductive graphene monolayers, *Adv Mater*, **21**, 4683–4686.

48. Emtsev K.V., Bostwick A., Horn K., et al. (2009) Towards wafer-size graphene layers by atmospheric pressure graphitization of silicon carbide, *Nat Mater*, **8**, 203–207.

49. Kim K.S., Zhao Y., Jang H., et al. (2009) Large-scale pattern growth of graphene films for stretchable transparent electrodes, *Nature*, **457**, 706–710.

50. Yu Q., Lian J., Siriponglert S., Li H., Chen Y.P., Pei S.S (2008) Graphene segregated on Ni surfaces and transferred to insulators, *Appl Phys Lett*, **93**.

51. Bae S., Kim H., Lee Y., et al. (2010) Roll-to-roll production of 30-inch graphene films for transparent electrodes, *Nat Nanotechnol*, **5**, 574–578.

52. Wei D., Liu Y., Zhang H., et al. (2009) Scalable synthesis of few-layer graphene ribbons with controlled morphologies by a template method and their applications in nanoelectromechanical switches, *J Am Chem Soc*, **131**, 11147–11154.

53. Zhang W., Cui J., Tao C.-a., et al. (2009) A strategy for producing pure single-layer graphene sheets based on a confined self-assembly approach, *Angew Chem Int Ed*, **48**, 5864–5868.

54. Chen Z., Ren W., Gao L., Liu B., Pei S., Cheng H.-M. (2011) Three-dimensional flexible and conductive interconnected graphene networks grown by chemical vapour deposition, *Nat Mater*, **10**, 424–428.

55. Yang X., Dou X., Rouhanipour A., Zhi L., Räder H.J., Müllen K. (2008) two-dimensional graphene nanoribbons, *J Am Chem Soc*, **130**, 4216–4217.

56. Deng D., Pan X., Yu L., et al. (2011) Toward N-doped graphene via solvothermal synthesis, *Chem Mater*, **23**, 1188–1193.

57. Li X., Wang X., Zhang L., Lee S., Dai H. (2008) Chemically derived, ultrasmooth graphene nanoribbon semiconductors, *Science*, **319**, 1229–1232.

58. Hirsch A. (2009) Unzipping carbon nanotubes: A peeling method for the formation of graphene nanoribbons, *Angew Chem Int Ed*, **48**, 6594–6596.

59. Kosynkin D.V., Higginbotham A.L., Sinitskii A., Lomedaet al. (2009) Longitudinal unzipping of carbon nanotubes to form graphene nanoribbons, *Nature*, **458**, 872–876.

60. Jiao L., Zhang L., Wang X., Diankov G., Dai H. (2009) Narrow graphene nanoribbons from carbon nanotubes, *Nature*, **458**, 877–880.

61. Kim K., Sussman A., Zettl A. (2010) Graphene nanoribbons obtained by electrically unwrapping carbon nanotubes, *ACS Nano*, **4**, 1362–1366.

62. Segal M. (2009) Selling graphene by the ton, *Nat Nanotechnol*, **4**, 612–614.

63. Brodie B.C. (1859) On the atomic weight of graphite, *Philos Trans R Soc London*, **149**, 249–259.

64. Staudenmaier L. (1898) Verfahren zur Darstellung der Graphitsäure, *Ber Dtsch Chem Ges*, **31**, 1481–1487.

65. Dreyer D.R., Park S., Bielawski C.W., Ruoff R.S. (2010) The chemistry of graphene oxide, *Chem Soc Rev*, **39**, 228–240.

66. He H., Klinowski J., Forster M., Lerf A. (1998) A new structural model for graphite oxide, *Chem Phys Lett*, **287**, 53–56.

67. He H., Forster M., Klinowski J. (1998) Structure of graphite oxide revisited, *J Phys Chem B*, **102**, 4477–4482.

68. Cai W., Piner R.D., Stadermann F.J., et al. (2008) Synthesis and solid-state NMR structural characterization of 13C-labeled graphite oxide, *Science*, **321**, 1815–1817.

69. Moon I.K., Lee J., Ruoff R.S., Lee H. (2010) Reduced graphene oxide by chemical graphitization, *Nat Commun*, **1**, 73.

70. Gao W., Alemany L.B., Ci L., Ajayan P.M. (2009) New insights into the structure and reduction of graphite oxide, *Nat Chem*, **1**, 403–408.

71. Li D., Muller M.B., Gilje S., Kaner R.B., Wallace G.G. (2008) Processable aqueous dispersions of graphene nanosheets, *Nat Nano*, **3**, 101–105.

72. Paredes J.I., Villar-Rodil S., Fernandez-Merino M.J., Guardia L., Martinez-Alonso A., Tascon J.M.D. (2011) Environment friendly approaches toward the mass production of processable graphene from graphite oxide, *J Mater Chem*, **21**, 298–306.

73. Zhang J., Yang H., Shen G., Cheng P., Zhang J., Guo S. (2010) Reduction of graphene oxide vial-ascorbic acid, *Chem Commun*, **46**, 1112–1114.

74. Gao J., Liu F., Liu Y., Ma N., Wang Z., Zhang X. (2010) Environment-friendly method to produce graphene that employs vitamin C and amino acid, *Chem Mater*, **22**, 2213–2218.

75. Zhu C., Guo S., Fang Y., Dong S. (2010) Reducing sugar: New functional molecules for the green synthesis of graphene nanosheets, *ACS Nano*, **4**, 2429–2437.

76. Zhou Y., Bao Q., Tang L.A.L., Zhong Y., Loh K.P. (2009) Hydrothermal dehydration for the "green" reduction of exfoliated graphene oxide to graphene and demonstration of tunable optical limiting properties, *Chem Mater*, **21**, 2950–2956.

77. Fan X., Peng W., Li Y., et al. (2008) Deoxygenation of exfoliated graphite oxide under alkaline conditions: A green route to graphene preparation, *Adv Mater*, **20**, 4490–4493.

78. Wang H., Robinson J.T., Li X., Dai H. (2009) Solvothermal reduction of chemically exfoliated graphene sheets, *J Am Chem Soc*, **131**, 9910–9911.

79. Chen W., Yan L. (2010) Preparation of graphene by a low-temperature thermal reduction at atmosphere pressure, *Nanoscale*, **2**, 559–563.

References | 239

80. Park S., An J., Jung I., et al. (2009) Colloidal suspensions of highly reduced graphene oxide in a wide variety of organic solvents, *Nano Lett*, **9**, 1593–1597.

81. Zhu Y., Murali S., Stoller M.D., Velamakanni A., Piner R.D., Ruoff R.S. (2010) Microwave assisted exfoliation and reduction of graphite oxide for ultracapacitors, *Carbon*, **48**, 2118–2122.

82. Zhang M., Liu S., Yin X.M., et al. (2011) Fast synthesis of graphene sheets with good thermal stability by microwave irradiation, *Chem – An Asian J*, **6**, 1151–1154.

83. Janowska I., Chizari K., Ersen O., et al. (2010) Microwave synthesis of large few-layer graphene sheets in aqueous solution of ammonia, *Nano Res*, **3**, 126–137.

84. Chen W., Yan L., Bangal P.R. (2010) Preparation of graphene by the rapid and mild thermal reduction of graphene oxide induced by microwaves, *Carbon*, **48**, 1146–1152.

85. Ang P.K., Wang S., Bao Q., Thong J.T.L., Loh K.P. (2009) High-throughput synthesis of graphene by intercalation–exfoliation of graphite oxide and study of ionic screening in graphene transistor, *ACS Nano*, **3**, 3587–3594.

86. Compton O.C., Jain B., Dikin D.A., Abouimrane A., Amine K., Nguyen S.T. (2011) Chemically active reduced graphene oxide with tunable C/O ratios, *ACS Nano*, **5**, 4380–4391.

87. Schniepp H.C., Li J.-L., McAllister M.J., et al. (2006) Functionalized single graphene sheets derived from splitting graphite oxide, *J Phys Chem B*, **110**, 8535–8539.

88. McAllister M.J., Li J.-L., Adamson D.H., et al. (2007) Single sheet functionalized graphene by oxidation and thermal expansion of graphite, *Chem Mater*, **19**, 4396–4404.

89. Lv W., Tang D.-M., He Y.-B., et al. (2009) Low-temperature exfoliated graphenes: Vacuum-promoted exfoliation and electrochemical energy storage, *ACS Nano*, **3**, 3730–3736.

90. Subrahmanyam K.S., Panchakarla L.S., Govindaraj A., Rao C.N.R. (2009) Simple method of preparing graphene flakes by an arc-discharge method, *J Phys Chem C*, **113**, 4257–4259.

91. Wu Z.-S., Ren W., Gao L., et al. (2009) Synthesis of graphene sheets with high electrical conductivity and good thermal stability by hydrogen arc discharge exfoliation, *ACS Nano*, **3**, 411–417.

92. Li Z., Yao Y., Lin Z., Moon K.-S., Lin W., Wong C. (2010) Ultrafast, dry microwave synthesis of graphene sheets, *J Mater Chem*, **20**, 4781–4783.

93. Park S.-H., Bak S.-M., Kim K.-H., et al. (2010) Solid-state microwave irradiation synthesis of high quality graphene nanosheets under hydrogen containing atmosphere, *J Mater Chem*, **21**, 680–686.

94. Luo D., Zhang G., Liu J., Sun X. (2011) Evaluation criteria for reduced graphene oxide, *J Phys Chem C*, **115**, 11327–11335.

95. Chandra V., Park J., Chun Y., Lee J.W., Hwang I.-C., Kim K.S. (2010) Water-dispersible magnetite-reduced graphene oxide composites for arsenic removal, *ACS Nano*, **4**, 3979–3986.

96. Chen S., Zhu J., Wu X., Han Q., Wang X. (2010) Graphene oxide-MnO_2 nanocomposites for supercapacitors, *ACS Nano*, **4**, 2822–2830.

97. Zhang H., Lv X., Li Y., Wang Y., Li J. (2009) P25-Graphene composite as a high performance photocatalyst, *ACS Nano*, **4**, 380–386.

98. Williams G., Seger B., Kamat P.V. (2008) TiO_2-Graphene nanocomposites. UV-assisted photocatalytic reduction of graphene oxide, *ACS Nano*, **2**, 1487–1491.

99. Cao A., Liu Z., Chu S., et al. (2010) A facile one-step method to produce graphene–CdS quantum dot nanocomposites as promising optoelectronic materials, *Adv Mater*, **22**, 103–106.

100. Zhang J., Xiong Z., Zhao X.S. (2011) Graphene–metal-oxide composites for the degradation of dyes under visible light irradiation, *J Mater Chem*, **21**, 3634–3640.

101. Zhu C., Guo S., Wang P., et al. (2010) One-pot, water-phase approach to high-quality graphene/TiO_2 composite nanosheets, *Chem Commun*, **46**, 7148–7150.

102. Wang H., Robinson J.T., Diankov G., Dai H. (2010) Nanocrystal growth on graphene with various degrees of oxidation, *J Am Chem Soc*, **132**, 3270–3271.

103. Wang X., Tabakman S.M., Dai H. (2008) Atomic layer deposition of metal oxides on pristine and functionalized graphene, *J Am Chem Soc*, **130**, 8152–8153.

104. Wang D., Choi D., Li J., et al. (2009) Self-assembled TiO_2-graphene hybrid nanostructures for enhanced Li-ion insertion, *ACS Nano*, **3**, 907–914.

105. Wang D., Kou R., Choi D., et al. (2010) Ternary self-assembly of ordered metal oxide-graphene nanocomposites for electrochemical energy storage, *ACS Nano*, **4**, 1587–1595.

106. Vickery J.L., Patil A.J., Mann S. (2009) Fabrication of graphene–polymer nanocomposites with higher-order three-dimensional architectures, *Adv Mater*, **21**, 2180–2184.

107. Liang J., Huang Y., Zhang L., et al. (2009) Molecular-level dispersion of graphene into poly(vinyl alcohol) and effective reinforcement of their nanocomposites, *Adv Funct Mater*, **19**, 2297–2302.

108. Salavagione H.J., Martinez G., Gomez M.A. (2009) Synthesis of poly(vinyl alcohol)/reduced graphite oxide nanocomposites with improved thermal and electrical properties, *J Mater Chem*, **19**, 5027–5032.

109. Jeong H.-K., Jin M., Ra E.J., et al. (2010) Enhanced electric double layer capacitance of graphite oxide intercalated by poly(sodium 4-styrensulfonate) with high cycle stability, *ACS Nano*, **4**, 1162–1166.

110. Liu J., Tao L., Yang W., et al. (2010) Synthesis, characterization, and multilayer assembly of pH sensitive graphene-polymer nanocomposites, *Langmuir*, **26**, 10068–10075.

111. Wang D.-W., Li F., Zhao J., et al. (2009) Fabrication of graphene/polyaniline composite paper via in situ anodic electropolymerization for high-performance flexible electrode, *ACS Nano*, **3**, 1745–1752.

112. Yan J., Wei T., Shao B., et al. (2010) Preparation of a graphene nanosheet/polyaniline composite with high specific capacitance, *Carbon*, **48**, 487–493.

113. Zhang K., Zhang L.L., Zhao X.S., Wu J. (2010) Graphene/polyaniline nanofiber composites as supercapacitor electrodes, *Chem Mater*, **22**, 1392–1401.

114. Liu A., Li C., Bai H., Shi G. (2010) Electrochemical deposition of polypyrrole/sulfonated graphene composite films, *J Phys Chem C*, **114**, 22783–22789.

115. Zhang L.L., Zhao S., Tian X.N., Zhao X.S. (2010) Layered graphene oxide nanostructures with sandwiched conducting polymers as supercapacitor electrodes, *Langmuir*, **26**, 17624–17628.

116. Mini P.A., Balakrishnan A., Nair S.V., Subramanian K.R.V. (2011) Highly super capacitive electrodes made of graphene/poly(pyrrole), *Chem Commun*, **47**, 5753–5755.

117. Xu Y., Wang Y., Liang J., et al. (2009) A hybrid material of graphene and poly (3,4-ethyldioxythiophene) with high conductivity, flexibility, and transparency, *Nano Res*, **2**, 343–348.

118. Kuilla T., Bhadra S., Yao D., Kim N.H., Bose S., Lee J.H. (2010) Recent advances in graphene based polymer composites, *Prog Polym Sci*, **35**, 1350–1375.

119. Conway B.E., *Electrochemical Supercapacitor and Technological Applications*, 1999, Kluwer Academia/Plenum Publishers, NY.

120. Largeot C., Portet C., Chmiola J., Taberna P.-L., Gogotsi Y., Simon P. (2008) Relation between the ion size and pore size for an electric double-layer capacitor, *J Am Chem Soc*, **130**, 2730–2731.

121. Chmiola J., Yushin G., Gogotsi Y., Portet C., Simon P., Taberna P.L. (2006) Anomalous increase in carbon capacitance at pore sizes less than 1 nanometer, *Science*, **313**, 1760–1763.

122. Huang J., Sumpter B.G., Meunier V. (2008) A universal model for nanoporous carbon supercapacitors applicable to diverse pore regimes, carbon materials, and electrolytes, *Chem Euro J*, **14**, 6614–6626.

123. Zheng J.P., Cygan P.J., Jow T.R. (1995) Hydrous ruthenium oxide as an electrode material for electrochemical capacitors, *J Electrochem Soc*, **142**, 2699–2703.

124. Zhang J.T., Ma J.Z., Zhang L.L., Guo P., Jiang J., Zhao X.S. (2010) Template synthesis of tubular ruthenium oxides for supercapacitor applications, *J Phys Chem C*, **114**, 13608–13613.

125. Toupin M., Brousse T., Belanger D. (2004) Charge storage mechanism of MnO_2 electrode used in aqueous electrochemical capacitor, *Chem Mater*, **16**, 3184–3190.

126. Zhang J.T., Chu W., Jiang J.W., Zhao X.S. (2011) Synthesis, characterization and capacitive performance of hydrous manganese dioxide nanostructures, *Nanotechnology*, **22**, 25703.

127. Snook G.A., Kao P., Best A.S. (2011) Conducting-polymer-based supercapacitor devices and electrodes, *J Power Sources*, **196**, 1–12.

128. Long J.W., Bélanger D., Brousse T., Sugimoto W., Sassin M.B., Crosnier O. (2011) Asymmetric electrochemical capacitors—Stretching the limits of aqueous electrolytes, *MRS Bull*, **36**, 513–522.

129. Hu C.-C., Chang K.-H., Lin M.-C., Wu Y.-T. (2006) Design and tailoring of the nanotubular arrayed architecture of hydrous RuO_2 for next generation supercapacitors, *Nano Lett*, **6**, 2690–2695.

130. Toupin M., Brousse T., Bélanger D. (2004) Charge storage mechanism of MnO_2 electrode used in aqueous electrochemical capacitor, *Chem Mater*, **16**, 3184–3190.

131. Shukla A.K., Sampath S., Vijayamohanan K. (2000) Electrochemical supercapacitors: Energy storage beyond batteries, *Curr Sci*, **79**, 1656–1661.

132. Miller J.R., Simon P. (2008) Fundamentals of electrochemical capacitor design and operation, *Electrochem Soc Interface*, **17**, 31–32.

133. Abruña H.D., Kiya Y., Henderson J.C. (2008) Batteries and electrochemical capacitors, *Phys Today*, **61**, 43–47.

134. Stoller M.D., Ruoff R.S. (2010) Best practice methods for determining an electrode material's performance for ultracapacitors, *Energy Environ Sci*, **3**, 1294–1301.

135. Denisa H.-J., Puziy A.M., Poddubnaya O.I., Fabian S.-G., Tascón J.M.D., Lu G.Q. (2009) Highly stable performance of supercapacitors from phosphorus-enriched carbons, *J Am Chem Soc*, **131**, 5026–5027.

136. Portet C., Taberna P.L., Simon P., Flahaut E. (2005) Influence of carbon nanotubes addition on carbon–carbon supercapacitor performances in organic electrolyte, *J Power Sources*, **139**, 371–378.

137. Taberna P.L., Simon P., Fauvarque J.F. (2003) Electrochemical characteristics and impedance spectroscopy studies of carbon–carbon supercapacitors, *J Electrochem Soc*, **150**, A292–A300.

138. Pech D., Brunet M., Durou H., et al. (2010) Ultrahigh-power micrometre-sized supercapacitors based on onion-like carbon, *Nat Nano*, **5**, 651–654.

139. Fabio A.D., Giorgi A., Mastragostino M., Soavi F. (2001) Carbon-poly(3-methylthiophene) hybrid supercapacitors, *J Electrochem Soc*, **148**, A845–A850.

140. Robert H.N., Michael T.B., Bruce C.B., (2011) Modeling the electrochemical impedance spectra of electroactive pseudocapacitor materials, *J Electrochem Soc*, **158**, A678–A688.

141. Yan J., Wei T., Shao B., et al. (2010) Electrochemical properties of graphene nanosheet/carbon black composites as electrodes for supercapacitors, *Carbon*, **48**, 1731–1737.

142. Niu Z., Zhou W., Chen J., et al. (2011) Compact-designed supercapacitors using free-standing single-walled carbon nanotube films, *Energy Environ Sci*, **4**, 1440–1446.

143. Ates M. (2011) Review study of electrochemical impedance spectroscopy and equivalent electrical circuits of conducting polymers on carbon surfaces, *Prog Org Coat*, **71**, 1–10.

144. Nian Y.-R., Teng H. (2002) Nitric acid modification of activated carbon electrodes for improvement of electrochemical capacitance, *J Electrochem Soc*, **149**, A1008–A1014.

145. Izadi-Najafabadi A., Yasuda S., Kobashi K., et al. (2010) Extracting the full potential of single-walled carbon nanotubes as durable supercapacitor electrodes operable at 4 V with high power and energy density, *Adv Mater*, **22**, E235–E241.

146. Miller J.R., Outlaw R.A., Holloway B.C. (2010) Graphene double-layer capacitor with ac line-filtering performance, *Science*, **329**, 1637–1639.

147. Zhao X., Tian H., Zhu M., et al. (2009) Carbon nanosheets as the electrode material in supercapacitors, *J Power Sources*, **194**, 1208–1212.

148. Stoller M.D., Park S., Zhu Y., An J., Ruoff R.S., (2008) Graphene-based ultracapacitors, *Nano Lett*, **8**, 3498–3502.

149. Wang Y., Shi Z., Huang Y., et al. (2009) Supercapacitor devices based on graphene materials, *J Phys Chem C*, **113**, 13103–13107.

150. Niu C., Sichel E.K., Hoch R., Moy D., Tennent H. (1997) High power electrochemical capacitors based on carbon nanotube electrodes, *Appl Phys Lett*, **70**, 1480–1482.

151. Vivekchand S., Rout C., Subrahmanyam K., Govindaraj A., Rao C. (2008) Graphene-based electrochemical supercapacitors, *J Chem Sci*, **120**, 9–13.

152. Du X., Guo P., Song H., Chen X. (2010) Graphene nanosheets as electrode material for electric double-layer capacitors, *Electrochim Acta*, **55**, 4812–4819.

153. Du Q., Zheng M., Zhang L., et al. (2010) Preparation of functionalized graphene sheets by a low-temperature thermal exfoliation approach and their electrochemical supercapacitive behaviors, *Electrochim Acta*, **55**, 3897–3903.

154. Zhu Y., Murali S., Stoller M.D., et al. (2011) Carbon-based supercapacitors produced by activation of graphene, *Science*, **332**, 1537–1541.

155. Burke A. (2007) R&D considerations for the performance and application of electrochemical capacitors, *Electrochim Acta*, **53**, 1083–1091.

156. Yoo J.J., Balakrishnan K., Huang J., et al. (2011) Ultrathin planar graphene supercapacitors, *Nano Lett*, **11**, 1423–1427.

157. Jeong H.M., Lee J.W., Shin W.H., et al. (2011) Nitrogen-doped graphene for high-performance ultracapacitors and the importance of nitrogen-doped sites at basal planes, *Nano Lett*, **11**, 2472–2477.

158. Xu B., Yue S., Sui Z., et al. (2011) What is the choice for supercapacitors: Graphene or graphene oxide?, *Energy Environ Sci*, **4**, 2826–2830.

159. Li D., Müller M.B., Gilje S., Kaner R.B., Wallace G.G. (2008) Processable aqueous dispersions of graphene nanosheets, *Nat Nanotechnol*, **3**, 101–105.

160. Lei Z., Christov N., Zhao X.S. (2011) Intercalation of mesoporous carbon spheres between reduced graphene oxide sheets for preparing high-rate supercapacitor electrodes, *Energy Environ Sci*, **4**, 1866–1873.

161. Loh K.P., Bao Q., Ang P.K., Yang J. (2010) The chemistry of graphene, *J Mater Chem*, **20**, 2277–2289.

162. Zhu Y., Murali S., Cai W., et al. (2010) Graphene and graphene oxide: synthesis, properties, and applications, *Adv Mater*, **22**, 3906–3924.

163. Xu Y., Sheng K., Li C., Shi G. (2010) Self-assembled graphene hydrogel via a one-step hydrothermal process, *ACS Nano*, **4**, 4324–4330.

164. Fan Z., Yan J., Zhi L., (2010) A three-dimensional carbon nanotube/graphene sandwich and its application as electrode in supercapacitors, *Adv Mater*, **22**, 3723–3728.

165. Wu Z.S., Ren W., Wang D.W., Li F., Liu B., Cheng H.M. (2010) High-energy MnO_2 nanowire/graphene and graphene asymmetric electrochemical capacitors, *ACS Nano*, **4**, 5835–5842.

166. Yang X., Zhu J., Qiu L., Li D. (2011) Bioinspired effective prevention of restacking in multilayered graphene films: Towards the next generation of high-performance supercapacitors, *Adv Mater*, **23**, 2833–2838.

167. Palacin M.R. (2009) Recent advances in rechargeable battery materials: A chemist's perspective, *Chem Soc Rev*, **38**, 2565–2575.

168. Yuan C.Z., Gao B., Shen L.F., et al. (2011) Hierarchically structured carbon-based composites: Design, synthesis and their application in electrochemical capacitors, *Nanoscale*, **3**, 529–545.

169. Ma J., Zhang J., Xiong Z., Yong Y., Zhao X.S. (2011) Preparation, characterization and antibacterial properties of silver-modified graphene oxide, *J Mater Chem*, **21**, 3350–3352.

170. Wang H., Casalongue H.S., Liang Y., Dai H. (2010) $Ni(OH)_2$ nanoplates grown on graphene as advanced electrochemical pseudocapacitor materials, *J Am Chem Soc*, **132**, 7472–7477.

171. Lv W., Sun F., Tang D.-M., et al. (2011) A sandwich structure of graphene and nickel oxide with excellent supercapacitive performance, *J Mater Chem*, **21**, 9014–9019.

172. Wu Z.-S., Wang D.-W., Ren W., et al. (2010) Anchoring hydrous RuO_2 on graphene sheets for high-performance electrochemical capacitors, *Adv Funct Mater*, **20**, 3595–3602.

173. Zhang J.T., Jiang J.W., Zhao X.S. (2011) Synthesis and capacitive properties of manganese oxide nanosheets dispersed on functionalized graphene sheets, *J Phys Chem C*, **115**, 6448–6454.

174. Yan J., Fan Z., Wei T., Qian W., Zhang M., Wei F. (2010) Fast and reversible surface redox reaction of graphene–MnO_2 composites as supercapacitor electrodes, *Carbon*, **48**, 3825–3833.

175. Zhang J.T., Ma J.Z., Jiang J.W., Zhao X.S. (2010) Synthesis and capacitive properties of carbonaceous sphere@MnO_2 rattle-type hollow structures, *J Mater Res*, **25**, 1476–1484.

176. Wang H., Hao Q., Yang X., Lu L., Wang X. (2009) Graphene oxide doped polyaniline for supercapacitors, *Electrochem Commun*, **11**, 1158–1161.

177. Zhang J., Zhao X.S. (2012) Conducting polymers directly coated on reduced graphene oxide sheets as high-performance supercapacitor electrodes, *J Phys Chem C*, **116**, 5420–5426.

178. Murugan A.V., Muraliganth T., Manthiram A. (2009) Rapid, facile microwave-solvothermal synthesis of graphene nanosheets and their polyaniline nanocomposites for energy strorage, *Chem Mater*, **21**, 5004–5006.

179. Xu J., Wang K., Zu S.-Z., Han B.-H., Wei Z. (2010) Hierarchical nanocomposites of polyaniline nanowire arrays on graphene oxide sheets with synergistic effect for energy storage, *ACS Nano*, **4**, 5019–5026.

180. Xu C., Du H., Li B., Kang F., Zeng Y. (2009) Asymmetric activated carbon-manganese dioxide capacitors in mild aqueous electrolytes containing alkaline-earth cations, *J Electrochem Soc*, **156**, A435–A441.

181. Fan Z., Yan J., Wei T., et al. (2011) Asymmetric supercapacitors based on graphene/MnO_2 and activated carbon nanofiber electrodes with high power and energy density, *Adv Funct Mater*, **21**, 2366–2375.

182. Zhang J., Jiang J., Li H., Zhao X.S. (2011) A high-performance asymmetric supercapacitor fabricated with graphene-based electrodes, *Energy Environ Sci*, **4**, 4009–4015.

183. Wang H., Liang Y., Mirfakhrai T., Chen Z., Casalongue H., Dai H. (2011) Advanced asymmetrical supercapacitors based on graphene hybrid materials, *Nano Res*, **4**, 729–736.

184. Yu G., Hu L., Vosgueritchian M., et al. (2011) Solution-processed graphene/MnO_2 nanostructured textiles for high-performance electrochemical capacitors, *Nano Lett*, **11**, 2905–2911.

185. Hu L., Choi J.W., Yang Y., et al. (2009) Highly conductive paper for energy-storage devices, *Proc Natl Acad Sci USA*, **106**, 21490–21494.

Chapter 9

Chemical Synthesis of Graphene and Its Applications in Batteries

Xufeng Zhou and Zhaoping Liu
Ningbo Institute of Material Technology and Engineering,
Chinese Academy of Sciences, Ningbo, Zhejiang 315201, P. R. China
liuzp@nimte.ac.cn

9.1 Introduction

Graphene is undoubtedly one of the most significant discoveries in the first decade of the 21st century, and has received continuous and increasing attentions all over the world. The preparation of free-standing and high quality single-layer graphene and the revealing of its amazing physical properties brought Prof. A. Geim and Prof. K. Novoselov a Nobel Prize only six years after their pioneering works [1]. Comparing with its relatively simple atomic structure, the physicochemical properties of graphene is astonishing and attractive to scientists worldwide [2]. Graphene has ultrahigh intrinsic mobility, high Young's modulus, excellent electrical and thermal conductivity, as well as large theoretical specific surface area. Meanwhile, the robust yet flexible 2D nanostructure of graphene

---

Two-Dimensional Carbon: Fundamental Properties, Synthesis, Characterization, and Applications
Edited by Yihong Wu, Zexiang Shen, and Ting Yu
Copyright © 2014 Pan Stanford Publishing Pte. Ltd.
ISBN 978-981-4411-94-3 (Hardcover), 978-981-4411-95-0 (eBook)
www.panstanford.com

provides infinite possibilities in construction of graphene-based materials with various structures and morphologies for diverse applications. So far, graphene has exhibited broad application potentials in next-generation transistors, energy storage materials, functional composite materials, flexible and transparent display, etc. The attractive prospect of graphene thus requires facile and mass preparation of this novel carbon material to meet increasing demands.

In this chapter, the development and current status of chemical synthesis of graphene will be introduced. Then, the application of chemically-derived graphene in energy-storage devices, mainly Li-ion batteries and supercapacitors, will be briefly discussed.

9.2 Chemical Synthesis of Graphene

The attractive application potentials of graphene demand highly efficient, low cost, and large scale preparation of this novel carbon material. Basically, the preparation process can be divided into two pathways, the top-down process and the bottom-up process. The former adopts exfoliation of graphitic carbon sources (graphite or its derivatives in most cases) into graphene sheets, while the latter synthesizes graphene from carbonaceous molecules. The synthesis methods introduced in this chapter focus on the exfoliation of graphite or its derivatives by a chemical process and mainly in solutions, which is considered to be the approach that can most likely realize the mass production of graphene in the near future.

9.2.1 Direct Exfoliation of Graphite

Though the π–π interaction between adjacent graphene layers in graphite is relatively weak comparing with covalent sp^2 hybridized carbon–carbon bonds within each graphene sheet, large forces are always needed to conquer the barrier to obtain graphene. The micromechanical exfoliation of pristine graphite, such as using scotch tape to exfoliate graphite as Prof. Geim did in his discovery of graphene [1], can produce high quality samples. However, the yield is ultra-low, which hinders its broad application. The direct exfoliation of graphite into graphene nanosheets can also be achieved in liquid phase. When graphite powder is dispersed in suitable organic solvents, such as N-methyl-pyrrolidone (NMP), single-layer graphene sheets can be obtained and steadily dispersed in a concentration up to 0.01 mg ml^{-1} [3]. It was pointed out that solvents whose surface energy matched that of graphene could minimize the energy cost of exfoliation to achieve better exfoliation effects. Though the graphene products are almost defect free, the low monolayer yield of ~1 wt% is far from the demand of large scale production.

Recently, Behabtu et al. reported a spontaneous exfoliation of graphite into single-layer graphene in chlorosulfonic acid without functionalization or sonication [4]. In their experiments, graphite was stirred in the acid for a minimum of 2 days, and then the dispersion was centrifuged for 12 h. Graphene sheets were obtained in the supernatant. The isotropic concentration of graphene could reach ~2 mg ml^{-1}. Nevertheless, chlorosulfonic acid is a superacid with ultra-strong oxidizability, and decomposes immediately when it contacts moisture. Therefore, the whole experiment needs to be operated in a glove box with extreme care. Thus, the scale up of this method still faces enormous obstacles.

9.2.2 Graphene from Graphite Oxide

Preparation of graphene by direct exfoliation of graphite either has an ultra-low yield or requires harsh experimental conditions, which can be attributed to the difficulty in breaking the π–π interaction between graphene layers. Therefore, the decrease in the interlayer interaction in graphite is critical to acquire a high throughput production of graphene, and chemical treatment of graphite is considered to be an effective way.

Oxidation of graphite is the most commonly used method. Since the first report in 1860 by Brodie [5], several oxidant systems have been proposed [6, 7]. Graphite oxide possesses a structure as illustrated in Fig. 9.1 [8]. Electron rich C=C bonds in graphene sheets can react with strong oxidants to form oxygenous groups covalently bonded to carbon atoms and distributed on both sides and fringes of graphene sheets. Hence, the planar structure of graphene sheets and the π–π interaction in between are destroyed, and the layer-to-layer distance is expanded. Moreover, the hydroxyl, carbonyl, and ether groups convert intrinsically hydrophobic graphene sheets into hydrophilic ones. Consequently, the exfoliation of graphite oxide is much easier than exfoliation of pristine graphite, especially in polar solvents.

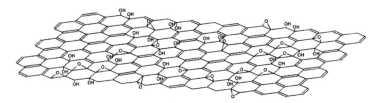

Figure 9.1 Structural model of a single-layer graphite oxide [8]. © 1998 Elsevier

In fact, the first attempt to exfoliate graphite oxide into ultra-thin sheets was carried out in as early as 1962 by Boehm et al. [9]. Transmission electron microscopy, X-ray diffraction, and methylene-blue-based surface area determination gave an estimated average thickness of the products in the range of 2–3 layers, and certain amounts of single layers were also observed. This research has not been paid much attention until the landmark scientific paper on graphene was published in 2004 by Prof. Geim and his colleagues [1]. Scientists soon turned this original discovery into a general way to prepare single-layer graphene. In 2006, Stankovich and co-workers applied sonication assisted exfoliation to produce single-layer graphite oxide, structurally equaling to graphene oxide, with a thickness of ~1 nm (Fig. 9.2) [10]. The following reduction of graphene oxide by dimethylhydrazine resulted in graphene. Numerous reports were then published on the synthesis and applications of graphene using similar reaction route.

The structural parameters, including thickness, lateral size, defects, and possible functional groups of the graphene product are closely related to the oxidation–exfoliation–reduction process, which needs to be carefully adjusted to control the physicochemical properties of as-prepared graphene. Graphite oxide is extremely hydrophilic, thus, the exfoliation of graphite oxide is mainly conducted in aqueous solutions. Sonication is considered to be effective to separate graphite oxide into ultra-thin graphene oxide sheets. Nevertheless, the strong energy derived from sonication treatment not only exfoliates graphite oxide along the c direction, but also breaks the basal plane of graphene oxide, usually resulting in small lateral size (several hundreds of nanometers to several micrometers), even the graphite raw material used has a large size of up to several hundreds of micrometers.

Figure 9.2 (a) SEM image and digital image (inset) of natural graphite raw material and (b) AFM image of graphene oxide. The inset is the photo of aqueous graphene oxide dispersion [10]. © 2006 Nature Publishing Group.

Our group has developed a modified exfoliation pathway using mild shaking instead of sonication to prevent graphene oxide sheets from fragmentation [11]. As shown in Fig. 9.3, using 50 mesh (~300 μm) natural graphite as the raw material, the graphene oxide obtained after 12 h of shaking in water with a tabletop shaker had a mean lateral size of ~100 μm, and some sheets even exceeded 200 μm, which was apparently larger than ultrasonically treated samples. The characterization of over 100 graphene oxide sheets using atomic force microscopy revealed that >95% of the sample had a thickness of ~1 nm, characteristic of single-layer graphene oxide, which implied that the shearing force induced by shaking is strong enough to separate graphene oxide sheets from each other in water. Moreover, the size of graphene oxide could be simply tuned by changing the size of graphite. For instance, using 50 μm graphite as the raw material, graphene oxide sheets with a mean size of ~10 μm could be prepared.

The exfoliation can also be realized through a solid-state process. Schnieppet et al. firstly reported a simultaneous exfoliation and reduction of graphene oxide by rapid heating to 1050 °C under inert atmosphere [12]. The rapid decomposition of oxygenous groups on graphene oxide at high temperatures generated large amounts of gaseous species which instantly expanded graphite oxide into ultrathin sheets. Meanwhile, the removal of oxygenous groups gave rise to the reduction of graphene oxide, and the electrical conductivity could be partially restored in the graphene products. Later on, it was discovered that similar process could also be achieved in vacuum yet at much lower temperatures of 200–300 °C, which could apparently reduce the cost of graphene production [13].

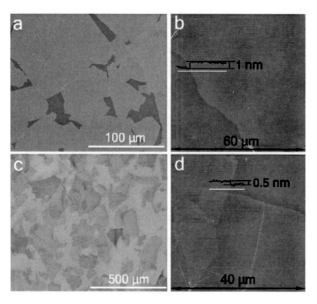

Figure 9.3 (a) SEM and (b) AFM images of ultra-large graphene oxide sheets prepared by mild shaking. (c) SEM and (d) AFM images of ultra-large graphene sheets obtained by chemical reduction [11]. © 2010 Royal Society of Chemistry.

The reduction process which finally turns graphene oxide into graphene also plays an important role in the whole preparation course. Similar to the exfoliation process, reduction can also be achieved either in solutions or in a solid state. Strong chemical reducing agents, such as hydrazine [14], dimethylhydrazine [10], and sodium borohydride [15], have been commonly used to remove oxygenous functional groups. However, they are usually toxic. Therefore, environment friendly reducing agents have been broadly investigated. Ascorbic acid (vitamin C) was the first green reductant introduced [16–18], which could achieve effective deoxygenation and match the efficiency of hydrazine in terms of reduction ability. Saccharides [19] and bovine serum albumin [20] were also successfully applied to reduce graphene oxide.

It was also discovered that strong alkaline solutions were capable of reducing graphene oxide [21]. Deoxygenation occurred when alkaline graphene oxide suspensions were heated to 50–90 °C. Hydrothermal treatment is another facile method to remove oxygenous groups via dehydration reactions [22]. This method does not introduce any impurities to the product and can be easily scaled up for practical use. Similarly, microwave assisted solvothermal reduction of graphene oxide has been carried out in various solvents, including tetraethyleneglycol, *N,N*-dimethylformamide, ethanol, 1-butanol, and water [23].

Electrochemical reduction of graphene oxide has been proposed as an alternative green route to graphene [24]. The reaction was carried out in a standard three-electrode cell. Graphene oxide was electrochemically reduced and deposited on the working electrode.

The reducing degree of graphene from solution-based reduction process, however, is limited. It is reported that in most cases, the C/O molar ratio of the graphene products ranges from 5/1 to 10/1 [17]. Therefore, the electrical conductivity of chemically reduced graphene is several magnitudes lower than that of pristine graphene. Graphene can also be produced via a thermal reduction route in vacuum or inert atmosphere [12]. The higher the reaction temperature, the longer the reaction time, and this results in higher reducing degree. Comparing with solution-based reducing process, thermally treated graphene has lower oxygen content and higher electrical conductivity [25]. In addition, thermal treatment of graphene oxide under reductive atmosphere, such as H_2, is a more effective way to restore the planar structure of graphene [26].

Even though most of the oxygenous groups on graphene oxide can be removed by either chemical or thermal reduction, the restoration to a perfect graphene structure cannot be achieved. In depth characterization of reduced graphene sheets using atomic resolution TEM reveals a large number of defects [27, 28]. As shown in Fig. 9.4, intact graphene islands of variable size between 3 and 6 nm are interspersed with defect areas dominated by clustered pentagons and heptagons. These disordered areas induce strain as well as in-plane and out-of-plane deformations. Therefore, the presence of a remarkable amount of topological defects enormously deteriorates the electron mobility and conductivity of reduced graphene.

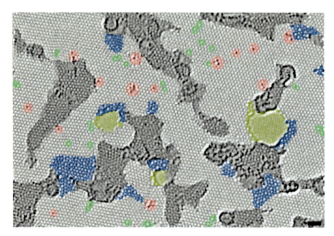

Figure 9.4 Aberration-corrected TEM image of a single layer reduced graphene oxide membrane with color added to highlight the different features. The defect free crystalline graphene area is displayed in the original light gray color. Contaminated regions are shaded in dark gray. Blue regions are the disordered single-layer carbon networks, or extended topological defects. Red areas highlight individual ad-atoms or substitutions. Green areas indicate isolated topological defects, that is, single bond rotations or dislocation cores. Holes and their edge reconstructions are colored in yellow. Scale bar 1 nm [27]. © 2010 American Chemical Society.

9.2.3 Graphene from Graphite Intercalated Compounds

The drawbacks of graphite oxide derived graphene drive the need of other methods that generate fewer defects in the graphene sheets. The insertion of ions or molecules into adjacent graphene layers can break the interlayer π–π interaction in graphite and expands the layer-to-layer distance, while reserves the structure perfection, to a large extent, of graphene layers. Such graphite intercalated compounds (GICs) have been discovered and broadly studied for a long period of time [29]. Expandable graphite is one of the representative GICs. The intercalated species in expandable graphite decompose and generate great amount of gas at high temperatures, which instantly expands scaly graphite into worm-like products with an apparent volume expansion over hundreds of times. Graphite thus is exfoliated along *c*-direction into nanosheets with a mean thickness of tens of nanometers in most cases. Such nanosheets can only be regarded as graphite nanoplates rather than graphene. Therefore, efforts have been made to optimize the intercalation and exfoliation conditions to reduce the thickness of nanosheets as much as possible.

A homogeneous suspension of graphene sheets or ribbons was produced by stirring potassium GIC, $K(THF)_xC_{24}$ (x=1–3) in NMP (THF stands for tetrahydrofuran) [30]. As-prepared graphene sheets have a thickness of 0.36 nm measured by scanning tunneling microscopy, which is characteristic of one-atom thick graphene. Alternatively, thin graphene sheets could also be prepared using ClF_3 as the inorganic volatile intercalating agent [31]. Comparing with common GICs, the thermal treatment of fluorinated GIC $C_2F \bullet nCF_3$ not only generated high vapor pressure that expands the graphite matrix, but also formed gaseous fluorocarbons (mainly CF_4) and other gaseous products by interaction between the graphite matrix and the intercalants, which made a large number of defects in the exfoliated graphite that was essential for easy dispersion. The expanded product had a surface area of 250–280 m^2g^{-1}, ~5 times higher than the conventional expanded graphite, which indicated a more highly expanded state and smaller thickness of graphene sheets.

An approach of exfoliation–reintercalation–expansion of graphite was taken to prepare high quality graphene as illustrated in Fig. 9.5 [32]. The thermally expanded graphite was first reintercalated with oleum, and then further inserted by tetrabutylammonium hydroxide (TBA) in N,N-dimethylformamide (DMF). The sample was sonicated in a DMF solution of 1,2-distearoyl-sn-glycero-3-phosphoethanolamine-N-[methoxy(polyethyleneglycol)-5000] (DSPE-mPEG) for 60 min to form a homogeneous suspension. After the removal of large pieces using centrifugation, the supernatant with large amounts of graphene sheets was obtained and could be transferred to other solvents including water and organic solvents.

In another report, thermally expanded graphite was mixed with 7,7,8,8-tetracyanoquinodimethane by grinding and several drops of dimethyl sulfoxide were added. After maintaining at room temperature for 12 h and drying under vacuum, the mixture was dispersed in aqueous solution of KOH and sonicated for 90 min. After centrifugation, the supernatant containing single or few-layer graphene sheets was obtained [33].

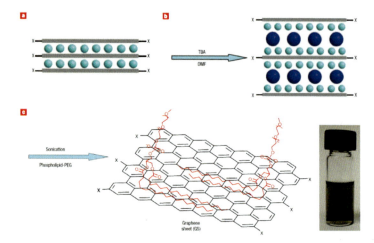

Figure 9.5 (a) Schematic representation of the exfoliated graphite reintercalated with sulfuric acid molecules (teal spheres) between the layers. (b) Schematic representation of TBA (blue spheres) insertion into the intercalated graphite. (c) Schematic representation of graphene sheets coated with DSPE-mPEG molecules and a photograph of a DSPE-mPEG/DMF solution of graphene [32]. © 2008 Nature Publishing Group.

9.3 Applications of Graphene in Batteries

9.3.1 Li ion Batteries

Lithium ion batteries (LIB) have nowadays been widely used in daily life, mainly as power supplies for portable and small-scale electronic equipments. It is anticipated that LIB will be applicable to electric vehicles and large-scale energy-storage systems in the near future due to its high energy density, long life, and safety. However, the charge/discharge performance of LIB needs to be improved to achieve such goals. Development of high-performance electrode materials is always a major task in this area. Carbon materials, such as graphite, carbon black, carbon nanotubes, and amorphous pyrolytic carbon, have been widely used in electrode materials for LIB. As a new member of the carbon materials, graphene with excellent conductivity, large surface area, and robust yet flexible sheet-like structure has exhibited attractive perspectives in the application in LIB.

Graphene-based electrode material for LIB was firstly reported in 2008 (Fig. 9.6) [34]. The specific capacity of graphene nanosheets was found to be 540 mAh g^{-1}, which was much larger than that of graphite. Soon, graphene anode materials with various synthetic routes and structures were successively reported [35–38]. Traditional graphite anode has a theoretical capacity of 372 mAh g^{-1}, corresponding to the atomic ratio of Li/C = 6/1 in the fully inserted Li/graphite composite. In addition, closely stacked graphene sheets in graphite may kinetically restrict the insertion and extraction of Li ions, inducing fast capacity fade at high charge/discharge rates. The electrochemical reaction of Li ion with graphene is different from the insertion/extraction mechanism for graphite. The large surface area and single-layer feature of graphene favors a fast surface reaction. With no spatial restriction, graphene is able to accommodate larger amounts of Li ions in a shorter period of time. Therefore, the reported graphene anodes exhibit much higher reversible capacities than graphite, even up to ~1000mAh g^{-1} [38], and their rate performance also excels graphite. It is important to note that the unique electrochemical reaction process of graphene causes a totally different charge/discharge curve from that of graphite. No typical potential plateau at ~0.2 V (vs. Li/Li$^+$) of graphite anode can

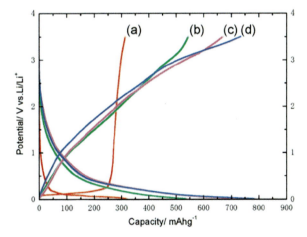

Figure 9.6 Charge/discharge profiles of (a) graphite, (b) graphene nanosheet, (c) graphene nanosheet + carbon nanotube, and (d) graphene nanosheet + C$_{60}$ at a current density of 0.05 A g^{-1} [34] © 2008 American Chemical Society.

be observed in the case of graphene. Instead, a continuous potential rise or fall is obtained in the whole potential range of the charge or discharge curves, respectively, as shown in Fig. 9.6. Such high charge potential, however, is not appropriate for an anode material, as it will dramatically reduce the voltage of the battery. Therefore, new structural design or modification needs to be carried out on graphene to make a balance of the capacity and the potential.

Comparing with graphene anode, graphene-based composite electrode materials in which graphene serves as a functional component to enhance the electrochemical performance is believed to be a more significant application direction for graphene in LIB. Following are some examples of how graphene improves the charge/discharge performance through its unique structure and outstanding physicochemical properties. The first graphene-based composite electrode material was graphene/SnO_2 composite reported in 2009 (Fig. 9.7) [39]. Structural characterizations showed that graphene nanosheets were distributed homogeneously between the loosely packed SnO_2 nanoparticles and an anoporous structure was formed. The obtained SnO_2/graphene composite material exhibited a reversible capacity of 810 mAh g^{-1}, which was higher than 550 mAh g^{-1} of bare SnO_2 nanoparticle. The cycling stability of graphene modified SnO_2 material was also enormously improved. After 30 charge/discharge cycles, the capacity of SnO_2/graphene composite still remained 570 mAh g^{-1}, while the capacity of bare SnO_2 nanoparticle dropped rapidly to 60 mAh g^{-1} only after 15 cycles. In this composite material, the volume expansion of SnO_2 nanoparticles was limited due to their dimensional confinement by surrounding graphene nanosheets, and the nanopores between the two components could be used as buffering space during charge/discharge, resulting in superior cycling performance. SnO_2/graphene composites were also prepared by our group using a simple solution-based oxidation–reduction reaction between graphene oxide and $SnCl_2$ [40]. In the composite material, uniform SnO_2 nanocrystals with diameters of 3–5 nm are homogeneously clung to the graphene matrix. With an optimum SnO_2/graphene molar ratio of 3.2/1, the composite electrode material can deliver a charge capacity of 840 mAh g^{-1} (with capacity retention of 86%) after 30 charge/discharge cycles at a current density of 67 mA g^{-1}. At even higher current densities of 400 and 1000 mA g^{-1}, it still delivers the charge capacities of 590 and 270 mAh g^{-1} up to 50 cycles, respectively.

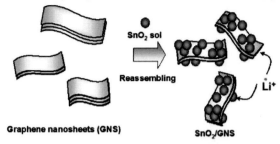

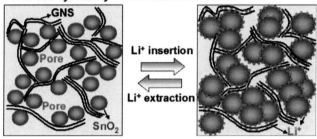

Figure 9.7 Illustration for the synthesis and the structure of SnO$_2$/graphene nanosheet composites [39]. © 2009 American Chemical Society.

The poor cycling stability of other metal oxide anode materials, such as Mn$_3$O$_4$, Fe$_3$O$_4$, and Co$_3$O$_4$ can also be greatly improved by graphene. For example, a flexible interleaved composite of graphene decorated with Fe$_3$O$_4$ nanoparticles was synthesized by an in situ reduction of iron hydroxide between graphene nanosheets [41]. The composite material delivered a reversible specific capacity approaching 1026 mAh g^{-1} after 30 cycles at 35mA g^{-1}, and 580 mAh g^{-1} after 100 cycles at 700 mA g^{-1}. In this case, graphene nanosheets also buffered the volume change of Fe$_3$O$_4$ nanoparticles during charge/discharge processes, and prevented them from agglomeration. Meanwhile, graphene acted as an excellent conductive agent that facilitated the electron transportation within the active materials. In another case, Mn$_3$O$_4$ nanoparticles loaded on reduced graphene oxide sheets were prepared by a two-step solution phase reaction [42]. In contrast to free Mn$_3$O$_4$ particles, the intimate interaction between Mn$_3$O$_4$ and graphene improved the electrical conducting of insulating Mn_3O_4. The composite showed a high specific capacity of ~900 mAh g^{-1}, near its theoretical capacity, with good rate capability and cycling stability. Similarly, Co_3O_4 nanoparticles anchored on conducting graphene were reported as a high-performance anode material [43]. The composite material exhibited a large reversible capacity of 935 mAh g^{-1} after 30 cycles, high Coulombic efficiency above 98%, and good rate capability.

Graphene modification is also successfully introduced in metal anodes, such as silicon, which have even larger volume change ratio than metal oxide anode materials. Si/graphene composite anode materials were first reported in 2010 [44]. A self-supporting Si–graphene composite paper was prepared by filtration of an aqueous mixture of commercial Si nanoparticles and graphene oxide, and heat treatment afterwards. The Si nanoparticles were well dispersed between graphene, while the reduced graphene sheets formed a continuous, highly conducting 3D network that served as a structural scaffold to anchor Si. Such sandwich-like structure induced a remarkably improved cycle stability of Si. The anode material electrode showed a capacity of >2200 mAh g^{-1} after 50 cycles and >1500 mAh g^{-1} after 200 cycles. The authors from the same group further optimized the structural design of such Si/graphene composite by generating nanometer-sized in-plane vacancies on graphene sheets using acid-sonication treatment as illustrated in Fig. 9.8 [45]. Such vacancies benefited fast ion transport in 3D graphene scaffold, resulting in unachievable combination power capability and energy storage capacity. The Si/graphene composite electrode achieved an ultra-high reversible capacity of 3200 mAh g^{-1} at current density of 1 A g^{-1}, and the capacity loss was only 0.14% per cycle within 150 charge/discharge cycles. When the current density was increased to 8 A g^{-1}, an unprecedentedly high capacity of 1100 mAh g^{-1} could be reached, and degraded at only ~0.34% per cycle for 150 cycles.

Some cathode materials, typically olivine phosphate materials, including $LiFePO_4$ and $LiMnPO_4$ have relatively low electron conductivity. Carbon coating is commonly adopted to raise their conductivity and thus improve the charge/discharge performance. Comparing with pyrolytic carbon coatings derived from carbonaceous molecules, graphene with a much higher conductivity and unique 2D sheet-like nanostructure is expected to have better effects. Recently,

Applications of Graphene in Batteries | 261

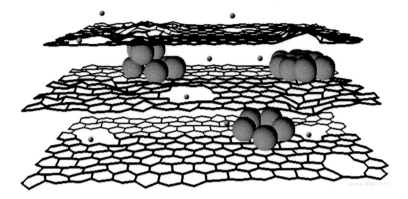

Figure 9.8 A schematic drawing of a section of a composite electrode material constructed with a graphene scaffold with in-plane carbon vacancy defects and Si nanoparticles [45]. © 2011 Wiley.

graphene modified LiFePO$_4$ cathode materials having excellent rate capability and cycling stability was prepared in our lab (Fig. 9.9) [46]. The composite was prepared with LiFePO$_4$ nanoparticles and graphene oxide nanosheets by spray-drying and annealing processes. The careful SEM and TEM characterizations revealed that the LiFePO$_4$ primary nanoparticles embedded in micro-sized spherical secondary particles were wrapped homogeneously and loosely with a graphene 3D network. The continuous graphene network had a lower oxygen content and a higher graphitization degree than conventional pyrolytic carbon coating layers, which provided a fast transportation pathway for electrons within the secondary particles. Meanwhile, considerable amounts of voids between the LiFePO$_4$ nanoparticles and the graphene sheets benefited the diffusion of Li ions. The composite cathode material could deliver a capacity of 70 mAh g^{-1} at an ultra-high discharge rate of 60 C, while the pyrolytic carbon coated LiFePO$_4$ material only reached a discharge capacity of 54 mAh g^{-1} at 30 C. The cycling stability of LiFePO$_4$ was also notably improved by graphene modification. A capacity decay rate of <15% was achieved when cycled under 10 C charging and 20 C discharging for 1000 times. Consequently, such LiFePO$_4$/graphene composite cathode materials with their controlled microstructure and excellent rate performance and cycling stability offer great potential for applications in high-power electrical sources.

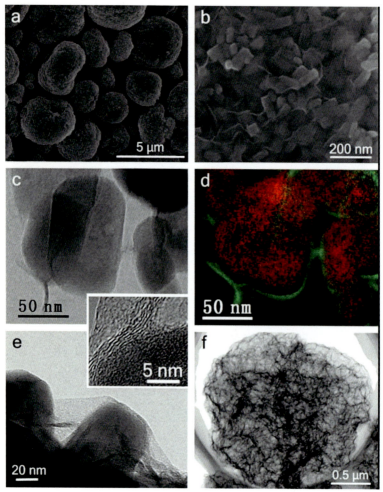

Figure 9.9 (a, b) SEM images of LiFePO$_4$/Graphene composites at different magnifications. (c) TEM image and (d) corresponding elemental map using electron energy loss spectroscopy (EELS) at the same area showing graphene sheets wrapping on LiFePO$_4$ nanoparticles, where red represents LiFePO$_4$ nanoparticles, and green represents graphene sheets. (e) TEM image of the edge of individual composite microsphere. The inset is the high resolution TEM image illustrating the 3–5 layers of graphene sheets on the surface of an LiFePO$_4$ nanoparticle. (f) TEM image of a 3D graphene network obtained by removing LiFePO$_4$ nanoparticles with an HCl solution [46]. © 2011 Royal Society of Chemistry.

Similar olivine-type lithium transition metal phosphates with a composition of $LiMn_{0.75}Fe_{0.25}PO_4$ grown on graphene sheets were reported to have ultrahigh rate performance [47]. Using a two-step solution-based synthetic route, $LiMn_{0.75}Fe_{0.25}PO_4$ nanorods with ultra-small size were deposited on highly conducting mildly oxidized graphene sheets. The electrical conductivity measured from pellets of the hybrid material was 0.1–1 S cm^{-1}, which is 10^{13}–10^{14} times higher than that of pure $LiMnPO_4$. Specific capacities of 132 mAh g^{-1} and 107 mAh g^{-1} were obtained at high discharge rates of $20C$ and $50C$, which is 85% and 70% of the capacity at $C/2$, respectively, and better than the rate performance of any pure or doped $LiMnPO_4$ cathode material ever reported.

Though electrode active material is the majority component in the electrode, conductive additives that usually have a weight ratio less than 10% in the electrode are also essential in LIBs. Conductive additives form a conducting network that ensures transportation of electrons from inside the electrode material to the current collector and vice versa. Carbon materials, including graphitic carbon, carbon black, carbon fiber, and carbon nanotube, are major conductive additives for LIBs. Recently, graphene conductive additives were applied substituting the commercial carbon conductive additives [48]. It showed that the charge/discharge performance of cells with 2 wt% of graphene excelled even that with 20 wt% of carbon black. With the same loading amount, graphene also prevails over carbon nanotube, a high-end conductive additive in the current LIB industry. The outstanding performance of graphene conductive additive can be ascribed to its high conductivity, large contact area between graphene and active materials, and better ability to form conductive networks. It should be mentioned that the graphene used in this study was prepared by a chemical oxidation–reduction method, thus its conductivity is limited. It can be expected that using graphene with higher electrical conductivity can further improve the performance of LIBs. Currently, few-layer defect-free graphene sheets as high-performance conductive additives are under investigation in our lab. Preliminary results show that the resistance of the electrode can be enormously reduced using such highly conductive graphene instead of carbon black, carbon nanotube, or chemically reduced graphene, and the rate capability, cycling stability as well as the thermal effects of the cells are remarkably improved. We believe that graphene-based conductive additives will be widely used in LIBs.

9.3.2 Supercapacitors

Though the energy-storage mechanism of supercapacitor is different from the traditional electrochemical batteries, the developments of high-performance supercapacitors with the energy density approaching batteries, and the emergence of hybrid cells combining supercapacitor and battery, gradually blur the boundaries between two areas. For a classical supercapacitor, which is also called electrochemical capacitor, charges are stored in electric double layers formed at the interface of the electrode materials and the electrolytes. Therefore, the capacity of a supercapacitor strongly depends on the surface area of the electrode materials. Activated carbon with a large specific surface area of up to $\sim$2000 m^2 g^{-1} is the most commonly used electrode material for supercapacitors. However, majority of the pores in the activated carbon are micropores which are not accessible by electrolytes. Thus the effective surface that can form electric double layers is limited. Compared with activated carbon, graphene has an even larger theoretical surface area of 2630 m^2 g^{-1}, which can be effectively utilized through a proper structure construction of the graphene electrode. The first graphene-based supercapacitor was reported in 2008 (Fig. 9.10) [49]. Agglomerate of chemically reduced graphene which had a Brunauer, Emmett and Teller (BET) surface area of 705 m^2 g^{-1} was used as the electrode material and assembled into a symmetric supercapacitor. Specific capacitances of 135 and 99 F g^{-1} in aqueous and organic electrolytes were obtained, respectively. In addition, high electrical conductivity of graphene gave consistently good performance over a wide range of voltage scan rates.

Besides the high surface area, functional groups, such as hydroxyl, carbonyl, and carboxyl groups attached to the chemically synthesized graphene sheets give additional pseudocapacitance. Lin et al. reported the superior capacitance of functionalized graphene prepared by controlled reduction of grapheneoxide [50]. The density of functionalities on the graphene sheets could be finely tuned by a thermal treatment of graphene oxide dispersed in dimethylformamide at a moderate temperature. A specific capacitance of up to 276 F g^{-1} at 0.1 A g^{-1} in a 1 M H_2SO_4 electrolyte was obtained, which was much higher than that of highly reduced graphene with less functional groups. A recent research even suggested the direct use of non-reduced graphene oxide as the electrode material for supercapacitor rather than the reduced one due to its higher capacitance [51].

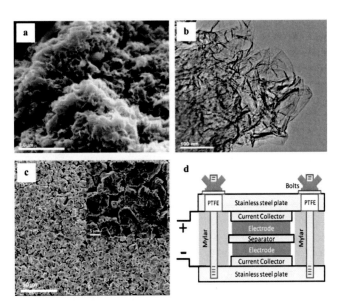

Figure 9.10 (a) SEM and (b) TEM images showing the morphology and structure of chemically reduced graphene, (c) low and high (inset) magnification SEM images of the surface of graphene electrode in the supercapacitor, and (d) schematic representation of the configuration of the supercapacitor cell for test [49]. © 2008 American Chemical Society.

Therefore, it is important to control the surface chemistry of graphene for supercapacitor applications. Some efforts have been made to graft functional groups onto graphene by post treatment. In one case, nitrogen-doped graphene was prepared using a plasma doping process [52]. It not only increased the specific capacitance to 280 F g^{-1}, four times higher than that of pristine graphene, but also preserved excellent cycling stability (>200000 times).

Similar to traditional carbon-based materials for supercapacitors, the pore structure of graphene electrode material also directly affects its capacitance because of the surface dominated reactions in supercapacitors. An optimum structure should ensure the exposure of the entire surface of graphene to the electrolytes, and provide suitable pore size which is easily accessible by ions and preserves the bulk density of graphene as high as possible. In a recently published paper, Prof. Ruoff and his colleagues applied a chemical activation process which was commonly used in the production of activated carbon to produce porous graphene with high gravimetric capacitance and energy density [53]. As shown in Fig. 9.11, microwave exfoliated graphene oxide was mixed with KOH and activated at 800 °C for 1 h. The activated graphene had an ultra-high BET specific surface area of $\sim$3100 m^2 g^{-1}, which was even larger than the theoretical value of graphene, and pore width in the range of 0.6–5 nm. High resolution TEM characterization showed the presence of a dense pore structure, and indicated that the walls of the pores were composed of highly curved carbon sheets, of predominantly single layer thickness. The activated graphene yielded a specific capacitance above 150 F g^{-1} in the tetraethylammonium tetrafluoroborate in acetonitrile (TEA BF_4/AN) electrolyte with a current density of 0.8 A g^{-1}, and 166 F g^{-1} in the 1-Butyl-3-methylimidazoliumtetrafluoro borate,98% in acetonitrile (BMIM BF4/AN) electrolyte at a current density of 5.7 A g^{-1}, corresponding to an energy density of $\sim$70 Wh kg^{-1}. Based on a weight ratio of 30% for the electrode active material in a packaged supercapacitor, the estimated energy density for an activated graphene-based supercapacitor device was $\sim$20 Wh kg^{-1}, which was four times higher than that of commercial activated carbon-based supercapacitors, and almost equaled to that of lead acid batteries. Such novel electrode material also exhibited excellent cycling stability. After 10000 cycles at a current density of 2.5 A g^{-1}, 97% of the initial capacitance was retained. The authors believed that the activation process could be easily scaled-up and graphene-based energy storage devices might be realized in a short period of time.

The combination of graphene with pseudocapacitive materials is another important aspect for the application of graphene in supercapacitors. MnO_2 is one of the most extensively studied active materials due to its relatively high capacitance, environmental compatibility, and low cost. However, its poor electrical conductivity causes low specific capacitance and poor capacitance retention ratio over a wide range of scan rates. Thus, incorporation of MnO_2 with high surface area conductive supports, such as mesoporous carbon and carbon nanotubes, has been proposed to optimize its electrochemical performance. The advantages in both conductivity and surface area impart graphene with a great potential in modification of MnO_2. Chen et al. in 2010 reported for the first time graphene oxide–MnO_2 nanocomposites as high-performance supercapacitive materials [54]. Employing an isopropyl alcohol–water solution as the reaction media, a simultaneous formation of needle-like 1D MnO_2 nanocrystals and exfoliation of graphene oxide sheets was realized.

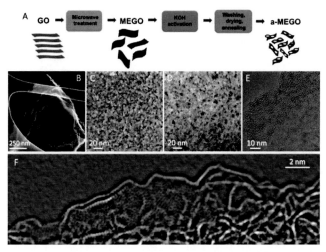

Figure 9.11 (A) A schematic drawing showing the process of activation of graphene. (B) Low and (C) high magnification SEM images of activated graphene. (D) Annular dark field scanning transmission electron microscopy image of the same area as (C) showing micro- and mesopores with a distribution of sizes between 1 and 10 nm. (E) High-resolution phase contrast electron micrograph of the thin edge of the graphene sample, revealing the presence of a dense network of nanometer-scale pores surrounded by highly curved, predominantly single layer carbon. (F) Exit wave reconstructed high resolution TEM image from the edge of the sample, in which the in-plane carbon atoms are clearly resolved, and a variety of *n*-membered carbon rings can be seen [53]. © 2011 Science.

The composite had specific capacitances of 216.0 F g^{-1} and 111.1 F g^{-1} at current densities of 150 mA g^{-1} and 1000 mA g^{-1}, respectively, in 1 M Na$_2$SO$_4$ aqueous solution. It also retained 84.1% (165.9 F g^{-1}) of the initial capacitance after 1000 cycles, while that of the pure MnO$_2$ retained only 69.0%. In another report, Wu et al. designed an asymmetric electrochemical capacitor, using graphene as the negative electrode and a MnO$_2$ nanowire/graphene composite as the positive electrode [55]. This novel type of supercapacitor exhibited an energy density of 30.4 Wh kg^{-1}, which was much higher than that of the symmetric ones based on either graphene electrodes or MnO$_2$ nanowire/graphene electrodes, and those of other MnO$_2$-based asymmetric supercapacitors reported previously. Recently, a graphene/MnO$_2$ nanostructured textile for high-performance electrochemical capacitors was also reported [56]. In the experiment, graphene nanosheets were firstly coated onto 3D porous textiles, and MnO + nanomaterials were then electrodeposited on graphene. The hybrid textile yielded a high specific capacitance of 315 F g^{-1}. An asymmetric electrochemical capacitor was then assembled using graphene/MnO$_2$ hybrid as the positive electrode and single-walled carbon nanotube as the negative electrode. The device exhibited a power density of 110 kW kg^{-1}, an energy density of 12.5 Wh kg^{-1}, and a cycling performance of ~95% capacitance retention over 5000 cycles, which had promising potential in energy-storage applications.

Besides metal oxides, conducting polymers are another type of pseudocapacitive materials with high capacitance. Their poor stability during charge/discharge cycles calls for suitable modification, where graphene also finds its positive influence in this area. Since 2009, graphene/conducting polymer composites for supercapacitors with various compositions and structures have been successively reported. Wang et al. prepared a freestanding paper-like graphene/polyaniline composite using an in situ anodic electropolymerization method [57]. The flexible composite paper had a tensile strength of 12.6 MPa, and a large electrochemical capacitance of 233 F g^{-1}. Similar graphene/polyaniline composite paper was prepared by vacuum filtration of a mixture of chemically reduced graphene and polyaniline nanofibers [58]. Polymer nanofibers were uniformly distributed between graphene sheets, and such structural features induced improved mechanical property, and conductivity 10 times higher than that of pure polyaniline nanofiber films. The composite showed large specific capacitance of 210 F g^{-1} at a current density of 0.3 A g^{-1}, and high capacitance of 155 F g^{-1} could be maintained after 800 charge/discharge cycles at a current density of 3 A g^{-1}. In another case, polyaniline nanowires aligned vertically on graphene oxide sheets were produced [59]. This hierarchically structured composite reached a capacitance of 555 F g^{-1} at a current density of 0.2 A g^{-1}, which was much higher than randomly connected polyaniline nanowires with no graphene oxide substrate. The results implied that a significant synergetic effect could be achieved by controlling the composition and structure at the nanoscale of graphene-based composite supercapacitive materials. The combination of graphene with other conducting polymers, such as polypyrrole was also reported [60].

Most recently, a novel surface-enabled Li ion-exchanging cell using graphene electrodes was proposed to be the next-generation high-power energy storage device [61]. As illustrated in Fig. 9.12, both the cathode and the anode are composed of porous graphene electrode materials with their surfaces in direct contact with electrolytes used in LIB, which enables fast surface adsorption and desorption of Li ions. Li particles of foils are implemented at the anode during cell assembling, which can be ionized at the first discharge cycle to ensure enough Li ions that can be absorbed on the whole surface of graphene electrodes. Meanwhile, oxygenous functional groups on the edge or surface of chemically synthesized graphene sheets provide an additional pseudocapacitance. This new type of energy-storage device actually combines the concept of supercapacitor and LIB, which gives rise to a superb charge/discharge performance. The device is capable of storing an energy density of 160 Wh g^{-1}, which is ~30 times higher than that of conventional symmetric supercapacitors, and comparable to that of LIBs. It also delivered a power density of 100 kW kg^{-1}, which is ~100 times higher than that of LIB.

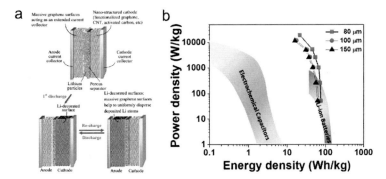

Figure 9.12 (a) The structure of a surface-enabled Li ion-exchanging cell, containing porous graphene and pieces of Li foil or surface-stabilized Li powder at the anode, and porous graphene at the cathode. The lower left and lower right portions show the structure of this cell after its first discharge and after being recharged, respectively. (b) The Ragone plots of graphene surface-enabled Li ion-exchanging cells with different electrode thicknesses, comparing with electrochemical capacitors and Li-ion batteries [61]. © 2011 American Chemical Society.

9.3.3 Other types of Batteries

Though LIB and supercapacitors are being paid most of the attention currently, the development of novel battery systems with higher energy density for next-generation energy-storage devices, such as lithium–sulfur batteries and Li–air batteries, have also attracted much interest. The application of graphene in these areas has also been investigated and some intriguing results have been obtained.

The exceptionally high specific capacity (1672 mAh g^{-1}), low cost, and environmental benignity make sulfur a promising cathode material. However, the low electrical conductivity, the volume variation during charge/discharge, and the dissolution of polysulfides in electrolyte cause poor cycle life and limited capacity of sulfur, which hinder the practical application of Li–S batteries. The modification of sulfur cathode thus becomes a critical task in this area. In 2011, Wang et al. reported the synthesis of graphene–sulfur composite materials by wrapping poly(ehthylene glycol) coated submicrometer sulfur particles with mildly oxidized graphene oxide sheets decorated by carbon black nanoparticles as shown in Fig. 9.13 [62]. The carefully designed structure raised the electrical conductivity of the sulfur cathode, entrapped the polysulfide intermediates, and accommodated some of the volume expansion of sulfur during discharging. At a rate of $C/5$, the composite had an initial capacity of ~750 mAh g^{-1} that decreased gradually to ~600 mAh g^{-1} after 10 conditioning cycles. Within the next 90 cycles, the capacity decreased by only 13%, showing good cycling stability. At a higher rate of $C/2$, even better cycling performance (9% of decay from the 10th to the 100th cycle) was achieved. For comparison, the specific capacity of sulfur particles without graphene wrapping decreased from ~700 mAh g^{-1} to ~330 mAh g^{-1} after 20 cycles. In another paper, graphene/sulfur composite with a different structure was proposed [63]. Using a simple chemical reaction–deposition reaction strategy and a low-temperature thermal treatment process, a uniform coating of sulfur with a thickness of tens of nanometers on graphene oxide sheets was obtained. The functional groups on graphene oxide played important roles in immobilizing S species and prevented the dissolution of polysulfides. Meanwhile, after heat treatment, partially reduced graphene oxide raised the conductivity of the composite. Accordingly, the graphene–sulfur composite showed excellent cycling stability that the capacity retention ratio >95% (from 1000 mAh g^{-1} to 954 mAh g^{-1}) was observed after 50 cycles at 0.1C. The rate capability of up to 2C was also achieved.

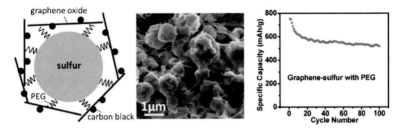

Figure 9.13 The structure model, the SEM image, and the cycling performance of the graphene–sulfur composite from left to right [62]. © 2011 American Chemical Society.

Lithium–air battery whose specific capacity is ten times higher than LIB is considered to be one of the next-generation energy-storage device with ultra-high energy density. The air electrode in a Li–air battery is usually composed of noble metals, such as Pt/Au, or metal oxides supported by carbon materials. However, the cycling efficiency and cycle life need great improvement. A recent research on graphene-based metal free catalyst for the air electrode showed some inspiring progress in this area [64]. At a current density of 0.5 mA cm^{-1}, the graphene sheets showed a high discharge voltage that was near that of the 20 wt% Pt/carbon black. Moreover, the difference in the charge voltage between the first and the 50th cycle was only 0.16 V, while that of the discharge voltage was as little as 0.07 V. The improved cycle stability could be attributed to the removal of adsorbed functional groups and crystallization of the graphene surface into a graphitic structure on heat treatment.

The applications of graphene in other battery systems, such as vanadium redox flow batteries [65, 66] and zinc–air batteries [67], have also been reported. Through the efforts from scientists all over the world, more graphene-based energy-storage devices with excellent performance can be expected.

9.4 Summary

Undoubtedly, graphene has now become a superstar in the whole world. Few materials, in the history, have received so much attention and are being developed so fast in such a short period of time as graphene. It is expected that graphene may bring a revolution in the diverse areas of science and technology in the near future. The

Chemical Synthesis of Graphene and Its Applications in Batteries

potential applications of graphene need an urgent solution for its mass production, which has not been resolved as yet. At this stage, chemical synthesis is regarded as the most efficient method of graphene production to meet its potentially vast demand in energy and chemical industries. Attempts for scaling up graphene production by chemical routes are being carried out worldwide, and mature technologies may emerge in a few years.

Batteries, which play important roles in our daily lives, provide a broad space for the application of graphene. Graphene-based electrode materials and additives for energy-storage devices have now become hot research spots. With continuous emerging reports on graphene-assisted improvement on performance of various types of batteries, people can foresee the bright future of graphene-based batteries running in every corner of life.

References

1. Novoselov K.S., Geim A.K., Morozov S.V., et al. (2004) Electric field effect in atomically thin carbon films, *Science*, **306**, 666–669.

2. Geim A.K. and Novoselov, K.S. (2007) The rise of graphene, *Nat Mater*, **6**, 183–191.

3. Hernandez Y., Nicolosi V., Lotya M., et al. (2008) High-yield production of graphene by liquid-phase exfoliation of graphite, *Nat Nanotechnol*, **3**, 563–568.

4. Behabtu N., Lomeda J.R., Green M.J., et al. (2010) Spontaneous high-concentration dispersions and liquid crystals of graphene, *Nat Nanotechnol*, **5**, 406–411.

5. Brodie B.C. (1860) Sur le poids atomique du graphite, Ann Chim Phys, **59**, 466.

6. Staudenmaier L. (1898) Verfahren zur Darstellung der Graphitsaure, *Ber Deut Chem Ges*, **31**, 1481.

7. Hummers W.S., Offeman R.E. (1958) Preparation of graphitic oxide, *J Am Chem Soc*, **80**, 1339–1339.

8. He H.Y., Klinowski J., Forster M., Lerf A. (1998) A new structural model for graphite oxide, *Chem Phys Lett*, **287**, 53–56.

9. Boehm V.H.P., Clauss A., Fischer G.O., Hofmann U. (1962) Dünnste Kohlenstoff-Folien, *Z Naturforschg*, **17b**.

10. Stankovich S., Dikin D.A., Dommett G.H.B., et al. (2006) Graphene-based composite materials, *Nature*, **442**, 282–286.

11. Zhou X.F., Liu Z.P. (2010) A scalable, solution-phase processing route to graphene oxide and graphene ultralarge sheets, *Chem Commun*, **46**, 2611–2613.

12. Schniepp H.C., Li J.L., McAllister M.J., et al. (2006) Functionalized single graphene sheets derived from splitting graphite oxide, *J Phys Chem* B, **110**, 8535–8539.

13. Lv W., Tang D.M., He Y.B., et al. (2009) Low-temperature exfoliated graphenes: Vacuum-promoted exfoliation and electrochemical energy storage, *ACS Nano*, **3**,3730–3736.

14. Stankovich S., Piner R.D., Chen X.Q., Wu N.Q., Nguyen S.T., Ruoff R.S. (2006) Stable aqueous dispersions of graphitic nanoplatelets via the reduction of exfoliated graphite oxide in the presence of poly(sodium 4-styrenesulfonate), *J Mater Chem*, **16**, 155–158.

15. Shin H.J., Kim K.K., Benayad A., et al. (2009) Efficient reduction of graphite oxide by sodium borohydrilde and its effect on electrical conductance, *Adv Funct Mater*,**19**, 1987–1992.

16. Zhang J.L., Yang H.J., Shen G.X., Cheng P., Zhang J.Y., Guo S.W. (2010) Reduction of graphene oxide via L-ascorbic acid, *Chem Commun*, **46**, 1112–1114.

17. Fernandez-Merino M.J., Guardia L., Paredes J.I., et al. (2010) Vitamin c is an ideal substitute for hydrazine in the reduction of graphene oxide suspensions, *J Phys Chem C*, **114**, 6426–6432.

18. Gao J., Liu F., Liu Y.L., Ma N., Wang Z.Q., Zhang X. (2010) Environment-friendly method to produce graphene that employs vitamin c and amino acid, *Chem Mater*, **22**, 2213–2218.

19. Zhu C.Z., Guo S.J., Fang Y.X., Dong S.J. (2010) Reducing sugar: New functional molecules for the green synthesis of graphene nanosheets, *ACS Nano*, **4**, 2429–2437.

20. Liu J.B., Fu S.H., Yuan B., Li Y.L., Deng Z.X. (2010) Toward a universal "adhesive nanosheet" for the assembly of multiple nanoparticles based on a protein-induced reduction/decoration of graphene oxide, *J Am Chem Soc*, **132**, 7279–+.

21. Fan X.B., Peng W.C., Li Y., et al. (2008) Deoxygenation of exfoliated graphite oxide under alkaline conditions: A green route to graphene preparation, *Adv Mater*, **20**, 4490–4493.

22. Zhou Y., Bao Q.L., Tang L.A.L., Zhong Y.L., Loh K.P. (2009) Hydrothermal dehydration for the "green" reduction of exfoliated graphene oxide to graphene and demonstration of tunable optical limiting properties, *Chem Mater*, **21**, 2950–2956.

23. Murugan A.V., Muraliganth T., Manthiram A. (2009) Rapid, facile microwave-solvothermal synthesis of graphene nanosheets and their polyaniline nanocomposites for energy strorage, *Chem Mater*, **21**, 5004–5006.

24. Guo H.L., Wang X.F., Qian Q.Y., Wang F.B., Xia X.H. (2009) A green approach to the synthesis of graphene nanosheets, *ACS Nano*, **3**, 2653–2659.

25. Becerril H.A., Mao J., Liu Z., Stoltenberg R.M., Bao Z., Chen Y. (2008) Evaluation of solution-processed reduced graphene oxide films as transparent conductors, *ACS Nano*, **2**, 463–470.

26. Kaniyoor A., Baby T.T., Ramaprabhu S. (2010) Graphene synthesis via hydrogen induced low temperature exfoliation of graphite oxide, *J Mater Chem*, **20**, 8467–8469.

27. Gomez-Navarro C., Meyer J.C., Sundaram R.S., et al. (2010) Atomic structure of reduced graphene oxide, *Nano Lett*, **10**, 1144–1148.

28. Erickson K., Erni R., Lee Z., Alem N., Gannett W., Zettl A. (2010) Determination of the local chemical structure of graphene oxide and reduced graphene oxide, *Adv Mater*, **22**, 4467–4472.

29. Inagaki M., Kang F., Toyoda M. (2004) Exfoliation of graphite via intercalation compounds, **29**, 1–69.

30. Valles C., Drummond C., Saadaoui H., et al. (2008) Solutions of negatively charged graphene sheets and ribbons, *J Am Chem Soc*, **130**, 15802–+.

31. Lee J.H., Shin D.W., Makotchenko V.G., et al. (2009) One-step exfoliation synthesis of easily soluble graphite and transparent conducting graphene sheets, *Adv Mater*, **21**, 4383–+.

32. Li X.L., Zhang G.Y., Bai X.D., et al. (2008) Highly conducting graphene sheets and Langmuir-Blodgett films, *Nat Nanotechnol*, **3**, 538–542.

33. Hao R., Qian W., Zhang L.H., Hou Y.L. (2008) Aqueous dispersions of TCNQ-anion-stabilized graphene sheets, *Chem Commun*, 6576–6578.

34. Yoo E., Kim J., Hosono E., Zhou H., Kudo T., Honma I. (2008) Large reversible Li storage of graphene nanosheet families for use in rechargeable lithium ion batteries, *Nano Lett*, **8**, 2277–2282.

35. Wang G.X., Shen X.P., Yao J., Park J. (2009) Graphene nanosheets for enhanced lithium storage in lithium ion batteries, *Carbon*, **47**, 2049–2053.

36. Wang C.Y., Li D., Too C.O., Wallace G.G. (2009) Electrochemical properties of graphene paper electrodes used in lithium batteries, *Chem Mater*, **21**, 2604–2606.

37. Bhardwaj T., Antic A., Pavan B., Barone V., Fahlman B.D. (2010) Enhanced electrochemical lithium storage by graphene nanoribbons, *J Am Chem Soc*, **132**, 12556–12558.

38. Pan D.Y., Wang S., Zhao B., et al. (2009) Li storage properties of disordered graphene nanosheets, *Chem Mater*, **21**, 3136–3142.

39. Paek S.M., Yoo E., Honma I. (2009) Enhanced cyclic performance and lithium storage capacity of SnO_2/graphene nanoporous electrodes with three-dimensionally delaminated flexible structure, *Nano Lett*, **9**, 72–75.

40. Wang X.Y., Zhou X.F., Yao K., Zhang J.G., Liu Z.P. (2011) A SnO_2/graphene composite as a high stability electrode for lithium ion batteries, *Carbon*, **49**, 133–139.

41. Zhou G.M., Wang D.W., Li F., et al. (2010) Graphene-wrapped Fe_3O_4 anode material with improved reversible capacity and cyclic stability for lithium ion batteries, *Chem Mater*, **22**, 5306–5313.

42. Wang H.L., Cui L.F., Yang Y.A., et al. (2010) Mn_3O_4 –graphene hybrid as a high-capacity anode material for lithium ion batteries, *J Am Chem Soc*, **132**, 13978–13980.

43. Wu Z.S., Ren W.C., Wen L., et al. (2010) Graphene anchored with Co_3O_4 nanoparticles as anode of lithium ion batteries with enhanced reversible capacity and cyclic performance, *ACS Nano*, **4**, 3187–3194.

44. Lee J.K., Smith K.B., Hayner C.M., Kung H.H. (2010) Silicon nanoparticles–graphene paper composites for Li ion battery anodes, *Chem Commun*, **46**, 2025–2027.

45. Zhao X., Hayner C.M., Kung M.C., Kung H.H. (2011) In-plane vacancy-enabled high-power Si-graphene composite electrode for lithium-ion batteries, *Adv Energy Mater*, **1**, 1079–1084.

46. Zhou X.F., Wang F., Zhu Y.M., Liu Z.P. (2011) Graphene modified $LiFePO_4$ cathode materials for high power lithium ion batteries, *J Mater Chem*, **21**, 3353–3358.

47. Wang H.L., Yang Y., Liang Y.Y., et al. (2011) $LiMn_{1-x}Fe_xPO_4$ nanorods grown on graphene sheets for ultrahigh-rate-performance lithium ion batteries, *Angew Chem Int Edit*, **50**, 7364–7368.

48. Su F.Y., You C.H., He Y.B., et al. (2010) Flexible and planar graphene conductive additives for lithium-ion batteries, *J Mater Chem*, **20**, 9644–9650.

49. Stoller M.D., Park S.J., Zhu Y.W., An J.H., Ruoff R.S. (2008) Graphene-based ultracapacitors, *Nano Lett*, **8**, 3498–3502.

50. Lin Z.Y., Liu Y., Yao Y.G., et al. (2011) Superior capacitance of functionalized graphene, *J Phys Chem C*, **115**, 7120–7125.

51. Xu B., Yue S.F., Sui Z.Y., et al. (2011) What is the choice for supercapacitors: Graphene or graphene oxide?, *Energy Environ Sci*, **4**, 2826–2830.

52. Jeong H.M., Lee J.W., Shin W.H., et al. (2011) Nitrogen-doped graphene for high-performance ultracapacitors and the importance of nitrogen-doped sites at basal planes, *Nano Lett*, **11**, 2472–2477.

53. Zhu Y.W., Murali S., Stoller M.D., et al. (2011) Carbon-based supercapacitors produced by activation of graphene, *Science*, **332**, 1537–1541.

54. Chen S., Zhu J.W., Wu X.D., Han Q.F., Wang X. (2010) Graphene oxide-MnO_2 nanocomposites for supercapacitors, *ACS Nano*, **4**, 2822–2830.

55. Wu Z.S., Ren W.C., Wang D.W., Li F., Liu B.L., Cheng H.M. (2010) High-energy MnO_2 nanowire/graphene and graphene asymmetric electrochemical capacitors, *ACS Nano*, **4**, 5835–5842.

56. Yu G.H., Hu L.B., Vosgueritchian M., et al. (2011) Solution-processed graphene/MnO_2 Nanostructured textiles for high-performance electrochemical capacitors, *Nano Lett*, **11**, 2905–2911.

57. Wang D.W., Li F., Zhao J.P., et al. (2009) Fabrication of graphene/polyaniline composite paper via in situ anodic electropolymerization for high-performance flexible electrode, *ACS Nano*, **3**, 1745–1752.

58. Wu Q., Xu Y.X., Yao Z.Y., Liu A.R., Shi G.Q. (2010) Supercapacitors based on flexible graphene/polyaniline nanofiber composite films, *ACS Nano*, **4**, 1963–1970.

59. Xu J.J., Wang K., Zu S.Z., Han B.H., Wei Z.X. (2010) Hierarchical nanocomposites of polyaniline nanowire arrays on graphene oxide sheets with synergistic effect for energy storage, *ACS Nano*, **4**, 5019–5026.

60. Biswas S., Drzal L.T. (2010) Multi layered nanoarchitecture of graphene nanosheets and polypyrrole nanowires for high performance supercapacitor electrodes, *Chem Mater*, **22**, 5667–5671.

61. Jang B.Z., Liu C.G., Neff D., et al. (2011) Graphene surface-enabled lithium ion-exchanging cells: Next-generation high-power energy storage devices, *Nano Lett*, **11**, 3785–3791.

62. Wang H.L., Yang Y., Liang Y.Y., et al. (2011) Graphene-wrapped sulfur particles as a rechargeable lithium-sulfur battery cathode material with high capacity and cycling stability, *Nano Lett*, **11**, 2644–2647.

63. Ji L., Rao M., Zheng H., et al. (2011) Graphene oxide as a sulfur immobilizer in high performance lithium/sulfur cells, *J Am Chem Soc*, **133**, 18522–18525.

64. Yoo E., Zhou H.S. (2011) Li–air rechargeable battery based on metal-free graphene nanosheet catalysts, *ACS Nano*, **5**, 3020–3026.

65. Han P.X., Yue Y.H., Liu Z.H., et al. (2011) Graphene oxide nanosheets/multi-walled carbon nanotubes hybrid as an excellent electrocatalytic material towards $VO(2+)/VO(2)(+)$ redox couples for vanadium redox flow batteries, *Energy Environ Sci*, **4**, 4710–4717.

66. Han P.X., Wang H.B., Liu Z.H., et al. (2010) Graphene oxide nanoplatelets as excellent electrochemical active materials for $VO(2+)/VO(2)(+)$ and $V(2+)/V(3+)$ redox couples for a vanadium redox flow battery, *Carbon*, **49**, 693–700.

67. Lee J.S., Lee T., Song H.K., Cho J., Kim B.S. (2011) Ionic liquid modified graphene nanosheets anchoring manganese oxide nanoparticles as efficient electrocatalysts for Zn–air batteries, *Energy Environ Sci*, **4**, 4148–4154.

Chapter 10

Photonic Properties of Graphene Device

Hua-Min Li and Won Jong Yoo
SKKU Advanced Institute of Nano Technology, Department of Nanoscience and Technology, Sungkyunkwan University, Suwon 440-746, Korea
yoowj@skku.edu

In this review chapter, photonic properties of graphene device have been presented, posing potential advantages of high-speed electronic response and wide photodetection bandwidth. The high-speed electronic response of photo-generated carriers is attributed to the carrier mobility in graphene photodetectors, which is much higher than those in the conventional semiconductor photodetectors. Meanwhile, the wide photodetection bandwidth is attributed to linear energy–momentum dispersion relation of graphene as a two-dimensional (2D) electron gas that gives rise to no energy bandgap at the neutral Dirac point. To enlighten the photocurrent generation mechanism of graphene devices, in this chapter, band diagrams with voltage bias applied have been presented with the help of laser illumination technique with spatial resolution on graphene. The previously reported theories on change in band structure have also been explained here. Lastly, various techniques

---

Two-Dimensional Carbon: Fundamental Properties, Synthesis, Characterization, and Applications
Edited by Yihong Wu, Zexiang Shen, and Ting Yu
Copyright © 2014 Pan Stanford Publishing Pte. Ltd.
ISBN 978-981-4411-94-3 (Hardcover), 978-981-4411-95-0 (eBook)
www.panstanford.com

to enhance photocurrent from graphene devices, e.g., asymmetric device structure, graphene stack, and surface plasmonics, have been presented in this chapter.

10.1 Introduction

Graphene as a single-layer atomic carbon crystal with the 2D honeycomb lattice structure has recently attracted enormous attention due to its unique electronic properties [1,2]. In particular, the pioneering reports [3,4] on the quantum electronic properties of graphene obtained from exfoliation of highly oriented pyrolytic graphite (HOPG) inspired many scientists to explore the unknown interesting properties of graphene due to the ease of the fabrication of quantum electronic devices via the formation of graphene as a 2D gas. More fascinating in graphene device is the possibility to demonstrate the quantum electronic properties at room temperature.

The unique electronic properties of graphene are attributed to its linear energy–momentum dispersion relation of 2D Dirac electrons which can be controlled by electric and magnetic fields [1,2]. Let us review how the linear energy–momentum dispersion relation and thereby the high carrier mobility is obtained from graphene. Graphene has two atoms per unit cell: A and B. As shown in Fig. 10.1, a low-energy band structure consists of Dirac cones located at two Brillouin zone corners $\mathbf{K}$ and $\mathbf{K}'$.

$$H_K = \hbar v \sigma \cdot \mathbf{k} = \hbar v \begin{bmatrix} 0 & k_x - ik_y \\ k_x + ik_y & 0 \end{bmatrix} \tag{10.1}$$

where v is the graphene Fermi velocity [5]. For each wave vector $\mathbf{k}$, an eigen-energy of $E_k = \pm \hbar v |\mathbf{k}|$ as a linear energy–momentum dispersion relation is obtained. The energy dispersion resembles the energy of relativistic particles which are quantum mechanically described by the Dirac equation.

Meanwhile, we are very interested in carrier mobility as a figure of merit on materials property for high frequency device applications. According to the theoretical and experimental results, carrier mobility generally expressed as $\mu = \sigma/ne$ is found to be much higher in graphene than those of the conventional silicon and other semiconductor materials, varying in the range of 10^3 to 10^6 cm^2 V^{-1}s^{-1} depending on lattice and impurity scattering. When field-effect transistor (FET) structure is used for the graphene device, the charge carrier density, n, can be determined by $n = (7.2 \times 10^{10}$ cm^{-2}V$^{-1})V_G$, where V_G is the gate voltage applied to the device, usually in the range of 0–100 V. That is, the charge carrier density in graphene is <7.2×10^{12} cm^{-2}. According to the relation on effective mass of charge carriers to n, $m^* = h(n/\pi)^{1/2}/(2v_F)$, where v_F is the Fermi velocity of $\sim 10^6$ ms^{-1} [1,3], the effective mass in graphene is found to be very low, <$0.1m_0$. By applying $\mu = \sigma/ne$ and the results reported in 2005 by Novoselov et al. [3], carrier mobility of $\sim$2,600 cm^2 V^{-1}s^{-1} is obtained. This result shows the great potential for the mobility in graphene to be enhanced by increasing the mean free path of carriers, that is, by suppressing the scattering of carriers in graphene, as the conductivity of a material can be directly proportional to the mean free path of carriers (l): $\sigma = ne\mu = (e^2/h)kl$. In addition, it is very interesting to observe that the mobilities in graphene can remain unchanged near to the room temperature, whereas other materials show a drastic decrease in their mobilities with increasing temperature [6].

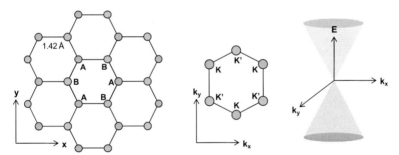

Figure 10.1 (a) Illustration of graphene lattice. (b) The first Brillouin zone of graphene with the Dirac points at the corners K and K'. (c) Energy bands of graphene at the Dirac points.

Considering the application of graphene to the electronic devices, particularly the digital electronic devices which are currently dominated by silicon semiconductor technology, the most important pre-requisite is a high ratio between on-current and off-current (I_{on}/I_{off}). According to the International Technology Roadmap for Semiconductors (ITRS), the usually required I_{on}/I_{off} for the high performance semiconductor devices to operate is in the range of 10^5–10^7. This high value of I_{on}/I_{off} can be obtained only when the electronic energy bandgap of the material is sufficiently large, > ~0.5 eV, because the electronic transport of charge carriers (electrons and holes) are impeded depending on the magnitude of energy bandgap which gives rise to an energy barrier for their charge transport.

However, the linear energy–momentum dispersion relation of graphene results in zero bandgap at the charge neutral points. This brings about graphene's metallic behavior, and becomes a major physical obstacle for graphene to be applied for the development of digital electronic devices. Due to this constraint, graphene has only been studied for the application to the analog devices such as radio frequency (RF) devices and photonic devices. Since the charge carriers in graphene can transport in the Fermi speed of $\sim 10^6$ ms^{-1}, as explained above, we can expect the future graphene devices operating in very high frequency for RF and photonic applications where the high I_{on}/I_{off} is not critical to realize their electronic functions.

In this chapter, we address the recent development of graphene devices to the photonic application. In addition to the advantage of high frequency of graphene photonic devices, the unique zero bandgap of graphene also gives rise to a broad absorption spectrum at least from the visible to the infrared range [7]. In fact, the infinitely long-wavelength light can be detected from the ideal graphene photonic devices, but the detection of long-wavelength light, including infrared light, is difficult due to the weakness of light signal and the unavailability of long-wavelength laser. The high Fermi velocity and carrier mobility can contribute to the high bandwidth potentially over 500 GHz for ultrafast light detection [8]. Although the graphene photodetectors are very promising to achieve the high frequency and wide band photodetection, it is inherently limited by a low photoabsorption as graphene is limited in depth as a 2D material. To overcome the limitation, a graphene stack photodetector has been introduced in this chapter. Other techniques to improve photocurrent of graphene by using asymmetric metallization and exciting surface plasmonics have also been introduced.

To understand the conversion of photon to electrical current in graphene, several mechanisms have been proposed [9]. A photovoltaic (PV) theory was proposed first to interpret the photocurrent generation in graphene. The photocurrent, which is generated only near the metal contacts, is considered to originate from the built-in electric field in graphene resulted from the band bending near the contacts and modulated by the electrical gating [10,11]. Meanwhile, a photothermoelectric (PTE) theory was proposed recently, which also contributes to the photocurrent generation in graphene where a thermoelectric voltage arises from a light-induced temperature difference and generates a heat current accompanied by electrical current [12]. A variety of other factors, e.g., single- and bi-layer interface junction [13], local chemical doping edges [14], hot carrier transport, and carrier multiplication [15,16] have also attracted attention and interest.

10.2 Energy Band Structure of Graphene

Here we use the p-type doped graphene as an example to discuss its energy band structure. The transfer characteristics of graphene FET always show the Dirac point (V_{Dirac}) at the positive V_G, indicating an unintentional p-type doping effect in graphene. This is attributed to (i) the formation of weak C–O bonds between graphene and SiO_2 surface [17,18] and (ii) external molecules, e.g., H_2O [19], CO_2, and O_2 [20], adsorbed on the channel surface in the ambient environment or between graphene and SiO_2 during the transfer process [21]. Assuming the graphene FET with V_{Dirac} of ~60 V is operated with the zero source voltage (V_S), constant drain voltage (V_D), and variable V_G, the interactions between the metal and graphene in both vertical (z-axis) and horizontal (x-axis) directions are illustrated by the energy band diagrams, as shown in Fig. 10.2. For the graphene covered by the electrode, its properties are controlled solely by the metal contact, since the gate field has a relatively negligible impact. The graphene in this region can be doped into n-type (by using Al, Ag, or Pd contact), or into p-type (by using Au or Pt contact), due to the charge transfer and chemical interaction between the graphene and the metal [22,23]. The doping induced Fermi level shift ($\Delta\varphi$) relative to the Dirac point energy level (E_{Dirac}) in this region can be estimated via a simple capacitor model [11,24]

$$\Delta\phi = \hbar v_F \sqrt{\pi\alpha \left| V_{\text{flat}} - V_{\text{Dirac}} \right|} \qquad (10.2)$$

where $\hbar$ is the reduced Planck's constant, v_F is the Fermi velocity, V_{flat} is the gate voltage for the flat band condition, and α is the gate coupling parameter that relates the electrostatically induced carrier density to V_G, which equals to ~7.2 × 10^{10} cm^{-2}V^{-1} for 300 nm-thick

SiO$_2$ dielectric [11,25]. For the graphene in the bulk channel, its carrier density is capacitively controlled by V_G. For the high negative (or positive) V_G, the graphene in the bulk channel region is strongly doped into p-type (or n-type), and the V_G-dependent Fermi level shift (ΔE) relative to E_{Dirac} in this region can be estimated as [24]

$$\Delta E = \hbar v_F \sqrt{\pi \alpha |V_G - V_{Dirac}|} \qquad (10.3)$$

Therefore, the entire graphene channel is divided into the metal doping region, the bulk channel region, and the transition region between the other two [10].

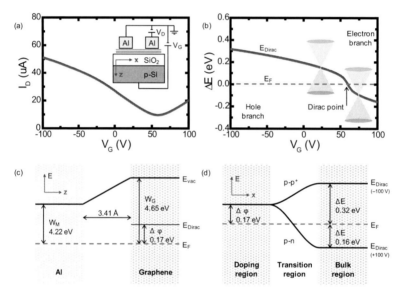

Figure 10.2 (a) Transfer characteristics of a single layer graphene FET shows the Dirac point at V_G of ~60 V. (b) The corresponding Fermi level shift of graphene in the bulk channel as a function of V_G. (c) Energy band diagram between the graphene and the metal contact (Al). (d) Energy band diagram of graphene along the channel at V_G of −100 and +100 V.

10.3 Photonic Absorption of Graphene

Graphene has unique optical properties [7,9,26] that give rise to a strong graphene–light interaction. For example, a single-layer graphene can absorb the same amount of 1.55 μm light as a 20 nm-thick InGaAs film. Here we review the wide and field-controllable photon absorption of graphene for its optoelectronic applications.

Assuming that only vertical (k-conserving) transitions are allowed for the normal light incidence, the optical transmittance (T) and reflectance (R) of the free-standing graphene can be derived using the Fresnel equations for a thin film with a fixed universal optical conductance of $G_0 = e^2/(4\hbar) \approx 6.08 \times 10^{-5}\ \Omega^{-1}$ as [27]

$$T \equiv (1 + 2\pi G_0/c) = (1 + 0.5\pi\alpha)^{-2} \approx 1 - \pi\alpha \approx 97.7\% \qquad (10.4)$$

$$R \equiv 0.25\pi^2\alpha^2 T \qquad (10.5)$$

where $\alpha = e^2/(4\pi\varepsilon_0\hbar c) = G_0/(\pi\varepsilon_0 c) \approx 1/137$ is the fine-structure constant and c is the speed of light. Therefore, the single-layer graphene should transmit ~97.7% of the incident light and absorb $\pi\alpha \approx 2.3\%$, being independent of the wavelength. Moreover, it is found that the optical properties of the graphene layers are proportional to the number of layers in the few-layer graphene structure. Each layer can be considered as a 2D electron gas with negligible perturbation from the adjacent layers, resulting in the few-layer graphene being optically equivalent to a superposition of single-layer graphene. In addition, the surface only reflects <0.1% of the incident light in the visible spectrum for the single-layer freestanding graphene [27], and this increases to ~2% for the ten layer structure [28].

The absorption spectrum of the single-layer graphene is quite flat from 300 to 2500 nm [27,29]. A slightly higher absorption at <500 nm range is probably due to the hydrocarbon contamination [27], and a peak in the ultraviolet region (~270 nm) is induced by the exciton-shifted van Hove singularity in the graphene density of states [7]. It is noted that the graphene has a very strong infrared (IR) response [24]. As one-atom-thick material, the single-layer graphene can absorb more than 2% of normally incident IR radiation, compared to ~1% absorption of a 10 nm-thick GaAs film. By using stacked graphene sheets, optical multipasses, or waveguiding technologies, the IR response of graphene can be enhanced considerably.

Another interesting property of graphene is that its optical absorption can be modulated by the external electrical gating [24,30]. The carrier density in the intrinsic graphene can be capacitively controlled by the gate field, leading to a shift of the Fermi level (ΔE_F) relative to the Dirac point energy level. In this case, the interband absorption of the photons only occurs when the photon energy is higher than $2\Delta E_F$, as shown in Fig. 10.3. Otherwise the transition is forbidden. Therefore, the optical properties of graphene in FET devices, e.g., absorbance and transmittance, can be changed by varying V_G. Owing to the tunability provided by electrical gating and charge injection, graphene can be a promising material for the novel optoelectronic devices such as the tunable IR detectors, modulators, and emitters [24].

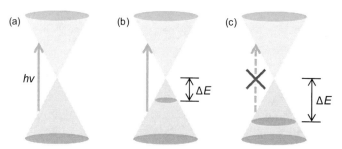

Figure 10.3 Band structure of graphene near the Dirac point illustrating the interband transition for (a) pristine graphene and (b) slightly p-type doped graphene, and the Pauli blocking for (c) the highly p-type doped graphene. Absorption is allowed only for the photon energy hv higher than $2\Delta E_F$ (solid arrow). Otherwise it is forbidden (dash arrow).

The high electrical conductivity and optical transmittance indicate graphene as an excellent conducting material for replacing the conventional electrodes in various optoelectronic applications, such as solar cells, flat panel displays, touch screens, and light-emitting diodes [26]. Graphene shows a better performance in terms of transmittance and sheet resistance compared to the conventional transparent conducting materials, including indium tin oxide (ITO), ZnO, and TiO_2. In addition, the conventional transparent conducting materials such as ITO are always brittle and expensive, whereas graphene can easily overcome these problems owing to its high flexibility and availability of materials for mass production.

10.4 Photocurrent Generation in Graphene

Considering the wide spectrum of potential applications of graphene, ranging from transistors and chemical sensors to nanoelectromechanical devices and composites, photonics and optoelectronics are believed to be one of the most promising fields

[31]. Owing to the combination of both unique optical and electronic properties of graphene, the novel photonic and optoelectronic devices which are fundamentally different from the conventional arrangements have been proposed for various applications, ranging from solar cells and light-emitting diodes to touch screens and ultrafast lasers [7,9,26]. Here we take the graphene photodetector as an example to demonstrate the light-to-current conversion which is one of the most important technologies for the rise of graphene in photonics and optoelectronics. A PV theory for the photocurrent generation in graphene has been reviewed, and several technologies for improving the performance of the graphene photodetector have been discussed in this chapter.

Photodetectors measure the photon flux or optical power by converting the absorbed photon energy into electrical current. A configuration of graphene FET photodetector similar to the back-gate graphene transistor [8] is shown in Fig. 10.4. Owing to the application of scanning photocurrent microscopy (SPCM) technology, which can provide the planar distribution of photocurrent profile, PV theory was proposed to understand the photocurrent generation in graphene. In the PV theory, the photocurrent generated only near the metal contacts is attributed to an internal (built-in) electric field in the graphene which is resulted from the band bending near the contacts and modulated by the electrical gating. An open circuit voltage V_{OC} is produced due to the separation of photon excited carriers under the open circuit condition, and a photocurrent $I_{ph} = V_{OC}/R_g$ is generated under the short circuit condition, where R_g is the resistance of the graphene channel [8]. The graphene channel with the metal contact can be divided into three regions [10]. The Fermi level of graphene in the metal controlled segment is pinned due to a relatively negligible impact of the gate field, whereas it is capacitively modulated by the electrical gating in the bulk channel segment. For the high negative (or positive) V_G, the graphene in the bulk channel region is strongly doped into p-type (or n-type), forming a transition segment located between the metal controlled segment and the bulk channel segment. The band-bending direction and potential drop of junction in the transition segment are determined by the difference between $\Delta\varphi$ and ΔE by Eqs. 10.2 and 10.3. The transition region separates the photon-excited electron and hole carriers, resulting in the photocurrent generation near the contacts. As a comparison, the bulk channel region does not contribute to the photocurrent generation since the electron and hole carriers will recombine immediately after the photon excitation.

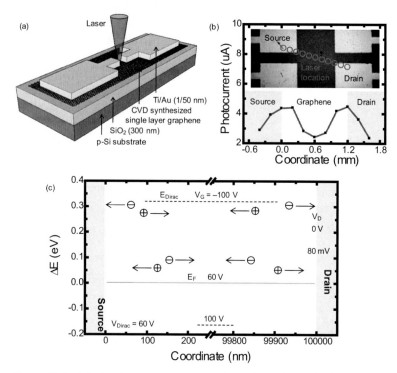

Figure 10.4 (a) A schematic representation of graphene back-gate FET photodetector under laser illumination. (b) Photocurrent generation as a function along the channel indicates that the photocurrent is only generated near the metal contacts. Inset: Microscopic image of laser scanning measurement. (c) Energy band diagrams of graphene along the channel at V_G of −100 and +100 V and V_D of 0 and 80 mV. The photon-excited electrons and holes are separated at the band bending region near the metal contact, contributing to the photocurrent generation.

At the same V_G, the photocurrent at the source and drain contacts always appear in pairs of opposite signs, indicating the opposite direction of the internal field at the two contacts. By varying V_G between positive and negative biases, the direction of the internal field at each contact is also switched [11]. Assuming that graphene at the source and drain is doped into p-type, the p-n-p type and p-p⁺-p type channels are formed for the positive and positive V_G, respectively. Moreover, it is found that the position of the maximum photocurrent at the p-n junction moves into the bulk channel (several hundred nanometers away from the contact) compared to the p-p$^+$ junction (at the contact edge) due to its relatively wider transition region. A higher photocurrent generation is also observed at the p-n junction compared to that at the p-p$^+$ junction due to its relatively higher internal electric field. This band-bending phenomenon in the transition region is analogous to a conventional metal–semiconductor or semiconductor–semiconductor junction in which the depletion region varies with the doping level [10].

More recently, a PTE theory was proposed which also contributes to the photocurrent generation in graphene. In the PTE theory, a thermoelectric voltage arises from a light-induced temperature difference, which generates a heat current accompanied by electric current [12,13,15,16]. Although the exact mechanism for the light to current conversion is still debated, it has been widely accepted that the photocurrent is only generated near the contacts where the internal electric field separates the photon excited carriers [31]. The photocurrent generation may be resulted from a combination of the direct PV and indirect PTE effects, and both effects are strongly enhanced at the band bending region [32].

10.5 Technology for Performance Improvement

There is a strong interaction between graphene and photons [26]. For example, a 20 nm-thick InGaAs film is required to absorb the same amount of light (1.55 µm wavelength) as a single-layer graphene [26], and 100% light absorption can be achieved in a single patterned sheet of doped graphene [33]. However, the performance of graphene photodetector is still limited by one potential drawback: the low photoresponse and very low external quantum efficiency (EQE) (0.1–0.2%) due to its intrinsically poor light absorption (~2.3% for single-layer graphene) compared to the traditional group III–V materials-based devices [31,34]. Therefore, various technologies have been proposed to improve the photocurrent generation in graphene, e.g., asymmetric metallization, layer-by-layer stacking, surface plasmonics, graphene bilayer, nanoribbon, p-i-n junction, microcavity, graphene–insulator stack, graphene–quantum dot hybrid, and graphene antenna sandwich structures. Here we select the first three methods as the examples for a brief discussion.

10.5.1 Asymmetric Metallization

In the conventional configuration of graphene photodetector, both the source and the drain electrodes consist of the same material, giving rise to the symmetric internal electric field profile in the channel. In this way, the photocurrents generated around these two electrodes have the same magnitude but opposite directions, and the total current is close to zero (for a small drain voltage). As a comparison, an asymmetric metallization which uses two different metallic materials can break the mirror symmetry of the internal electric field within the channel, allowing the photocurrents near both the electrodes to flow in the same direction at an appropriate gate voltage, and contributing to an enhanced total photocurrent [35].

10.5.2 Graphene Stack Channel

Being different from an HOPG multi-layer graphene [36] where an *ABAB*-ordered arrangement (or Bernal stacking that resembles the structure of graphite) provides a strong interlayer coupling between the adjacent layers [37,38], a large scale graphene stack structure can be easily formed by layer-by-layer transfer and overlap of the single-layer graphene synthesized via chemical vapor deposition (CVD) [39,40]. Graphene is polycrystalline, and only physically stacked via a transfer process, so that the single-layer graphene in the graphene stack structure is crystallographically misaligned to each other and the interlayer coupling is decoupled by breaking the *ABAB*-ordered arrangement [37,38]. The π bonds between each stacking layer are absent, and the graphene stack structure (up to tri-layer stacking) retains the electron–hole ambipolar transportation characteristics. Compared to the single-layer graphene devices, the graphene stack photodetector can significantly enhance the photocurrent generation, owing to (i) the improved carrier transport, such as the increased carrier mobility, and reduced sheet and contact resistances, and (ii) the enhanced photon absorption due to the increased thickness of graphene channel [41], as shown in Fig. 10.5.

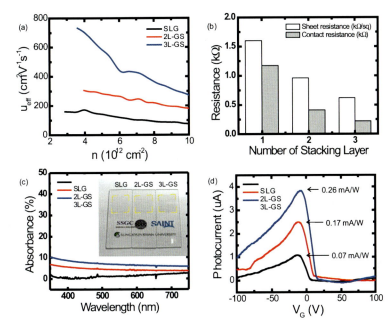

Figure 10.5 Improvement of electrical and optoelectronic performance of graphene stack structure, including (a) field effect mobility (holes), (b) sheet and contact resistances, (c) optical absorbance in the visible spectrum, and (d) photocurrent generation as a function of V_G with the maximum photoresponse. Inset of (c): Photograph of the graphene stack structure on a transparent glass slide, indicating their different optical transmittances.

10.5.3 Graphene Plasmonics

Surface plasmon resonance (SPR) can be excited on graphene by integrating with a thin layer of plasmonic nanostructures, in which the incident light is efficiently converted into plasmonic oscillations, giving rise to a dramatic enhancement of the local electric field. The plasmon resonance is appropriate for the graphene photodetector since the enhanced field only appears in the near-field region where the band bending is exactly located and responsible for the separation of photon excited electron and hole carriers [31,34]. Therefore, the enhanced electromagnetic energy can be guided directly into the band bending region via the plasmonic nanostructure, contributing to the photocurrent generation. It has been reported that the photocurrent can be enhanced by up to 15–20 times, and the EQE can be increased by up to ~1.5% [31,34,42,43]. In addition, since the plasmon resonance only occurs at certain frequency (plasmon resonance frequency) which is highly dependent on the size, shape, material, and periodicity of the nanostructure, the lack of spectral selectivity in graphene can be solved by coupling with the plasmonic nanostructures of designed plasmon resonance frequency, and a highly sensitive multi-color graphene photodetector can be achieved [34].

10.6 Summary

In this chapter, we have discussed the photonic properties of graphene device by focusing on its potential advantages in high-speed electronic response and wide photodetection bandwidth. These in turn are attributed to the unique band structures of graphene. The research on graphene-based photonics is advancing rapidly; the unique properties of graphene make it very promising in applications in some niche areas which graphene has significant advantages over other types of semiconductors.

References

1. Castro Neto A.H., Guinea F., Peres N.M.R., Novoselov K.S., Geim A.K. (2009) The electronic properties of graphene, *Rev Mod Phys*, **81**(1), 109–162.

2. Das Sarma S., Adam S., Hwang E.H., Rossi E. (2011) Electronic transport in two-dimensional graphene, *Rev Mod Phys*, **83**(2), 407–470.

3. Novoselov K.S., Geim A.K., Morozov S.V., et al. (2005) Two-dimensional gas of massless Dirac fermions in graphene, *Nature*, **438**, 197–200.

4. Zhang Y., Tan Y.-W., Stormer H.L., Kim P. (2005) Experimental observation of the quantum Hall effect and Berry's phase in graphene, *Nature*, **438**, 201–204.

5. Nomura K., MacDonald A.H. (2007) Quantum transport of massless Dirac Fermions, *Phys Rev Lett*, **98**(076602).

6. Chen J.-H., Jang C., Ishigami M., et al. (2009 Diffusive charge transport in graphene on SiO_2, *Solid State Commun*, **149**, 1080–1086.

7. Bonaccorso F., Sun Z., Hasan T., Ferrari A.C. (2010) Graphene photonics and optoelectronics, *Nat Photonics*, **4**, 611–622.

8. Xia F., Mueller T., Lin Y.-M., Valdes-Garcia A., Avouris P. (2009) Ultrafast graphene photodetector, *Nat Nanotechnol*, **4**, 839–843.

9. Freitag M. (2010) Optical and thermal properties of graphene field-effect transistors, *Phys Status Solidi B*, **247**, 2895–2903.

10. Xia F., Mueller T., Golizadeh-Mojarad R., et al. (2009) Photocurrent imaging and efficient photon detection in a graphene transistor, *Nano Lett*, **9**(3), 1039–1044.

11. Mueller T., Xia F., Freitag M., Tsang J., Avouris P. (2009) Role of contacts in graphene transistors: A scanning photocurrent study, *Phys Rev B*, **79**(245430).

12. Basko D. (2011) A photothermoelectric effect in graphene, *Science*, **334**, 610–611.

13. Xu X., Gabor N.M., Alden J.S., van der Zande A. M., McEuen P.L. (2009) Photo-thermoelectric effect at a graphene interface junction, *Nano Lett*, **10**, 562–566.

14. Peters E.C., Lee E.J.H., Burghard M., Kern K. (2010) Gate dependent photocurrents at a graphene p-n junction, *Appl Phys Lett*, **97**(193102).

15. Gabor N.M., Song J.C.W., Ma Q.,et al. (2011) Hot carrier-assisted intrinsic photoresponse in graphene, *Science*, **334**, 648–652.

16. Song J.C.W., Rudner M.S., Marcus C.M., Levitov L.S. (2011) Hot carrier transport and photocurrent response in graphene, *Nano Lett*, **11**, 4688–4692.

17. Kang Y.-J., Kang J., Chang K.J. (2008) Electronic structure of graphene and doping effect on SiO_2, *Phys Rev B*, **78**(115404).

18. Shi Y., Dong X., Chen P., Wang J., Li L.-J. (2009) Effective doping of single-layer graphene from underlying SiO_2 substrates, *Phys Rev B*, **79**(115402).

19. Leenaerts O., Partoens B., Peeters F.M. (2008) Adsorption of H_2O, NH_3, CO, NO_2, and NO on graphene: A first-principles study, *Phys Rev B*, **77**(125416).

20. Huang B., Li Z., Liu Z., et al. (2008) Adsorption of gas molecules on graphene nanoribbons and its implication for nanoscale molecule sensor, *J Phys Chem C*, **112**, 13442–13446.

21. Di Bartolomeo A., Giubileo F., Santandrea S., et al. (2011) Charge transfer and partial pinning at the contacts as the origin of a double dip in the transfer characteristics of graphene-based field-effect transistors, *Nanotechnology*, **22**(275702).

22. Giovannetti G., Khomyakov P.A., Brocks G., Karpan V.M., van den Brink J., Kelly P.J. (2008) Doping graphene with metal contacts, *Phys Rev Lett*, **101**,(026803).

23. Khomyakov P.A., Giovannetti G., Rusu P.C., Brocks G., van den Brink J., Kelly P.J. (2009) First-principles study of the interaction and charge transfer between graphene and metals, *Phys Rev B*, **79**(195425).

24. Wang F., Zhang Y., Tian C., et al. (2008) Gate-variable optical transitions in graphene, *Science*, **320**(5873), 206–209.

25. Huard B., Stander N., Sulpizio J.A., Goldhaber-Gordon D. (2008) Evidence of the role of contacts on the observed electron–hole asymmetry in graphene, *Phys Rev B*, **78**(121402(R)).

26. Avouris P. (2010) Graphene: Electronic and photonic properties and devices, *Nano Lett*, **10**, 4285–4294.

27. Nair R.R., Blake P., Grigorenko A.N., et al. (2008) Fine structure constant defines visual transparency of graphene, *Science*, **320**(5881), 1308.

28. Casiraghi C., Hartschuh A., Lidorikis E., et al. (2007). rayleigh imaging of graphene and graphene layers, *Nano Lett*, **27**(9), 2711–2717.

29. Mak K.F., Sfeir M.Y., Wu Y., Lui C.H., Misewich J.A., Heinz T.F. (2008) Measurement of the optical conductivity of graphene, *Phys Rev Lett*, **101**(196405).

30. Li Z.Q., Henriksen E.A., Jiang Z., et al. (2008) Dirac charge dynamics in graphene by infrared spectroscopy, *Nat Phys*, **4**, 532–535.

31. Echtermeyer T.J., Britnell L., Jasnos P.K., et al. (2011) Strong plasmonic enhancement of photovoltage in graphene, *Nat Commun*, **2**(458).

32. Lemme M.C., Koppens F.H.L., Falk A.L., et al. (2011) Gate-activated photoresponse in a graphene p-n junction, *Nano Lett*, **11**, 4134–4137.

33. Thongrattanasiri S., Koppens F.H.L., Javier Garcia de Abajo F. (2012) Complete optical absorption in periodically patterned graphene, *Phys Rev Lett*, **108**(047401).

34. Liu Y., Cheng R., Liao L., et al. (2011) Plasmon resonance enhanced multicolour photodetection by graphene, *Nat Commun*, **2**(579).

35. Mueller T., Xia F., Avouris P. (2010) Graphene photodetectors for high-speed optical communications, *Nat Photonics*, **4**, 297–301.

36. Novoselov K.S., Geim A.K., Morozov S.V., et al. (2004) Electric field effect in atomically thin carbon films, *Science*, **306**(5696), 666–669.

37. Yu T., Liang C.-W., Kim C., Song E.-S., Yu B. (2011) Three-dimensional stacked multilayer graphene interconnects, *IEEE Electron Device Lett*,**32**(8), 1110–1112.

38. Yu T., Kim E., Jain N., Xu Y., Geer R., Yu B. (2011) Carbon-based interconnect: Performance, scaling and reliability of 3D stacked multilayer graphene system, *IEEE IEDM Tech Dig*, 159–162.

39. Kim K.S., Zhao Y., Jang H., et al. (2009) Large-scale pattern growth of graphene films for stretchable transparent electrodes, *Nature*, **457**, 706–710.

40. Li X., Cai W., An J., et al. (2009) Large-area synthesis of high-quality and uniform graphene films on copper foils, *Science*, **324**, 1312–1314.

41. Li H.-M., Shen T.-Z., Lee D.-Y., Yoo W.J. (2012) High photocurrent and quantum efficiency of graphene photodetector using layer-by-layer stack structure and trap assistance, *IEEE IEDM Tech Dig,* 549–552.

42. Ju L., Geng B., Horng J., et al. (2011) Graphene plasmonics for tunable terahertz metamaterials, *Nat Nanotechnol*, **6**, 630–634.

43. Shi S.-F., Xu X., Ralph D.C., McEuen, P.L. (2011) Plasmon resonance in individual nanogap electrodes studied using graphene nanoconstrictions as photodetectors, *Nano Lett*, **11**, 1814–1818.

Chapter 11

Graphene Oxides and Reduced Graphene Oxide Sheets: Synthesis, Characterization, Fundamental Properties, and Applications

Shixin Wu and Hua Zhang

School of Materials Science and Engineering, Nanyang Technological University,
50 Nanyang Avenue, Singapore 639798, Singapore
hzhang@ntu.edu.sg

11.1 Introduction

Graphene, a recently emerging two-dimensional (2D) crystalline material, is a single layer of sp^2-hybridized carbon atoms bound into a hexagonal lattice [1]. It has shown various fascinating properties and great importance in the research and industrial studies of carbon nanostructures.

Recently, the unique electronic properties of graphene have been studied extensively. Unlike quantum properties of other materials understood using the Schrödinger equation, charge carriers in graphene are more naturally described by the Dirac equation with

---

Two-Dimensional Carbon: Fundamental Properties, Synthesis, Characterization, and Applications
Edited by Yihong Wu, Zexiang Shen, and Ting Yu
Copyright © 2014 Pan Stanford Publishing Pte. Ltd.
ISBN 978-981-4411-94-3 (Hardcover), 978-981-4411-95-0 (eBook)
www.panstanford.com

their zero effective mass [1,2]. Graphene, a semimetal with a tiny overlap between valence and conductance bands, demonstrates a remarkable ambipolar electric field effect with high charge carrier mobility (up to $10,000$ cm^{2}V^{-1}s^{-1}) under room temperature [3]. The charge carriers can even travel for hundreds of nanometers without scattering under room temperature [4].

Besides, single-layer graphene with an extraordinary large surface area of 2630 m^2 g^{-1} [5] and a low opacity of $\sim2.3\%$ towards visible light [6] exhibits its preponderance in non-electronic properties. It has a great breaking strength of 42 N m^{-1}, close to the intrinsic strength of a defect-free sheet, with a corresponding Young's modulus of 1.0 TPa [7]. It shows a superior room-temperature thermal conductivity of ~5000 Wm^{-1}K^{-1} which suggests the application of graphene for thermal management [8].

11.2   Methods of Production of Graphene

Efforts to produce graphene started from the preparation of graphitic oxide in 1859 by Brodie [9]. Until now, various methods have been developed to fabricate, grow, or synthesize graphene. Weak van der Waals interactions between the adjacent graphene layers exist to bind them in the bulk graphite [10]. Hence the single-layer graphene can be prepared from the well-known top-down approach of micromechanical exfoliation of graphite (also named as the "Scotch tape" method) [3]. Besides, versatile bottom-up approaches can be adopted to grow graphene sheets. Chemical vapor deposition (CVD) can grow single- or few-layer graphene sheets on metal substrates, such as Ni [11,12], Cu [13–15], and Ru [16]. Afterwards, the as-grown graphene can be transferred to arbitrary substrates. The epitaxial growth of graphene can also be achieved on insulating substrates, such as SiC [17,18]. Moreover, graphene can be prepared through the synthesis from organic precursors [19,20].

However, one of the most commonly used and developed method to acquire graphene is the chemical reduction of graphene oxide (GO), which is synthesized by oxidation and exfoliation of graphite [21–26]. Fabrication of GO by the Hummers method involves the treatment of graphite with strong acids and oxidizers, such as sulfuric acid, nitric acid, potassium permanganate, and potassium chlorate [27–35]. Graphene thus can be produced by chemical reduction of GO with hydrazine [21–26, 36–38], dimethyhydrazine

[39], hydroquinone [40], NaOH or KOH [41], vitamin C [42], or bovine serum albumin (BSA) [43]. Table 11.1 summarizes the advantages and disadvantages of the above mentioned four primary methods to produce graphene. In addition to chemical reduction of GO, there are many other methods to reduce GO, such as thermal reduction via the high-temperature deoxygenating process [44–46], hydrothermal dehydration of exfoliated GO using supercritical water as the reducing agent [47], electrochemical reduction method [48–50], photochemical reduction [51,52], etc.

Table 11.1 Advantages and disadvantages of four primary methods to produce graphene

Method	Advantages	Disadvantages
Micromechanical exfoliation of graphite	Produces high-quality graphene	Low throughput, low yield
CVD/Epitaxial growth	Produces high-quality and large-area graphene	Requires stringent conditions and careful control for uniform growth of graphene
Chemical reduction of GO	Low cost, large-amount production	Produces poor-quality graphene with extensive modification of graphene
Bottom-up synthesis from organic precursors	Size controllable synthesis of graphene	Difficult to yield large-scale graphene with narrow size distribution and almost impossible to avoid side reactions

11.3 Introduction to Graphene Oxide

As a precursor to synthesize graphene (also referred to as reduced graphene oxide, rGO), GO has a graphene-like structure but with a range of oxygen functional groups. The most accepted model for its structure is the Lerf–Klinowski model [53], which describes that GO has a layered structure containing abundant epoxy and hydroxyl groups on the basal planes and a few carbonyl and carboxyl groups at the edges. Subsequently, Gao et al. proved that there are five- and six-membered-ring lactols on the periphery of GO and added them into the structural model (Fig. 11.1) [54].

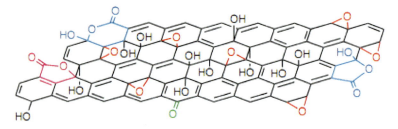

Figure 11.1 Structural model of graphene oxide. Reproduced with permission [54]. Copyright 2009, Nature Publishing Group.

GO is hydrophilic as the oxygen functionalities exist on its layered structure and water molecules are intercalated between the layers [55]. It can be readily exfoliated to produce a stable colloidal suspension in water [27,28,30,32,56] or some organic solvents [33,34,57] through sonication. Surface charge measurement showed that the zeta potential of GO in water with a concentration of 0.05 mg ml^{-1} and pH of 7 was about −40 mV [37]. It is believed that the ionization of carboxyl and hydroxyl groups on its surface leads to the negative charge on GO.

11.4 Methods to Produce Stable Dispersion of Reduced Graphene Oxide

Graphene (or rGO) is conductive and hydrophobic since most of the oxygen functional groups have been removed and the electrical conducting property of the conjugated graphitic network has been re-established during the process of reducing GO. However, rGO exhibits a much poorer conductivity than the pristine graphene due to the presence of residual oxygen functional groups and numerous defects that interrupt the conjugated graphitic network. Consequently, many methods have been developed to produce graphene with less modifications and higher conductivity, such as the liquid-phase exfoliation [58,59] and intercalation [60,61], electrochemical exfoliation [62], etc. In addition to the sheet morphology, graphene with other morphologies is also produced, like graphene quantum dots [63,64], graphene nanoribbons [28,65–68], and graphene nanomeshes [69–71].

Because of the hydrophobicity, aggregation of rGO is often observed in solutions. However, it is necessary to produce homogeneous and stable rGO dispersion in bulk quantity for the composite syntheses or device fabrications. The most simple method is to reduce GO using hydrazine without any surfactant in an alkaline solution with pH of 10 by adding ammonia [37]. With the optimal weight ratio of hydrazine to GO (7:10) for the reduction, a stable rGO aqueous solution with concentration of <0.5 mg ml^{-1} can be obtained [37]. Besides, hydrazine reduction of GO in a suspension with a volume ratio of dimethyl formamide (DMF):H$_2$O = 9:1 could also produce the homogeneous rGO suspension with pH of ~7 [72]. The as-obtained rGO suspension can be further dispersed in various organic solvents, e.g. ethanol, acetone, acetonitrile, DMF, *n*-methylpyrrolidone (NMP), tetrahydrofuran (THF), dimethyl sulfoxide (DMSO) [72].

In addition, the homogeneous rGO dispersion can be achieved through noncovalent functionalization and covalent functionalization of rGO. Noncovalent functionalization of rGO with some polymers, organic molecules, or biomolecules through the van der Waals force or the π–π interaction produces rGO suspension with tunable solubility. For example, stable rGO dispersion in water can be obtained by adding an amphiphilic polymer of poly(sodium-4 styrene sulfonate) [73], conducting polymer of sulfonated polyaniline [74], a pyrene derivative of 1-pyrenebutyrate [75], or single-strand DNA (ssDNA) [76] during the reduction of GO with hydrazine. Recently, our group reported the hydrazine reduction of GO with the presence of an amphiphilic conjugated polymer of PEG–OPE (PEG = poly(ethylene glycol), OPE = oligo(phenylene ethenylene)) (Fig. 11.2) [21]. As-obtained rGO had a strong π–π interaction with PEG–OPE and could be dispersed well in various organic solvents and water.

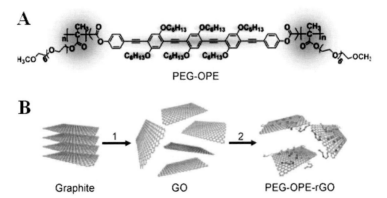

Figure 11.2 (A) Chemical structure of PEG–OPE, and (B) two-step synthesis of PEG–OPE-functionalized rGO sheets. Reproduced with permission [21]. Copyright 2010, Wiley-VCH.

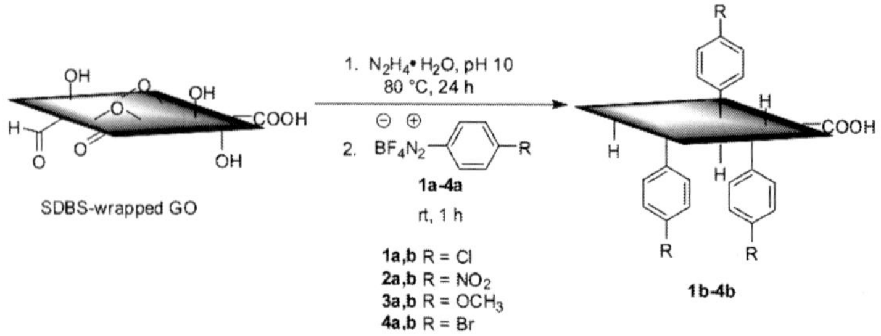

Figure 11.3 Schematic illustration of diazonium functionalization of SDBS-wrapped hydrazine reduced GO. Reproduced with permission [79]. Copyright 2008, American Chemical Society.

Covalent functionalization of rGO often involves reactions between functional molecules and oxygenated groups on GO or rGO. Hence the bonding between rGO and other functionalized material can be reinforced in the composite and homogeneous rGO dispersion can be achieved. Niyogi et al. demonstrated that rGO dissolved in tetrahydrofuran (THF), CCl_4, and 1,2-dichloroethane could be obtained by reactions between octadecylamine (ODA) and carboxylic acid groups present on oxidized graphite [77]. Besides, sulfonation of GO with aryl diazonium salt of sulfanilic acid reported by Si et al. could introduce negatively charged $-SO_3^-$ groups on GO and stabilize rGO in water upon the hydrazine reduction [78]. Moreover, Lomeda et al. functionalized surfactant (sodium dodecylbenzenesulfonate, SDBS)-protected rGO through the aryl diazonium treatment (Fig. 11.3) [79]. This method was developed from that of functionalization of carbon nanotubes (CNTs). The resulting rGO can be dispersed in N,N'-dimethylformamide (DMF), N,N'-dimethylacetamide (DMAc), and 1-methyl-2-pyrrolidinone (NMP) [79].

11.5 Characterizations and Fundamental Properties of Reduced Graphene Oxide

11.5.1 Characterization with Atomic Force Microscopy

Theoretically, monolayer pristine graphene has a van der Waals thickness of 0.34 nm [56]. However, the thickness of GO sheets is higher because of the presence of the oxygen functionalities and the displacement of sp³-hybridized carbon atoms [56]. The atomic force microscopy (AFM) images of GO and chemically converted graphene (or rGO) sheets showed a similar thickness of ~1 nm (Fig. 11.4), indicating that the complete exfoliation of GO and rGO down to single-layer sheets was achieved in the dispersion [37,57].

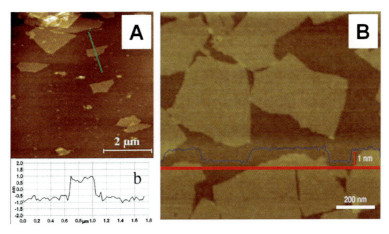

Figure 11.4 (A) AFM image of exfoliated GO sheets. Reproduced with permission [57]. Copyright 2008, American Chemical Society. (B) AFM image of rGO sheets with a height profile (blue curve; scale bar, 1 nm) taken along the red line. Reproduced with permission [37]. Copyright 2008, Nature Publishing Group.

11.5.2 Characterization with X-Ray Diffraction

Wang et al. studied the oxidation process from natural flake graphite to GO by X-ray diffraction (XRD) [40]. Figure 11.5A shows the XRD patterns of pristine graphite and GO obtained after different oxidation times. The XRD pattern of pristine graphite (black) contains a sharp diffraction peak at 26.23°, corresponding to the interlayer spacing of 0.34 nm of (002) planes. As the oxidation of graphite proceeds, the (002) diffraction peak intensity decreases sharply. Simultaneously a diffraction peak at 11.8° (d = 0.749 nm) appears and increases slightly with the oxidation time. Upon reduction, the (002) diffraction peak at about 26–27° reappears in the XRD pattern of rGO (Fig. 11.5B), indicating the restoration of the ordered stacking of the carbon sheets [40].

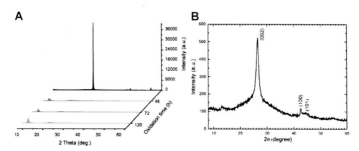

Figure 11.5 XRD patterns of (A) pristine graphite and GO after different oxidation durations, and (B) rGO. Reproduced with permission [40]. Copyright 2008, American Chemical Society.

11.5.3 Characterization with X-Ray Photoelectron Spectroscopy (XPS)

X-Ray Photoelectron Spectroscopy (XPS) spectrum (Fig. 11.6) can be used to analyze the different oxidation states of element C in both GO and rGO, from which we can distinguish between GO and rGO [46]. There are non-oxygenated ring C (C–C), single bonded C to oxygen (C–O), and carbonyl C (C=O) in the C1s XPS spectrum of GO. The XPS spectrum of rGO contains a strong signal for non-oxygenated ring C, while single bonded C to oxygen cannot be observed and carbonyl C exhibits smaller peak intensities compared to that in GO. The additional nitrogen bonded C (C–N) comes from the reduction process with hydrazine [46]. The comparison of XPS spectra between GO and rGO suggests that most of the oxygen functional groups in GO have been removed during the reduction process using hydrazine.

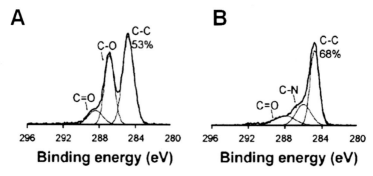

Figure 11.6 C1s XPS spectra of (A) GO and (B) rGO. Reproduced with permission [46].

11.5.4 Characterization with Raman Spectroscopy

It is known that defects in graphite can induce many special and meaningful Raman peaks. Figure 11.7A shows the Raman spectra of GO and rGO obtained using sodium borohydride (NaBH$_4$) of different molar concentrations. All the Raman spectra of GO and rGO illustrate the two characteristic D and G peaks corresponding to the disorder-induced mode and the first-order scattering of the E_{2g} mode, respectively [80,81]. The D peak suggests the reduction of in-plane sp^2 domain sizes due to oxidation. As seen from Fig. 11.7B, the D/G intensity ratio increases slightly after the GO is reduced and reaches the highest value when the GO is reduced using 150 mM of NaBH$_4$, indicating more disordering in the rGO structure. This is becausemany smaller sp^2 domains (i.e., ordering structure) are created during the reduction process, resulting in a decrease in the average size of sp^2 domains in the rGO [56]. Besides exhibiting the structural difference between GO and rGO, Raman spectra can also be used for identification of the layer number of graphene with less than five layers produced by the mechanical exfoliation method [82].

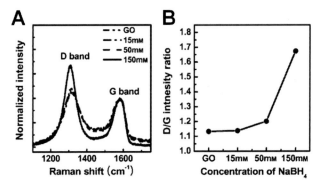

Figure 11.7 (A) Raman spectra of GO and rGO obtained by using NaBH$_4$ of different molar concentrations, and (B) dependence of D/G intensity ratio on the concentration of NaBH$_4$ for reducing GO. Reproduced with permission [80]. Copyright 2009, Wiley-VCH.

11.5.5 Conductivity

GO is insulating, but rGO is conductive due to the re-establishment of the conjugated graphitic network during the reduction process.

Conductivity of rGO obtained using 15 mM of $NaBH_4$ (~$1.5E-4$ S m^{-1}) is about 4 orders of magnitude higher than that of GO ($6.8E-8$ S m^{-1}), and the conductivity can increase to 45 S m^{-1} when reduced using 150 mM of $NaBH_4$ [80]. Figure 11.8 shows the increase in conductivity and C/O ratio of GO and rGO with molar concentration of $NaBH_4$ used for reducing GO, indicating that the conductivity of GO and rGO is closely related to their C/O ratio.

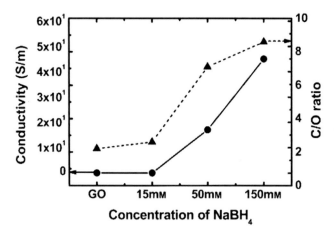

Figure 11.8 Dependence of conductivity and C/O ratio of GO and rGO on molar concentration of $NaBH_4$ used for reducing GO. Reproduced with permission [80].Copyright 2009, Wiley-VCH.

11.6 Sensing Applications of Reduced Graphene Oxide

11.6.1 Field-Effect Transistor Sensors

Electronic sensors based on field-effect transistors (FETs) are promising because of their small size, simple configuration, and real-time detection. Yet they are compromised in sensitivity and response time with the conventional semiconducting channels. The development of nanomaterials and nanotechnologies provides ideal channel materials with extremely high sensitivity and low noise. Graphene has attracted considerable interest as the active channel due to its high charge mobility and capacity [1], large detection area, facile and homogenous functionalization [83], relatively low

$1/f$ noise [84], and high biocompatibility [85]. Most importantly, the atomic thickness of graphene makes its electrical properties very sensitive towards the change of local environments, offering ultimate sensitivity [83]. Although the pristine graphene has been employed as both chemical [83,86] and biological sensors [87–89] and shown promising performance, it suffers from the low yield and incontrollable in size and device performance. Wafer-scale chemical vapor deposition (CVD)-grown graphene (CVD-graphene) is one good way to achieve practical electronic sensors with good reproducibility [90–92]. Lieber and co-workers reported an integrated electronic sensor array with high reproducibility based on CVD-graphene for the multiple pH detection [91]. Chemically derived rGO is also preferred for its massive production, consistent device performance, and reasonable compromising in sensitivity.

Compared to nanoelectronics based on individual rGO sheets [93–96], rGO based thin film electronics is more favored owing to its facile device fabrication and high reproducibility in device performance. Robinson et al. reported the first gas sensor based on the rGO thin film [97]. By monitoring the change in the conductance upon exposure to the target vapor, the rGO thin film based sensor demonstrated a better sensitivity towards various toxic gases than did the CNT thin-film based sensor. By patterning electrodes on the discontinuous single-layer rGO thin film, Fowler et al. demonstrated a practical gas sensing platform with detection limit down to 28 part per billion (ppb) towards the 2,4-dinitrotoluene, an explosive residue [98]. More recently, by taking advantage of the solution processability of rGO, Dua et al. reported a flexible vapor sensor based on the inject-printed rGO thin film, which exhibited a low detection limit at ppb level [99].

Besides the excellent performance in gas sensing, rGO is also favored in bio-sensing because of its good biocompatibility [85]. Specific detection of target DNA in real time was demonstrated by a series functionalization of rGO film [100]. Recently, our group also did several studies in realizing the potential of rGO thin film transistors in chemical and biological sensing. As shown in Fig. 11.9, a uniform-patterned rGO thin film was used as the active channel to detect hormonal catecholamine molecules and their dynamic secretion from living PC 12 cells in the buffer solution [101]. The real-time monitoring of living cells was realized by directly culturing cells on the patterned rGO thin film due to its large scale uniformity. Moreover, a real-time rGO thin film transistor for heavy metal detection based on the specific ion–protein interaction was realized, showing a far superior performance to the fluorescent heavy metal sensor with a detection limit down to nM [102]. Besides, it is possible to realize an all-rGO transistor that uses rGO film as both active channel and electrodes (drain and source) due to the unique thickness dependent electrical properties of the rGO film [103]. Such all-rGO transistors show perfect flexibility, transparency, and capability to specifically detect avidin in real time based on the biotin–avidin interaction.

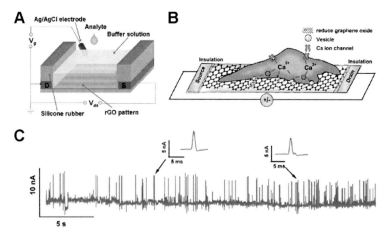

Figure 11.9 (A) Schematic illustration of the experimental setup for front-gate rGO based thin film transistor for sensing application. (B) Schematic illustration of the interface between a PC 12 cell and an rGO thin film transistor. (C) Real-time response of rGO/poly(ethylene terephthalate) (PET) transistor to the vesicular secretion of catecholamines from PC 12 cells stimulated by high K$^+$ concentration solution. Reproduced with permission [101]. Copyright 2010, American Chemical Society.

11.6.2 Electrochemical Sensors

The electrochemical property of rGO has attracted increasing attention. Recently, Zhou et al. proved that the rGO electrode has a large potential window of ~2.5 V, a large apparent electrode area of 0.092 cm^2, and a low charge-transfer resistance of 160.8 Ω [104], indicating its high electrochemical activity. In addition, the low peak-to-peak potential separation (ΔE_p) of the rGO electrode in the cyclic voltammograms of the representative redox systems, $[Fe(CN)_6]^{3-/4-}$ and $[Ru(NH_3)_6]^{3+/2+}$ [105–108], related to the electron transfer coefficient [109], suggests the fast electron transfer on rGO electrode, which can be explained by its high density of electronic states.

Due to its large specific surface area, high electrochemical activity, and fast electron transfer rate, rGO electrode is promising for electrochemical sensing. For example, the direct electrochemistry of glucose oxidase (GOD) on rGO electrode [49,109–111] shows an excellent electron transfer between the rGO electrode and the active center of GOD, hence rGO electrode can be used for glucose detections [49,110–117]. A glucose biosensor based on rGO which was fabricated by Wu et al. exhibited a low detection limit of 10 μM, a wide linear range from 0.1 mM to 10 mM, and a high sensitivity of ~110 $\mu AmM^{-1}cm^{-2}$ [111]. It is believed that the high adsorption capacity and unique electronic structure of rGO are the reasons for the superior performance of the rGO-based biosensor [111]. Besides, the surface defects present on rGO, which result in high electronic state density near Fermi level, may contribute to the fast electron transfer kinetics of GOD and the high sensitivity of the rGO-based biosensor [111]. Moreover, rGO based electrochemical sensors have been constructed for detecting H_2O_2 [105,118,119], reduced form of β-nicotinamide adenine dinucleotide [105,120,121], dopamine [108,122–127], and DNA bases [104,125, 128].

11.6.3 Matrices for Mass Spectrometry

The applications of the laser desorption/ionization mass spectrometry in fundamental researches started from 1970s [129]. Subsequently, the matrix-assisted laser desorption/ionization mass spectrometry with single laser was developed and awarded the Nobel Prize in 2002 [129]. Nowadays, the matrix-assisted laser desorption/ionization time-of-flight mass spectrometry (MALDI-TOF MS) has become an important tool for the rapid, accurate, and sensitive analysis of various materials, especially the high molecular mass biomolecules [130]. However, it is difficult to characterize the low-mass analytes (molecular mass < 500 Da) due to the matrix ion interference and detector saturation [130].

To overcome this problem, Dong et al. used rGO as a matrix for analysis of small molecules by MALDI-TOF MS because of its large surface area, monolayer structure, and unique electronic property

[130]. The MALDI-TOF MS with rGO as the matrix and an energy receptacle for laser radiation can detect small molecules such as amino acids, polyamines, steroids, nucleosides, and anticancer drugs [130]. It shows a higher signal with a lower laser power threshold for desorption/ionization, and no interference from the matrix background ions compared to the conventional matrices [130]. Our group also employed the rGO film in the MALDI-TOF MS to detect octachlorodibenzo-p-dioxin (OCDD) as little as 500 pg, while no signals of OCDD were observed from the conventional organic matrices of 2,5-dihydroxybenzoic acid (DHB), α-cyano-4-hydroxycinnamic acid (CHCA), and sinapic acid (SA) [131].

11.7 Device Applications of Reduced Graphene Oxide

11.7.1 Memory Devices

Over the last decade the dominant material used as the transparent and conductive layer for device fabrications is indium tin oxide (ITO). However, it has several drawbacks which limit its applications. It is brittle and easily cracks or fractures at low strains of 2–3%, which leads to degraded conductivity when microcracks propagate [132]. In addition, indium is limited on earth and hence expensive for industrial applications [133,134]. Therefore, the economical rGO with superior electronic, optical, and mechanical properties could be a promising substitute for ITO in device fabrications, for example, memory devices. Recently, our group constructed a polymer memory device using rGO (Fig. 11.10) [26,135]. The fabricated rGO film with a low sheet resistance of 160–500 Ω sq^{-1} was used as an electrode in the device with a structure of rGO/poly(3-hexylthiophene) (P3HT):phenyl-C61-butyric acid methyl ester (PCBM)/Al [26]. The as-constructed device exhibited a write-once-read-many-times (WORM) effect with a high ON/OFF current ratio of 10^6 [26].

11.7.2 Solar Cells

Besides memory devices, it is common to apply rGO film as window electrode in solar cells [133,136]. A dye-sensitized solid solar cell was fabricated using the rGO film on quartz as transparent anode, a blocking TiO$_2$ layer for electron transport, a spiro-OMeTAD layer for hole transport, and Au as cathode (Fig. 11.11) [133].

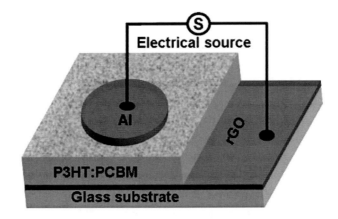

Figure 11.10 Schematic illustration of the fabricated rGO/P3HT:PCBM/Al memory device. Reproduced with permission [26]. Copyright 2010, Wiley-VCH.

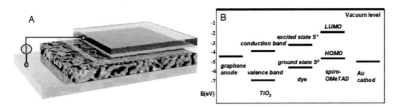

Figure 11.11 (A) Illustration of the dye-sensitized solar cell with rGO electrode; the four layers from bottom to top are Au, dye-sensitized heterojunction, TiO$_2$, and rGO. (B) Energy level diagram of the device in (A). Reproduced with permission [133]. Copyright 2008, American Chemical Society.

The current–voltage (I–V) characteristics of the solar cell fabricated with rGO film, referred to as rGO-device, and the one fabricated with fluorine tin oxide (FTO), referred to as FTO-device, are summarized in Table 11.2. From the comparison between the two solar cells, it can be found that the J_{sc} and PCE of rGO-device are lower than those of FTO-device, which can be attributed to the higher resistance and relatively lower light transmittance of rGO film, and the electronic interfacial change [133]. However, there is room for rGO-based device improvement by using rGO films of better quality or large-size (micrometer-scale) rGO sheet. The high-quality rGO film may be obtained through thermal treatment, other synthesis approaches (e.g., bottom-up approach), or more appropriate engineering procedures.

Table 11.2 Summary of *I–V* characteristics of rGO-device and FTO-device [133]

Device	Photocurrent density (J_{sc}) [mA cm^{-2}]	Open-circuit voltage (V_{oc}) [V]	Filling factor (FF)	Power conversion efficiency (PCE) [%]
rGO-device	1.01	0.7	0.36	0.26
FTO-device	3.02	0.76	0.36	0.84

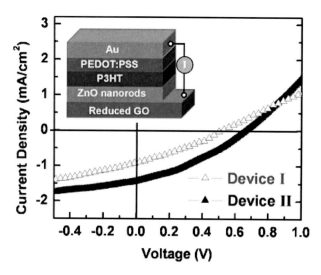

Figure 11.12 The obtained *I–V* curves for ZnO/P3HT hybrid solar cells by using (I) one-step and (II) two-step reduced GO films as electrodes. Inset: schematic illustration of the fabricated solar cell. Reproduced with permission [137]. Copyright 2009, Wiley-VCH.

Recently, our group has reported the application of rGO film as a transparent electrode in the hybrid solar cell fabrication [137]. The cell contains a layered structure of quartz/rGO/ZnO nanorods/P3HT/PEDOT:PSS/Au (inset of Fig. 11.12). The current density–voltage (*I–V*) curve of Device II (Fig. 11.12) made by two-step reduced GO film shows an open-circuit voltage (V_{oc}) of 0.66 V, a short-circuit current density (J_{sc}) of 1.43 mA cm^{-2}, a fill factor (FF) of 0.33, and an overall power conversion efficiency (PCE, η) of 0.31%. The two-step reduced GO film with a higher conductivity has a work function of ~4.7 eV, which showed a better performance because it produced a larger V_{oc} and provided a better match between the Fermi level of rGO and the conduction band of ZnO.

Furthermore, the rGO film can be transferred onto the flexible substrate of polyethylene terephthalate (PET) for fabrication of organic photovoltaic (OPV) device [138]. The performance of the fabricated device is dominated by the light transmission efficiency when the rGO transmittance is less than 65%, while it depends on the charge transport efficiency when the rGO transmittance is above 65% [138].

11.8 Applications of Graphene Oxide

11.8.1 Fluorescence Sensors

The superior electrical property of rGO makes it popular for various sensing and device applications. The insulating GO also shows great potentials owning to its unique fluorescence quenching capability, large surface area, and good solubility.

Fluorescence sensors are simple, rapid, cost-effective, and hence favorable in various biodetections. The unique fluorescence superquenching effect of GO, arising from its long range energy transfer properties [139], along with its excellent water solubility, makes GO an ideal fluorescence sensing platform for the detection of chemicals and biomolecules [140–143].

Recently, Lu et al. used GO as a sensing platform to selectively detect DNA and proteins (Fig. 11.13) [144]. Since the single-stranded DNA (ssDNA) has a strong binding with GO surface through the π–π interaction [145], the dye-labeled ssDNA probe can be completely quenched when binding to GO and forming the ssDNA-FAM–GO complex (step a in Fig. 11.13, FAM is the fluorescein-based fluorescent dye). In a typical detection process, the hybridization between the complementary target DNA and the dye-labeled probe DNA led to the formation and release of double-stranded DNA (dsDNA) from GO surface, resulting in the restoration of dye fluorescence (step b in Fig. 11.13). By replacing the dye-labeled ssDNA with dye-labeled aptamer, the sensing platform based on GO was able to detect the human thrombin with excellent selectivity.

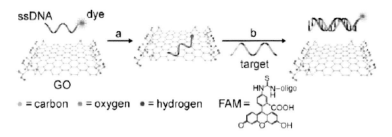

Figure 11.13 The schematic illustration of the target-induced fluorescence change of the ssDNA-FAM–GO complex. Reproduced with permission [144]. Copyright 2009, Wiley-VCH.

Moreover, He et al. demonstrated a GO-based multicolor fluorescence sensor for the multiplex, sequence-specific DNA detection [146]. By separately labeling three tumor-suppressor genes with different dyes, the simultaneous detection of multiple targets in a homogenous solution was realized. Such a homogeneous, mix-and-detect assay method is extremely rapid and can be finished within minutes. Similar fluorescent detection of other chemicals and biomolecules, including various enzymes [147,148] and metal ions [149], was also realized with different fluorescent probes such as quantum dots, pyrene, mercury-specific oligonucleotide, and peptide.

Besides the quenching effect, the near-UV to blue fluorescence of GO itself motivated the development of another type of GO-based fluorescence sensor [150,151]. For example, a GO-based immuno-biosensor for detection of the rotavirus as a pathogen model has been developed [151]. The GO array bound to the amino-modified glass surface was firstly modified with antibodies for rotavirus by the carbodiimide-assisted amidation reaction and subsequently captured target rotavirus cells by specific antigen–antibody interaction. Then the gold nanoparticles (AuNPs) modified antibodies could selectively bind to the previously formed GO–rotavirus composite. Such binding would lead to the reduction in the GO fluorescence due to the fluorescence resonance energy transfer between the AuNPs and GO sheets, enabling the detection of pathogenic target cells.

11.8.2 Solar Cells

Besides graphene, GO can also be used as a component in solar cells, such as hole transport layer (HTL) in solar cells. Li et al. fabricated a photovoltaic device with the layered structure of ITO/GO/P3HT:PCBM/Al (Fig. 11.14) [152]. Solar cells with a HTL of PEDOT:PSS and without any HTL were constructed to have a comparison. Table 11.3 shows the *I–V* characteristics of the fabricated devices.

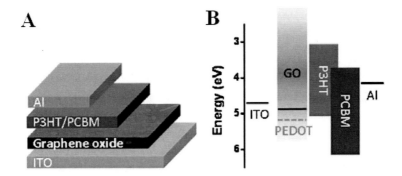

Figure 11.14 (A) Illustration of the polymer solar cell structure using GO as the hole transport layer. (B) Energy level diagram of the device in (A). Reproduced with permission [152]. Copyright 2010, American Chemical Society.

Table 11.3 *I–V* characteristics of GO-based solar cells and control devices

Solar cell	J_{sc} (mA cm^{-2})	V_{oc} (V)	FF (%)	PCE (%)
GO (2 nm)	11.4	0.57	54.3	3.5 ± 0.3
Control device with PEDOT:PSS	11.15	0.58	56.9	3.6 ± 0.2
Control device without any HTL	9.84	0.45	41.5	1.8 ± 0.2

Source: Reproduced with permission [152]. Copyright 2010, American Chemical Society.

The comparable PCE of GO-based polymer solar cell to PEDOT:PSS based device indicates the efficient hole transport in GO from P3HT to ITO electrode. The carrier transport mechanism in the as-constructed devices has been further investigated. It was

Graphene Oxides and Reduced Graphene Oxide Sheets

found that the recombination rate in the GO-based device was the lowest, leading to the effective separation of charge carriers and suppression of leakage current. In addition, the electron injection from PCBM LUMO to ITO could be blocked by the GO layer because of its large band gap (~3.6 eV). Moreover, an application of GO as the selective interfacial layer of hole in the inverted polymer solar cell has been reported by Gao et al. [153].

11.9 Conclusion

Graphene or rGO, derived chemically from GO, is being readily used in sensors, memory devices, and solar cells. The rGO-based FET sensors with high reliability and capability of massive production exhibit good sensitivity towards various gases, chemicals, and biomolecules. Such sensors can be even built on flexible substrates due to the easy solution-based manipulation of rGO. In addition, rGO electrode has shown excellent electrochemical activity, rapid charge transfer rate, and large specific surface area, and hence has achieved an exceptional performance in the electrochemical sensors. Besides, rGO with its unique structure and electronic properties can also be used as the matrix in mass spectrometry, showing high signal without interference from the matrix background ions. Moreover, memory devices and solar cells fabricated from rGO films have also shown superior performances. However, there is room for the device enhancement by improving the electrical conductivity and optical transmittance of rGO films. Furthermore, the insulating GO can be used as a fluorescence sensing platform to detect chemicals and biomolecules owing to its unique fluorescence quenching and emission properties, and a hole transport layer in solar cells owing to its specific energy band gap.

References

1. Geim A.K., Novoselov K.S. (2007) The rise of graphene, *Nat Mater*, **6**, 183–191.

2. Novoselov K.S., Geim A.K., Morozov S.V., et al. (2005) Two-dimensional gas of massless Dirac fermions in graphene, *Nature*, **438**, 197–200.

3. Novoselov K.S., Geim A.K., Morozov S.V., et al. (2004) Electric field effect in atomically thin carbon films, *Science*, **306**, 666–669.

4. Geim A.K. (2009) Graphene: Status and prospects, *Science*, **324**, 1530–1534.

5. Stoller M.D., Park S.J., Zhu Y.W., An J.H., Ruoff R.S. (2008) Graphene-based ultracapacitors, *Nano Lett*, **8**, 3498–3502.

6. Nair R.R., Blake P., Grigorenko A.N., et al. (2008) Fine structure constant defines visual transparency of graphene, **320**, 1308.

7. Lee C., Wei X., Kysar J.W., Hone J. (2008) Measurement of the elastic properties and intrinsic strength of monolayer graphene, *Science*, **321**, 385–388.

8. Balandin A.A., Ghosh S., Bao W., et al. (2008) Superior thermal conductivity of single-layer graphene, *Nano Lett*, **8**, 902–907.

9. Brodie B.C. (1859) On the atomic weight of graphite, *Philos Trans R Soc London*, **149**, 249–259.

10. Rozploch F., Patyk J., Stankowski J. (2007) Graphenes bonding forces in graphite, *Acta Phys Pol A*, **112**, 557 J.562.

11. Kim K.S., Zhao Y., Jang H., et al. (2009) Large-scale pattern growth of graphene films for stretchable transparent electrodes, *Nature*, **457**, 706–710.

12. Reina A., Jia X.T., Ho J., et al. (2009) Large area, few-layer graphene films on arbitrary substrates by chemical vapor deposition, *Nano Lett*, **9**, 30–35.

13. Lee S., Lee K., Zhong Z. (2010) Wafer scale homogeneous bilayer graphene films by chemical vapor deposition, *Nano Lett*, **10**, 4702–4707.

14. Wei D., Liu Y., Wang Y., Zhang H., Huang L., Yu G. (2009) Synthesis of *N*-doped graphene by chemical vapor deposition and its electrical properties, *Nano Lett*, **9**, 1752–1758.

15. Li X., Cai W., An J., et al. (2009) Large-area synthesis of high-quality and uniform graphene films on copper foils, *Science*, **324**, 1312–1314.

16. Sutter P.W., Flege J.-I., Sutter E.A. (2008) Epitaxial graphene on ruthenium, *Nat Mater*, **7**, 406–411.

17. Berger C., Song Z.M., Li X.B., et al. (2006) Electronic confinement and coherence in patterned epitaxial graphene, *Science*, **312**, 1191–1196.

18. Emtsev K.V., Bostwick A., Horn K., et al. (2009) Towards wafer-size graphene layers by atmospheric pressure graphitization of silicon carbide, *Nat Mater*, **8**, 203–207.

19. Markus M., Christian K., Klaus M. (1998) Giant polycyclic aromatic hydrocarbons, *Chem Eur J*, **4**, 2099–2109.

20. Yang X., Dou X., Rouhanipour A., Zhi L., Rader H.J., Mullen K. (2008) Two-dimensional graphene nanoribbons, *J Am Chem Soc*, **130**, 4216–4217.

21. Qi X.Y., Pu K.Y., Li H., et al. (2010) Amphiphilic graphene composites, *Angew Chem Int Ed*, **49**, 9426–9429.

22. Qi X.Y., Pu K.Y., Zhou X.Z., et al. (2010) Conjugated-polyelectrolyte-functionalized reduced graphene oxide with excellent solubility and stability in polar solvents, *Small*, **6**, 663–669.

23. Wu S.X., Yin Z.Y., He Q.Y., Huang X.A., Zhou X.Z., Zhang H. (2010) Electrochemical deposition of semiconductor oxides on reduced graphene oxide-based flexible, transparent, and conductive electrodes, *J Phys Chem C*, **114**, 11816–11821.

24. Wu S.X., Yin Z.Y., He Q.Y., Lu G., Zhou X.Z., Zhang H. (2011) Electrochemical deposition of Cl-doped n-type Cu_2O on reduced graphene oxide electrodes, *J Mater Chem*, **21**, 3467–3470.

25. Huang X.Y., Yin Z. Y., Wu S. X.et al. (2011) Graphene-based materials: Synthesis, characterization, properties and applications, *Small*, **7**, 1876–1902.

26. Liu J.Q., Lin Z.Q., Liu T.J., et al. (2010) Multilayer stacked low-temperature-reduced graphene oxide films: Preparation, characterization, and application in polymer memory devices, *Small*, **6**, 1536–1542.

27. Zhou X.Z., Huang X., Qi X.Y., et al. (2009) In situ synthesis of metal nanoparticles on single-layer graphene oxide and reduced graphene oxide surfaces, *J Phys Chem C*, **113**, 10842–10846.

28. Zhou X.Z., Lu G., Qi X.Y., et al. (2009) A method for fabrication of graphene oxide nanoribbons from graphene oxide wrinkles, *J Phys Chem C*, **113**, 19119–19122.

29. Lu G., Zhou X.Z., Li H., et al. (2010) Nanolithography of single-layer graphene oxide films by atomic force microscopy, *Langmuir*, **26**, 6164–6166.

30. Huang X., Li S.Z., Huang Y.Z., et al. (2011) Synthesis of hexagonal close-packed gold nanostructures, *Nat Commun*, **2**, 292.

31. Li B., Cao X., Ong H.G., et al. (2010) All-carbon electronic devices fabricated by directly grown single-walled carbon nanotubes on reduced graphene oxide electrodes, *Adv Mater*, **22**, 3058–3061.

32. Shi W.H., Zhu J.X., Sim D.H., et al. (2011) Achieving high specific charge capacitances in Fe_3O_4/reduced graphene oxide nanocomposites, *J Mater Chem*, **21**, 3422–3427.

33. Zhu J.X., Zhu T., Zhou X.Z., et al. (2011) Facile synthesis of metal oxide/reduced graphene oxide hybrids with high lithium storage capacity and stable cyclability, *Nanoscale*, **3**, 1084–1089.

34. Zhu J.X., Sharma Y.K., Zeng Z.Y., et al. (2011) Cobalt oxide nanowall arrays on reduced graphene oxide sheets with controlled phase, grain size, and porosity for Li-ion battery electrodes, *J Phys Chem C*, **115**, 8400–8406.

35. Xiao N., Dong X.C., Song L., et al. (2011) Enhanced thermopower of graphene films with oxygen plasma treatment, *ACS Nano*, **5**, 2749–2755.

36. Tung V.C., Allen M.J., Yang Y., Kaner R.B. (2009) High-throughput solution processing of large-scale graphene, *Nat Nanotechnol*, **4**, 25–29.

37. Li D., Muller M.B., Gilje S., Kaner R.B., Wallace G.G. (2008) Processable aqueous dispersions of graphene nanosheets, *Nat Nanotechnol*, **3**, 101–105.

38. Hu W., Peng C., Luo W., et al. (2010) Graphene-based antibacterial paper, *ACS Nano*, **4**, 4317–4323.

39. Stankovich S., Dikin D.A., Dommett G.H.B., et al. (2006) Graphene-based composite materials, *Nature*, **442**, 282–286.

40. Wang G., Yang J., Park J., et al. (2008) Facile synthesis and characterization of graphene nanosheets, *J Phys Chem C*, **112**, 8192–8195.

41. Xiaobin F., Wenchao P., Yang L., et al. (2008) Deoxygenation of exfoliated graphite oxide under alkaline conditions: A green route to graphene preparation. *Adv Mater*, **20**, 1–4.

42. Dua V., Surwade S.P., Ammu S., et al. (2010) All-organic vapor sensor using inkjet-printed reduced graphene oxide, *Angew Chem Int Ed*, **49**, 2154–2157.

43. Liu J., Fu S., Yuan B., Li Y., Deng Z. (2010) Toward a universal adhesive nanosheet for the assembly of multiple nanoparticles based on a protein-induced reduction decoration of graphene oxide, *J Am Chem Soc*, **132**, 7279–7281.

44. McAllister M.J., Li J.-L., Adamson D.H., et al. (2007) Single sheet functionalized graphene by oxidation and thermal expansion of graphite, *Chem Mater*, **19**, 4396–4404.

45. Schniepp H.C., Li J.-L., McAllister M.J., et al. (2006) Functionalized single graphene sheets derived from splitting graphite oxide, *J Phys Chem B*, **110**, 8535–8539.

46. Becerril H.A., Mao J., Liu Z., Stoltenberg R.M., Bao Z., Chen Y. (2008) Evaluation of solution-processed reduced graphene oxide films as transparent conductors, *ACS Nano*, **2**, 463–470.

47. Zhou Y., Bao Q., Tang L.A.L., Zhong Y., Loh K.P. (2009) Hydrothermal dehydration for the "green" reduction of exfoliated graphene oxide to

graphene and demonstration of tunable optical limiting properties, *Chem Mater*, **21**, 2950–2956.

48. Zhou M., Wang Y.L., Zhai Y.M., et al. (2009) Controlled synthesis of large-area and patterned electrochemically reduced graphene oxide films, *Chem Eur J*, **15**, 6116–6120.

49. Wang Z., Zhou X., Zhang J., Boey F., Zhang H. (2009) Direct electrochemical reduction of single-layer graphene oxide and subsequent functionalization with glucose oxidase, *J Phys Chem C*, **113**, 14071–14075.

50. Guo H.-L., Wang X.-F., Qian Q.-Y., Wang F.-B., Xia X.-H. (2009) A green approach to the synthesis of graphene nanosheets, *ACS Nano*, **3**, 2653–2659.

51. Huang X., Zhou X., Wu S., et al. (2010) Reduced graphene oxide-templated photochemical synthesis and in situ assembly of Au nanodots to orderly patterned Au nanodot chains, *Small*, **6**, 513–516.

52. Williams G., Seger B., Kamat P.V. (2008) TiO_2-graphene nanocomposites. UV-assisted photocatalytic reduction of graphene oxide, *ACS Nano*, **2**, 1487–1491.

53. Lerf A., He H., Forster M., Klinowski J. (1998) Structure of graphite oxide revisited, *J Phys Chem B*, **102**, 4477–4482.

54. Gao W., Alemany L.B., Ci L., Ajayan P.M. (2009) New insights into the structure and reduction of graphite oxide, *Nat Chem*, **1**, 403–408.

55. Buchsteiner A., Lerf A., Pieper J. (2006) Water dynamics in graphite oxide investigated with neutron scattering, *J Phys Chem B*, **110**, 22328–22338.

56. Stankovich S., Dikin D.A., Piner R.D., et al. (2007) Synthesis of graphene-based nanosheets via chemical reduction of exfoliated graphite oxide, *Carbon*, **45**, 1558–1565.

57. Paredes J.I., Villar-Rodil S., Martínez-Alonso A., Tascón J.M.D. (2008) Graphene oxide dispersions in organic solvents, *Langmuir*, **24**, 10560–10564.

58. Lotya M., Hernandez Y., King P.J., et al. (2009) Liquid phase production of graphene by exfoliation of graphite in surfactant/water solutions, *J Am Chem Soc*, **131**, 3611–3620.

59. Hernandez Y., Nicolosi V., Lotya M., et al. (2008) High-yield production of graphene by liquid-phase exfoliation of graphite, *Nat Nanotechnol*, **3**, 563–568.

60. Li X.L., Zhang G.Y., Bai X.D., et al. (2008) Highly conducting graphene sheets and Langmuir-Blodgett films, *NatNanotechnol*, **3**, 538–542.

61. Vallés C., Drummond C., Saadaoui H., et al. (2008) Solutions of negatively charged graphene sheets and ribbons, *J Am Chem Soc*, **130**, 15802–15804.

62. Liu N., Luo F., Wu H., Liu Y., Zhang C., Chen J. (2008) One-step ionic-liquid-assisted electrochemical synthesis of ionic-liquid-functionalized graphene sheets directly from graphite, *Adv Funct Mater*, **18**, 1518–1525.

63. Ponomarenko L.A., Schedin F., Katsnelson M.I., et al. (2008) Chaotic dirac billiard in graphene quantum dots, *Science*, **320**, 356–358.

64. Yan X., Cui X., Li B., Li L.S. (2010) Large, solution-processable graphene quantum dots as light absorbers for photovoltaics, *Nano Lett*, **10**, 1869–1873.

65. Li X., Wang X., Zhang L., Lee S., Dai H. (2008) Chemically derived, ultrasmooth graphene nanoribbon semiconductors, *Science*, **319**, 1229–1232.

66. Duan H., Xie E., Han L., Xu Z. (2008) Turning PMMA nanofibers into graphene nanoribbons by in situ electron beam irradiation, *Adv Mater*, **20**, 3284–3288.

67. Bai J., Duan X., Huang Y. (2009) Rational fabrication of graphene nanoribbons using a nanowire etch mask, *Nano Lett*, **9**, 2083–2087.

68. Johnson J.L., Behnam A., Pearton S.J., Ural A. (2010) Hydrogen sensing using Pd-functionalized multi-layer graphene nanoribbon networks, *Adv Mater*, **22**, 4877–4880.

69. Akhavan O. (2010) Graphene nanomesh by ZnO nanorod photocatalysts, *ACS Nano*, **4**, 4174–4180.

70. Kim M., Safron N.S., Han E., Arnold M.S., Gopalan P. (2010) Fabrication and characterization of large-area, semiconducting nanoperforated graphene materials, *Nano Lett*, **10**, 1125–1131.

71. Bai J., Zhong X., Jiang S., Huang Y., Duan X. (2010) Graphene nanomesh, *Nat Nanotechnol*, **5**, 190–194.

72. Park S., An J.H., Jung I.W., et al. (2009) Colloidal suspensions of highly reduced graphene oxide in a wide variety of organic solvents, *Nano Lett*, **9**, 1593–1597.

73. Stankovich S., Piner R.D., Chen X.Q., Wu N.Q., Nguyen S.T., Ruoff R.S. (2006) Stable aqueous dispersions of graphitic nanoplatelets via the reduction of exfoliated graphite oxide in the presence of poly(sodium 4-styrenesulfonate), *J Mater Chem*, **16**, 155–158.

74. Bai H., Xu Y.X., Zhao L., Li C., Shi G.Q. (2009) Non-covalent functionalization of graphene sheets by sulfonated polyaniline, *Chem Commun*, **13**, 1667–1669.

75. Xu Y., Bai H., Lu G., Li C., Shi G. (2008) Flexible graphene films via the filtration of water-soluble noncovalent functionalized graphene sheets, *J Am Chem Soc*, **130**, 5856–5857.

76. Patil A.J., Vickery J.L., Scott T.B., Mann S. (2009) Aqueous stabilization and self-assembly of graphene sheets into layered bio-nanocomposites using DNA, *Adv Mater*, **21**, 3159–3164.

77. Niyogi S., Bekyarova E., Itkis M.E., McWilliams J.L., Hamon M.A., Haddon R.C. (2006) Solution properties of graphite and graphene, *J Am Chem Soc*, **128**, 7720–7721.

78. Si Y., Samulski E.T. (2008) Synthesis of water soluble graphene, *Nano Lett*, **8**, 1679–1682.

79. Lomeda J.R., Doyle C.D., Kosynkin D.V., Hwang W.-F., Tour J.M. (2008) Diazonium functionalization of surfactant-wrapped chemically converted graphene sheets, *J Am Chem Soc*, **130**, 16201–16206.

80. Shin H.J., Kim K.K., Benayad A., et al. (2009) Efficient reduction of graphite oxide by sodium borohydrilde and its effect on electrical conductance, *Adv Funct Mater*, **19**, 1987–1992.

81. Reich S., Thomsen C. (2004) Raman spectroscopy of graphite, *Philos Trans R Soc London, Ser A*, **362**, 2271–2288.

82. Ferrari A.C., Meyer J.C., Scardaci V., et al. (2006) Raman spectrum of graphene and graphene layers, *Phys Rev Lett*, **97**, 187401.

83. Huang Y.X., Dong X.C., Shi Y.M., Li C.M., Li L.J., Chen P. (2010) Nanoelectronic biosensors based on CVD grown graphene, *Nanoscale*, **2**, 1485–1488.

84. Schedin F., Geim A.K., Morozov S.V., et al. (2007) Detection of individual gas molecules adsorbed on graphene, *Nat Mater*, **6**, 652–655.

85. Agarwal S., Zhou X.Z., Ye F., et al. (2010) Interfacing live cells with nanocarbon substrates, *Langmuir*, **26**, 2244–2247.

86. Dan Y.P., Lu Y., Kybert N.J., Luo Z.T., Johnson A.T.C. (2009) Intrinsic response of graphene vapor sensors, *Nano Lett*, **9**, 1472–1475.

87. Ohno Y., Maehashi K., Yamashiro Y., Matsumoto K. (2009) Electrolyte-gated graphene field-effect transistors for detecting pH protein adsorption, *Nano Lett*, **9**, 3318–3322.

88. Ohno Y., Maehashi K., Matsumoto K. (2010) Label-free biosensors based on aptamer-modified graphene field-effect transistors, *J Am Chem Soc*, **132**, 18012–18013.

89. Cohen-Karni T., Qing Q., Li Q., Fang Y., Lieber C.M. (2010) Graphene and nanowire transistors for cellular interfaces and electrical recording, *Nano Lett*, **10**, 1098–1102.

90. Dong X.C., Shi Y.M., Huang W., Chen P., Li L.J. (2010) Electrical detection of DNA hybridization with single-base specificity using transistors based on CVD-grown graphene sheets, *Adv Mater*, **22**, 1649–1653.

91. Park J.-U., Nam S., Lee M.-S., Lieber C.M. (2012) Synthesis of monolithic graphene–graphite integrated electronics, *Nat Mater*, **11**, 120–125.

92. Kwak Y.H., Choi D.S., Kim Y.N., et al. (2012) Flexible glucose sensor using CVD-grown graphene-based field effect transistor, *Biosens Bioelectron*, **37**, 82–87.

93. Lu G.H., Ocola L.E., Chen J.H. (2009) Reduced graphene oxide for room-temperature gas sensors, *Nanotechnology*, **20**, 445502.

94. Mohanty N., Berry V. (2008) Graphene-based single-bacterium resolution biodevice and DNA transistor: Interfacing graphene derivatives with nanoscale and microscale biocomponents, *Nano Lett*, **8**, 4469–4476.

95. Kurkina T., Sundaram S., Sundaram R.S., et al. (2012) Self-assembled electrical biodetector based on reduced graphene oxide, *ACS Nano*, **6**, 5514–5520.

96. Chen K., Lu G., Chang J., et al. (2012) Hg(II) ion detection using thermally reduced graphene oxide decorated with functionalized gold nanoparticles, *Anal Chem*, **84**, 4057–4062.

97. Robinson J.T., Perkins F.K., Snow E.S., Wei Z.Q., Sheehan P.E. (2008) Reduced graphene oxide molecular sensors, *Nano Lett*, **8**, 3137–3140.

98. Fowler J.D., Allen M.J., Tung V.C., Yang Y., Kaner R.B., Weiller B.H. (2009) Practical chemical sensors from chemically derived graphene, *ACS Nano*, **3**, 301–306.

99. Dua V., Surwade S.P., Ammu S., et al. (2010) All-organic vapor sensor using inkjet-printed reduced graphene oxide, *Angew Chem Int Ed*, **49**, 2154–2157.

100. Stine R., Robinson J.T., Sheehan P.E., Tamanaha C.R. (2010) Real-time DNA detection using reduced graphene oxide field effect transistors, *Adv Mater*, **22**, 5297–5300.

101. He Q.Y., Sudibya H.G., Yin Z.Y., et al. (2010) Centimeter-long and large-scale micropatterns of reduced graphene oxide films: Fabrication and sensing applications, *ACS Nano*, **4**, 3201–3208.

102. Sudibya H.G., He Q.Y., Zhang H., Chen P. (2011) Electrical detection of metal ions using field-effect transistors based on micropatterned reduced graphene oxide films, *ACS Nano*, **5**, 1990–1994.

103. He Q.Y., Wu S.X. , Gao S., et al. (2011) Transparent, flexible, all-reduced graphene oxide thin film transistors, *ACS Nano*, **5**, 5038–5044.

104. Zhou M., Zhai Y., Dong S. (2009) Electrochemical sensing and biosensing platform based on chemically reduced graphene oxide, *Anal Chem*, **81**, 5603–5613.

105. Lin W.J., Liao C.S., Jhang J.H., Tsai Y.C. (2009) Graphene modified basal and edge plane pyrolytic graphite electrodes for electrocatalytic oxidation of hydrogen peroxide and beta-nicotinamide adenine dinucleotide, *Electrochem Commun*, **11**, 2153–2156.

106. Yang S.L., Guo D.Y., Su L., et al. (2009) A facile method for preparation of graphene film electrodes with tailor-made dimensions with Vaseline as the insulating binder, *Electrochem Commun*, **11**, 1912–1915.

107. Tang L.H., Wang Y., Li Y.M., Feng H.B., Lu J., Li J.H. (2009) Preparation, structure, and electrochemical properties of reduced graphene sheet films, *Adv Funct Mater*, **19**, 2782–2789.

108. Shang N.G., Papakonstantinou P., McMullan M., et al. (2008) Catalyst-free efficient growth, orientation and biosensing properties of multilayer graphene nanoflake films with sharp edge planes, *Adv Funct Mater*, **18**, 3506–3514.

109. Shao Y.Y., Wang J., Wu H., Liu J., Aksay I.A., Lin Y.H. (2010) Graphene based electrochemical sensors and biosensors: A review, *Electroanalysis*, **22**, 1027–1036.

110. Shan C.S., Yang H.F., Song J.F., Han D.X., Ivaska A., Niu L. (2009) Direct electrochemistry of glucose oxidase and biosensing for glucose based on graphene, *Anal Chem*, **81**, 2378–2382.

111. Wu P., Shao Q.A., Hu Y.J., et al. (2010) Direct electrochemistry of glucose oxidase assembled on graphene and application to glucose detection, *Electrochim Acta*, **55**, 8606–8614.

112. Chen D., Tang L.H., Li J.H. (2010) Graphene-based materials in electrochemistry, *Chem Soc Rev*, **39**, 3157–3180.

113. Kang X.H., Wang J., Wu H., Aksay I.A., Liu J., Lin Y.H. (2009) Glucose oxidase-graphene-chitosan modified electrode for direct electrochemistry and glucose sensing, *Biosens Bioelectron*, **25**, 901–905.

114. Gu H., Yu Y.Y., Liu X.Q., Ni B., Zhou T.S., Shi G.Y. (2012) Layer-by-layer self-assembly of functionalized graphene nanoplates for glucose sensing in vivo integrated with on-line microdialysis system, *Biosens Bioelectron*, **32**, 118–126.

115. Jiang Y.Y., Zhang Q.X., Li F.H., Niu L. (2012) Glucose oxidase and graphene bionanocomposite bridged by ionic liquid unit for glucose biosensing application, *Sens Actuators, B*, **161**, 728–733.

116. Zhang Q., Wu S.Y., Zhang L., et al. (2011) Fabrication of polymeric ionic liquid/graphene nanocomposite for glucose oxidase immobilization and direct electrochemistry, *Biosens Bioelectron*, **26**, 2632–2637.

117. Liu S., Tian J.Q., Wang L., Luo Y.L., Lu W.B., Sun X.P. (2011) Self-assembled graphene platelet-glucose oxidase nanostructures for glucose biosensing, *Biosens Bioelectron*, **26**, 4491–4496.

118. Lu Q., Dong X.C., Li L.J., Hu X.A. (2010) Direct electrochemistry-based hydrogen peroxide biosensor formed from single-layer graphene nanoplatelet-enzyme composite film, *Talanta*, **82**, 1344–1348.

119. Fan L.S., Zhang Q.X., Wang K.K., Li F.H., Niu L. (2012) Ferrocene functionalized graphene: Preparation, characterization and efficient electron transfer toward sensors of H_2O_2, *J Mater Chem*, **22**, 6165–6170.

120. Liu H., Gao J., Xue M.Q., Zhu N., Zhang M.N., Cao T.B. (2009) Processing of graphene for electrochemical application: Noncovalently functionalize graphene sheets with water-soluble electroactive methylene green, *Langmuir*, **25**, 12006–12010.

121. Shan C.S., Yang H.F., Han D.X., Zhang Q.X., Ivaska A., Niu L. (2010) Electrochemical determination of NADH and ethanol based on ionic liquid-functionalized graphene, *Biosens Bioelectron*, **25**, 1504–1508.

122. Hou S.F., Kasner M.L., Su S.J., Patel K., Cuellari R. (2010) Highly sensitive and selective dopamine biosensor fabricated with silanized graphene, *J Phys Chem C*, **114**, 14915–14921.

123. Wang Y., Li Y.M., Tang L.H., Lu J., Li J.H. (2009) Application of graphene-modified electrode for selective detection of dopamine, *Electrochem Commun*, **11**, 889–892.

124. Alwarappan S., Erdem A., Liu C., Li C.Z. (2009) Probing the electrochemical properties of graphene nanosheets for biosensing applications, *J Phys Chem C*, **113**, 8853–8857.

125. Lim C.X., Hoh H.Y., Ang P.K., Loh K.P. (2010) Direct voltammetric detection of DNA and pH sensing on epitaxial graphene: An insight into the role of oxygenated defects, *Anal Chem*, **82**, 7387–7393.

126. Ma X.Y., Chao M.Y., Wang Z.X. (2012) Electrochemical detection of dopamine in the presence of epinephrine, uric acid and ascorbic acid using a graphene-modified electrode, *Anal Methods*, **4**, 1687–1692.

127. Mallesha M., Manjunatha R., Nethravathi C., et al. (2011) Functionalized-graphene modified graphite electrode for the selective determination of dopamine in presence of uric acid and ascorbic acid, *Bioelectrochemistry*, **81**, 104–108.

128. Huang K.J., Niu D.J., Sun J.Y., et al. (2011) Novel electrochemical sensor based on functionalized graphene for simultaneous determination of adenine and guanine in DNA, *Colloids Surf, B*, **82**, 543–549.

129. Wang H.L., Hu Y.J., Xing D. (2011) Recent progress of two-step laser desorption/laser ionization mass spectrometry and its application, *Chin J Anal Chem*, **39**, 276–282.

130. Dong X., Cheng J., Li J., Wang Y. (2010) Graphene as a novel matrix for the analysis of small molecules by MALDI-TOF MS, *Anal Chem*, **82**, 6208–6214.

131. Zhou X.Z., Wei Y.Y. He Q.Y., Boey F., Zhang Q.C., Zhang H. (2010) Reduced graphene oxide films used as matrix of MALDI-TOF-MS for detection of octachlorodibenzo-*p*-dioxin, *Chem Commun*, **46**, 6974–6976.

132. Cairns D.R., Witte R.P., Sparacin D.K., et al. (2000) Strain-dependent electrical resistance of tin-doped indium oxide on polymer substrates, *Appl Phys Lett*, **76**, 1425–1427.

133. Wang X., Zhi L., Mullen K. (2007) Transparent, conductive graphene electrodes for dye-sensitized solar cells, *Nano Lett*, **8**, 323–327.

134. Kalita G., Matsushima M., Uchida H., Wakita K., Umeno M. (2010) Graphene constructed carbon thin films as transparent electrodes for solar cell applications, *J Mater Chem*, **20**, 9713–9717.

135. Liu J., Yin Z., Cao X., et al. (2010) Bulk heterojunction polymer memory devices with reduced graphene oxide as electrodes, *ACS Nano*, **4**, 3987–3992.

136. Junbo W., Hector A.B., Zhenan B., Zunfeng L., Yongsheng C., Peter P. (2008) Organic solar cells with solution-processed graphene transparent electrodes, *Appl Phys Lett*, **92**, 263302.

137. Zongyou Y., Shixin W., Xiaozhu Z., et al. (2010) Electrochemical deposition of ZnO nanorods on transparent reduced graphene oxide electrodes for hybrid solar cells, *Small*, **6**, 307–312.

138. Yin Z.Y., Sun S.Y., Salim T., et al. (2010) Organic photovoltaic devices using highly flexible reduced graphene oxide films as transparent electrodes, *ACS Nano*, **4**, 5263–5268.

139. Li D., Kaner R.B. (2008) Graphene-based materials, *Science*, **320**, 1170–1171.

140. Chen J.L., Yan X.P., Meng K., Wang S.F. (2011) Graphene oxide based photoinduced charge transfer label-free near-infrared fluorescent biosensor for dopamine, *Anal Chem*, **83**, 8787–8793.

141. He Y., Wang Z.G., Tang H.W., Pang D.W. (2011) Low background signal platform for the detection of ATP: When a molecular aptamer beacon meets graphene oxide, *Biosens Bioelectron*, **29**, 76–81.

142. Xing X.J., Liu X.G., Yue H., Luo Q.Y., Tang H.W., Pang D.W. (2012) Graphene oxide based fluorescent aptasensor for adenosine deaminase detection using adenosine as the substrate, *Biosens Bioelectron*, **37**, 61–67.

143. Zhao X.H., Ma Q.J., Wu X.X., Zhu X. (2012) Graphene oxide-based biosensor for sensitive fluorescence detection of DNA based on exonuclease III-aided signal amplification, *Anal Chim Acta*, **727**, 67–70.

144. Lu C.-H., Yang H.-H., Zhu C.-L., Chen X., Chen G.-N. (2009) A graphene platform for sensing biomolecules, *Angew Chem Int Ed*, **48**, 4785–4787.

145. Wu M., Kempaiah R., Huang P.-J.J., Maheshwari V., Liu J. (2011) Adsorption and desorption of DNA on graphene oxide studied by fluorescently labeled oligonucleotides, *Langmuir*, **27**, 2731–2738.

146. He S., Song B., Li D., et al. (2010) A graphene nanoprobe for rapid, sensitive, and multicolor fluorescent DNA analysis, *Adv Funct Mater*, **20**, 453–459.

147. Feng D., Zhang Y., Feng T., Shi W., Li X., Ma H. (2011) A graphene oxide-peptide fluorescence sensor tailor-made for simple and sensitive detection of matrix metalloproteinase 2, *Chem Commun*, **47**, 10680–10682.

148. Li J., Lu C.-H., Yao Q.-H., et al. (2011) A graphene oxide platform for energy transfer-based detection of protease activity, *Biosens Bioelectron*, **26**, 3894–3899.

149. Liu X., Miao L., Jiang X., Ma Y., Fan Q., Huang W. (2011) Highly sensitive fluorometric Hg^{2+} biosensor with a mercury(II)-specific oligonucleotide (MSO) probe and water-soluble graphene oxide (WSGO), *Chin J Chem*, **29**, 1031–1035.

150. Liu F., Choi J.Y., Seo T.S. (2010) Graphene oxide arrays for detecting specific DNA hybridization by fluorescence resonance energy transfer, *Biosens Bioelectron*, **25**, 2361–2365.

151. Jung J.H., Cheon D.S., Liu F., Lee K.B., Seo T.S. (2010) A graphene oxide based immuno-biosensor for pathogen detection, *Angew Chem Int Ed*, **49**, 5708–5711.

152. Li S.-S., Tu K.-H., Lin C.-C., Chen C.-W., Chhowalla M. (2010) Solution-processable graphene oxide as an efficient hole transport layer in polymer solar cells, *ACS Nano*, **4**, 3169–3174.

153. Gao Y., Yip H.-L., Hau S.K., et al. (2010) Anode modification of inverted polymer solar cells using graphene oxide, *Appl Phys Lett*, **97**, 203306-3.

Index

ACN, *see* activated carbon nanofibers

activated carbon 2, 4, 189, 207, 214, 230, 264–65

activated carbon nanofibers (ACN) 230–31

AFM, *see* atomic force microscopy

ALD, *see* atomic layer deposition

amorphous carbons 2, 97, 112, 128, 138

aqueous electrolytes 190, 209, 229–30, 233

asymmetric electrochemical capacitors 267–68

asymmetric supercapacitors 233

atomic force microscopy (AFM) 40, 73, 99, 251, 302–3

atomic layer deposition (ALD) 204

batteries 134, 187–89, 211, 247–48, 250, 252, 254, 256–70, 272

lithium-ion 189

vanadium redox flow 271

Bernal stacking graphene 155–56

biaxial strain 173–74

bilayer graphene 26, 39, 73, 75, 156–57

BLG, *see* bilayer graphene

C-face graphene 50–53

capacitors

conventional 183, 187–89

electrolytic 188

carbon

disordered 98, 126, 128

hybridized 164, 183–84, 194, 248

porous 189, 218–19, 224

carbon allotropes 1–3, 79

carbon nanosheets 99, 103–5, 194, 216–17

carbon nanostructures 2, 4, 96, 122, 124, 140, 297

carbon nanotube, unzipping 185, 195

carbon nanotubes 2, 79, 122, 142, 184, 256–57, 263, 266, 302

carbon nanowalls 5, 95–102, 104, 106–12, 114–16, 121–46

potential applications of 121, 134

structural characterization of 121–22, 124

synthesis of 96–97, 99

carbon segregation process 70, 72, 83, 88

catalysts 70, 79, 81, 84, 99, 115, 122–23, 134, 140, 144–46

CCS, *see* confinement controlled sublimation

charged impurities 25, 27–29

chemical vapor deposition (CVD) 4, 37, 68, 70–71, 84, 96, 99, 185, 191–92, 216, 224, 288, 290, 298, 307

chemically modified graphene (CMGs) 185–86, 195, 197–98, 204, 206

CMGs, *see* chemically modified graphene

conducting polymers 189, 202, 206, 209, 228–29, 232, 268, 301 conductive additives 139, 227, 263

conductivity 24–25, 27–29, 196, 200, 219, 227–28, 253, 260, 263, 266, 268, 270, 281, 300, 305–6

confinement controlled sublimation (CCS) 42, 54–55

copper 70, 78–84, 86–88, 100, 123, 143, 298

copper foils 80–82, 86, 192

copper grain boundaries 84–85

copper substrates 79, 81, 84–85, 100–1, 111

CVD, see chemical vapor deposition

cyclic voltammetry 135, 137, 145, 210, 215, 222

cyclic voltammograms 135–36, 144

cycling stability 258, 260–61, 263

deoxygenation 198–200, 253

diamond 2, 4, 36, 79, 104, 184

diamond-like carbon (DLC) 2, 4

Dirac electrons 55, 280

Dirac fermions 3, 36

Dirac point 3, 12, 17, 26–28, 30, 161–62, 170, 175, 281, 283–84, 286

DLC, see diamond-like carbon

edge chirality 154–55, 176

EDL, see electrical double layer

EELS, see electron energy loss spectroscopy

EIS, see electrochemical impedance spectroscopy

electrical conductivity 4, 168, 201, 221–22, 229, 251, 253, 263, 270, 316

electrical double layer (EDL) 207, 210, 216

electrical gating 283, 286–87

electrical transport 17, 27–29, 101–2

electrochemical capacitors 187, 189, 226, 264, 268–69

electrochemical impedance spectroscopy (EIS) 210, 212–13, 215

electrochemical polymerization 206

electrode materials 101, 188–89, 207, 211–13, 217–20, 224, 227–28, 233, 256, 263–64
  composite 258, 261
  graphene-based composite 258
  porous graphene 269

electrodes, superconducting 102

electrolyte decomposition 136

electrolytes 135–36, 139, 145, 188–90, 207, 215–16, 219, 225, 229, 232, 264–66, 269
  ionic liquid 218

electron doping 162

electron energy loss spectroscopy (EELS) 262

electronic coupling 156

electronic devices 95, 107, 121, 280–81

energy dispersion, linear 17, 30

energy storage 4–5, 95, 153, 184, 187, 189

energy-storage devices 248, 269, 272

epitaxial graphene, growth of 35–36, 38, 40, 42, 44, 46, 48, 50, 52, 54, 56

EQE, see external quantum efficiency

equivalent series resistance (ESR) 212, 215

ESR, see equivalent series resistance

exfoliation 191, 196, 198, 200, 218, 221, 223, 248–51, 255, 266, 280

external quantum efficiency (EQE)
289, 292

few-layer graphene (FLG) 4, 75,
156, 158
few-layer graphene films 70–71,
73–74, 83
FLG
see few-layer graphene
turbostratic 156–57
fullerenes 2, 122, 184–85

geometric surface areas (GSA)
144–46
GFs, *see* graphene foams
GICs, *see* graphite intercalated
compounds
glass substrates 73–74
glassy carbon 126–28
glassy carbon electrode 144
glucose 197
glucose oxidase 309
graphene
activated 266–67
activation of 267
applications of 250, 271
band structure of 13, 38
chemical synthesis of 247–56,
258, 260, 262, 264, 266,
268, 270, 272
conductivity of 26, 28, 159,
217
density of states of 26, 113
disordered 155, 166
doped 160, 283, 286, 289
electronic band structure of
15, 17, 29, 176
electronic properties of 4,
154, 280, 287
electronic structure of 155
epitaxial 26, 35–38, 40, 42,
44, 46, 48, 50, 52, 54, 56,
68, 191
epitaxial growth of 191, 298

exfoliated 69, 76
Fermi level of 21, 160, 287
fluorinated 167–68
fluorination of 167
free-standing 38, 190, 285
high electrical conductivity of
13, 264
high-quality 46–47, 49, 56,
75, 88, 193, 216, 299
hydrogenated 167–68
large-area 67–68, 70, 72, 74,
76, 78, 80, 82, 84, 86, 88,
299
multilayer 4, 38–39, 44–45,
50, 55,81
nitrogen-doped 265
photocurrent generation in
286–87
photonic absorption of
284–85
porous 265, 269
pristine 154, 159, 162, 164,
168, 174, 185–86, 196,
202, 204, 219, 253, 265,
300, 307
Raman spectroscopy of 155
single-crystal 77, 84–85, 87
single layer 4, 18, 23, 98,
155–59, 172–73, 291
single-layer 2–3, 68, 75,
78–82, 185, 190–91,
247–50, 284–85, 289–90,
298
graphene-based composite
materials 186, 202, 233
graphene-based electrode material
for LIB 257
graphene-based electrode
materials 272
graphene-based materials,
development of 232–33
graphene-based supercapacitor
264
graphene devices 279–82, 284,
286, 288, 290, 292

graphene electrodes 264–65, 267, 269

graphene electronics 50

graphene films
pre-grown CVD multi-layer 86–87
single-layer 70, 79, 81, 85
stacked multi-layer 82
uniform 57
wafer-scale bilayer 83

graphene flakes 69

graphene foams (GFs) 194

graphene grain boundaries 85–86

graphene grains 80, 84–87
single-crystal 85–86

graphene growth 4, 39, 42, 44–48, 50, 54, 56–57, 77, 79–81, 83, 86, 193
bilayer 83
high-quality 56, 81

graphene islands 40, 51–53, 55

graphene nanoribbons 23, 169, 195–96, 300

graphene nanosheets 230, 248, 257–59, 268
freestanding 192

graphene networks 186, 262

graphene nucleation 45, 85–86

graphene oxide 5, 69, 167–68, 174, 186, 223, 250–53, 258, 260, 266, 270, 298–300, 313, 315
electrochemical reduction of 253
reduced 185, 202, 216, 270, 299–303, 305–7, 309–11
reduction of 197, 250–51
thermal treatment of 253, 264

graphene oxide sheets 251, 261 266, 268, 270

graphene photodetectors 279, 282, 287, 289–91

graphene plasmonics 291

graphene quantum dots 300

graphene ribbons 194

graphene scaffold 260–61

graphene sheets 95–96, 98, 100–1, 107–8, 112, 121–22, 191, 193–94, 196, 218–19, 224, 232–33, 248–49, 254–56, 260–64
exfoliated 112
few-layer 71, 84-85, 112, 192, 255, 298
free-standing single layer 98
oxidized 198, 221–22, 263
pristine 216
reduced 253, 260

graphene stack photodetector 282, 290

graphene synthesis 73, 190, 194, 270

graphene synthesis methods 68–69

graphite 2, 4, 68–69, 112, 114, 126, 134, 136, 138, 166–67, 184–85, 191, 248–49, 254–57, 298–99
direct exfoliation of 248–49
exfoliated 255–56
pristine 98, 248–49, 303–4

graphite crystals 68

graphite flakes, multilayered 221–22

graphite intercalated compounds (GICs) 254

graphite oxide 4, 68–69, 167, 186, 195, 199, 218, 221–24, 249–50, 254
exfoliation of 199, 249–50
reduction of 69, 199, 218
single-layer 250

graphite sheets, two-dimensional 121–22

graphitic carbons 136, 184

graphitization
degree of 128

high degree of 128, 133–34, 144, 146

GSA, *see* geometric surface areas

high resolution transmission electron microscopy (HRTEM) 44–45, 98, 157

highly oriented pyrolytic graphite (HOPG) 69, 81, 107–8, 112, 127–28, 190, 280

hole transport layer (HTL) 315–16

HOPG, *see* highly oriented pyrolytic graphite

HRTEM, *see* high resolution transmission electron microscopy

HTL, *see* hole transport layer

hydrazine 69, 197, 202, 252, 298, 301, 304

hydrazine reduction 301–2

hydrocarbons 68, 70, 79, 88
polycyclic aromatic 194

indium tin oxide (ITO) 76, 82, 286, 310

ITO, *see* indium tin oxide

JJA, *see* Josephson junction array

Josephson junction array (JJA) 103–5

Klein tunneling 3, 22–24, 174

Kohn anomalies (KAs) 160, 162

LEEM, *see* low-energy electron microscopy

LIBs, *see* lithium-ion batteries

lithium-ion batteries (LIBs) 5, 121–23, 134–35, 146, 189, 225, 256, 258, 263, 269, 271

low-energy electron microscopy (LEEM) 41–43, 84

magnetoresistance 105–6

manganese oxide 209–10, 226, 230

MCS, *see* mesoporous carbon spheres

mesoporous carbon spheres (MCS) 224

nanoelectronics, graphene-based 191

nanographite grains 99

OCV, *see* open-circuit voltage

open-circuit voltage (OCV) 212, 312–13

optoelectronics 82, 88, 286–87

PAH, *see* polycyclic aromatic hydrocarbons

PCE, *see* power conversion efficiency

PECVD, *see* plasma-enhanced chemical vapor deposition

photocurrent generation 282–83, 287–91

plasma-enhanced chemical vapor deposition (PECVD) 99, 122, 134

plasmon resonance 291–92

polyaniline 229–31, 268

polycyclic aromatic hydrocarbons (PAH) 194

polymerization 206, 228

porous carbon architecture, 3D hierarchical 224

power conversion efficiency (PCE) 311–13, 315

pseudocapacitance 208–9, 221–22, 225, 228, 264, 269

pseudospin 16

Pt catalysts 140–42

Pt loading 141, 144–45

Pt nanoparticles 141–43, 146

QIC, *see* quantum interference corrections
quantum Hall effect 55
quantum interference corrections (QIC) 24–25

Raman scattering 154–55
reversible hydrogen electrode (RHE) 144–45
RHE, *see* reversible hydrogen electrode

scanning electron microscope (SEM) 98, 107, 122, 252, 265
scanning tunnelling microscopy (STM) 40–42
SEI, *see* solid electrolyte interface
SEM, *see* scanning electron microscope
semiconductors 37, 289, 292
silicon 43, 46–47, 49, 54, 56, 191, 260
silicon carbide 35, 69, 191
solid electrolyte interface (SEI) 135
step bunching 43, 45, 49, 56
STM, *see* scanning tunnelling microscopy
sulfur 198, 203, 270–71
supercapacitor applications 184, 206, 208, 223, 228, 233, 265
supercapacitor devices 184, 211, 220, 225, 231–32

activated graphene-based 266
supercapacitor electrodes 207–9, 211, 213, 215–17, 219, 221, 223, 225, 227, 229, 231
supercapacitors 5, 37, 183–84, 187–90, 207, 210–13, 215–20, 225, 228–29, 232–33, 248, 264–69

TEM, *see* transmission electron microscopy
thermal conductivity 36, 247
thin film transistors 307–8
transmission electron microscopy (TEM) 5, 121–22, 124, 129, 143–44, 250
transparent electrodes 68, 74, 82, 193, 312
trigonal warping of the electronic spectrum 15, 17, 25–26
trilayer graphene, intercalated 163–64

weak localization 24–26

X-ray diffraction (XRD) 141, 143–45, 250, 303
X-ray photoelectron spectroscopy 77, 304
XRD, *see* X-ray diffraction

zero-bias resistance (ZBR) 109–10, 112–13
zigzag edge 21, 169–72